# BIOTECHNOLOGY

## Scientific Advancement versus Public Safety

# BIOTECHNOLOGY

## Scientific Advancement versus Public Safety

**CONRAD B QUINTYN**

*Bloomsburg University of Pennsylvania, USA*

NEW JERSEY • LONDON • SINGAPORE • BEIJING • SHANGHAI • HONG KONG • TAIPEI • CHENNAI • TOKYO

*Published by*

World Scientific Publishing Co. Pte. Ltd.
5 Toh Tuck Link, Singapore 596224
*USA office:* 27 Warren Street, Suite 401-402, Hackensack, NJ 07601
*UK office:* 57 Shelton Street, Covent Garden, London WC2H 9HE

**Library of Congress Cataloging-in-Publication Data**
Names: Quintyn, Conrad B., author.
Title: Biotechnology : scientific advancement versus public safety /
Conrad B Quintyn, Bloomsburg University of Pennsylvania, USA.
Description: First edition. | Hackensack, NJ : World Scientific Publishing Co. Pte. Ltd., [2023] |
Includes bibliographical references and index.
Identifiers: LCCN 2022025058 | ISBN 9789811259258 (hardcover) |
ISBN 9789811259265 (ebook for institutions) | ISBN 9789811259272 (ebook for individuals)
Subjects: LCSH: Biotechnology.
Classification: LCC TP248.2 .Q56 2023 | DDC 660.6--dc23/eng/20220723
LC record available at https://lccn.loc.gov/2022025058

**British Library Cataloguing-in-Publication Data**
A catalogue record for this book is available from the British Library.



For any available supplementary material, please visit
https://www.worldscientific.com/worldscibooks/10.1142/12916#t=suppl

Desk Editor: Vanessa Quek

Typeset by Stallion Press
Email: enquiries@stallionpress.com

Printed in Singapore

To my dad, who took us out of ignorance and never abandoned us.
(1930–2018)

# Preface

The impetus for writing this book came after reading Daniel J. Kevles's book *In the Name of Eugenics: Genetics and the Uses of Human Heredity*, first published by Harvard University Press in 1985. Although all the chapters of Kevles's book were quite stimulating, Chapter 17, "A New Eugenics," really piqued my interest. In this chapter, Kevles presented a short history of the research that led to genetic engineering (such as cloning amphibians, in vitro fertilization, and recombinant DNA) and the determination of geneticists to "reduce the incidence of genetic disease in the population."[1] In addition, Kevles made it clear that cloning, germinal choice, and sperm banks — echoes of the "old" eugenics — were intertwined with genetic engineering. Outside of Kevles's book, genetic engineering and medical genetics have been an interest (and concern) of mine since my postgraduate studies in biological anthropology. As a student of evolutionary biology, I praise the scientific breakthroughs in biotechnology leading to cures for some diseases. But, I worry about the impact of genetic engineering on future evolutionary change and the *Anthropocene* — the period of greatest human impact on Earth's biotic and abiotic resources. Humans' impact on the natural environment over

[1] Daniel Kevles makes a distinction between the "old" eugenics where the 'powers that be' forcibly sterilized people without power (i.e., poor; mentally ill, etc.) and the "new" eugenics where post-WWII enlightened officials (reeling from the atrocities committed by the Nazis) were committed to ridding *all* their people of diseases.

the past 12,000 years — through the agricultural revolution, civilization, the Industrial Revolution, and the atomic revolution — has been so great that the negative effects are still apparent today in terms of infectious and noninfectious diseases (such as respiratory diseases and cancers) and global warming. The long-term evolutionary impact on human biology and, as far as global warming is concerned, the evolutionary impact on other organisms remains unknown.

Returning to genetic engineering, biotechnology is advancing globally at an incredible rate, and as with computer technology and the internet, there is no strong regulation. Or, put another way, the current regulations are being stretched beyond their limits. For example, an important question posed today is whether genome editing should be regulated as a technology or a drug. In the 1970s, the important question was whether recombinant retroviruses, with their propensity for mutagenesis, should be used in gene therapy. This concern prompted the US National Academy of Sciences and National Institutes of Health (NIH) to ask concerned scientists to organize a meeting, leading to the Asilomar Conference on Recombinant DNA, held in February 1975 at the Asilomar Conference Center in Pacific Grove, California, which proposed guidelines to ensure the safe use of recombinant DNA. Forty years later in January 2015, another Asilomar-like conference was held in Napa, California, to discuss the path forward for genomic engineering. The questions raised at this conference surrounding the use of gene editing tools required a global discussion that resulted in an international summit on human gene editing, held in Washington, DC, in December of the same year. Three years later in November 2018, a second international conference held in Hong Kong addressed the claims of a Chinese scientist who had supposedly used genome editing to make genetic alterations in early embryos and implant them. Moreover, scientists are still operating in the Wild West, with mostly good intentions mixed in with hubris, lucrative patents for their host institutions, fame, and Nobel Prizes. Infused with these potent stimuli, they propel themselves blindly ahead with no consideration of the consequences. Mary Shelley's *Frankenstein* and H. G. Wells's *The Island of Dr. Moreau* are just two prominent fictional tales written to reach a wider audience and emphasize philosophical themes such as the violation of nature, cruelty, moral responsibility, and respecting the boundaries of humankind in the context of experiments with animal-human hybrids and animal chimeras. Taking the baton from Kevles, I used his Chapter 17 as a foundation on which to build a discussion of bioengineering in the 21st century.

My definition of bioengineering includes the modification of any biological life form (genetically modified organism or GMO), not just *Homo sapiens*, to replace or repair its "defects." One can think of this as an unintended consequence of the old eugenics. This definition is not based on any religious belief of spirituality encompassing all nature. The definition is based on science. There would be no genetic engineering if DNA was not practically universal. Genes can be transferred from one organism to another and subsequently expressed in the new host organism. For example, a transgenic piglet engineered to express a gene from jellyfish that encodes green fluorescent protein had a yellow nose and hoofs; inactivation of a gene for the p53 tumor suppressor resulted in transgenic mice that exhibited increased incidence of tumors in several tissues; a lentivirus was used to deliver an erythropoietin (EPO) gene, which builds the EPO hormone that controls red blood cell production; the gene encoding the *Bacillus thuringiensis* (Bt) toxin, which causes paralysis and death when consumed by insects, was inserted in plants; and three genes inserted in the rice genome allowed expression of the pathway for β-carotene in an endosperm (Golden Rice).[2]

Animal, plant, and microbe models are constantly being used to understand processes in human biology. For instance, the mechanics of DNA replication could not be understood without studying the bacterium *Escherichia coli.* DNA to DNA hybridization occurs readily between humans and nonhuman primates (an average difference of two chromosomes), and somatic cell hybridization between humans and mice (which have 40 chromosomes compared to 46 in humans) have made these animals the traditional animal models for humans. Add the numerous other biological life-forms used every day in research to understand human biology and promote human survival: malaria parasites, influenza virus subtypes, mosquitoes, fruit flies, guinea pigs, ferrets, corn, potatoes, and

[2]The universality of DNA has led to the proliferation of responsible and questionable research in the modification of biological lifeforms.

tobacco, to name a few. Additionally, the UK Human Fertilisation and Embryology Authority (HFEA) in 2007 authorized the research into the feasibility of using human-animal embryos as an alternative source of stem cells.[3] In short, humankind invades, manipulates, and patents human, animal, insect, and plant genomes in a way that is analogous to European colonialism, which led to the coining of the term *biocolonialism* to capture this particular behavior that is tied to the instinct to control life.[4] It would be human hubris to exclude nonhumans from this discussion of modifying life, because of the universality of DNA discussed above. (It is interesting that scientists have avoided legal liability and government regulation when using nonhuman genomes.)

As Francis Galton phrased it, the idea of eugenics was to promote "improving the stock . . . [via] judicious mating" in order "to give the more suitable races or strains of blood a better chance of prevailing speedily over the less suitable."[5] Later in his life, however, he surrendered to the proposition of sterilization to check the reproduction of the unfit. Nevertheless, the key change to this definition after World War II — which required the addition of the word *new* before the word *eugenics* — was to advance genetic research to root out defective genes from the human gene pool. For example, pre- and postnatal screening were initiated to remove mutant genes (or "bad" genes in most cases) from the human gene pool (e.g., those for phenylketonuria or PKU, sickle cell anemia, Tay-Sachs disease, cystic fibrosis, and trisomy 21 or Down syndrome) rather than promoting the genetics of a respective social economic class and restricting the genetics of another class. Today, the goal remains rooting out defective genes to help humans, which indirectly includes

---

[3] Pro-life proponents believe that harvesting the blastocyst (which becomes the embryo later in development) for stem cell research is abortion. Life begins at concept for them and destruction of the early human embryo is morally unacceptable. If animal embryos are injected with human stem cells, then one has a source for stem cells (i.e., animal embryos with human cells).

[4] Biotechnology is the new frontier where biotech companies are harvesting and patenting biological materials (e.g., Myriad Genetics hold patents for the tumor suppressor genes BRCA-1 and 2).

[5] Francis Galton's terms "suitable" and less "suitable" were largely based on social economic status and education of individuals.

their ecology (by improving livestock, improving crop yields, and protecting crops from insects and environmental stresses). In my elaboration of biotechnology, I also discuss GMOs, gene therapy (including induced pluripotent stem cells), cloning, artificial reproductive technologies (such as in vitro fertilization), nanobiotechnology and nanomedicine, xenobiology, genetic editing, isolation of embryonic stem (ES) cells, and synthetic biology, which involves "facilitating and accelerating the design, manufacture and modification of genetic materials in living organisms."[6]

My motivation for writing this book is rooted in several important concerns within biotechnology. For instance, although the mice in which the gene for the p53 tumor suppressor was inactivated were born normal, they exhibited an increased incidence of tumors during growth and development. This problem highlights the complexity of gene function and the varied reactions when one gene is inactivated: genes rarely work in isolation. In another example, the first DNA vaccine spliced an influenza gene encoding a nucleoprotein into a plasmid, which was then injected into mice, where it caused an immune response to influenza. Viruses have an RNA genome replicated by a viral RNA polymerase that lacks proofreading ability. Over time, mutations accumulate in surface proteins resulting in different strains of the virus. Thus, the DNA vaccine could create a new strain of virus or an allergic reaction. In a more alarming example, an interrupting RNA vector was used to suppress the mutant gene responsible for the X-linked form of severe combined immunodeficiency disease (X-SCID) and replace it with a working copy. During clinical trials, several patients developed the same rare type of leukemia. The engineered retrovirus used to introduce the X-SCID gene activated a proto-oncogene (tumor-causing) in these patients. Although some patients survived long after this gene therapy, the critical factors that biologists must consider when performing genetic engineering are unintended effects of genes, immunogenic properties of viral vectors used, pleiotropic effects (the individual's genetic background, population history, physiology, diet, and environment), and their own tunnel vision in treating genes as separate from environmental context. In another worrying example, to produce

[6]When accused of "playing God," synthetic biologists respond by stating that they are "engineers" or "designers" not creators.

plants resistant to insects, a gene for Bt toxin was inserted into crop plants, which caused insects to die when they ingested the plants. Enzymes within the insects' systems convert Bt toxin into an insect-specific toxin. These enzymes have not been found in other organisms yet; consequently, Bt toxin is harmless to them. But for how long will this be true? Finally, what are the long-term effects of genetically modified crops on human health? Will introduced genes in crops spread into wild relatives, impacting biodiversity? These questions are unanswered right now, and I believe this is a major problem.

The questions listed above, particularly the concerns regarding long-term, heritable effects, can also be applied to cloning technology, in which natural fertilization is bypassed. The possibility of human cloning captures the public imagination. For example, cloning could replace a dead child or a spouse, solve infertility, provide a spare self from which to obtain tissue or organs after trauma or disease, or resurrect recently extinct animal species. In reality, scientists believe that cloning mammals will result in genetic improvement of livestock and produce human proteins to treat diseases. One problem, however, is that cloning — whether by blastomere isolation, splitting of preimplantation embryos, or nuclear transplantation — results in animals with birth defects that could modify the resultant human protein. Generally, developing animals spontaneously abort at various stages. Dolly the sheep, made famous because she was the first mammal to be cloned from adult cells, was the only success out of 434 attempted fusions of donor nucleus and egg.[7] If the clone survives birth, it may not survive infancy. More importantly, the clone may have a combination of normal and abnormal chromosomes in its cells, which will be inherited and have implications for species survival. Other problems include the accumulation of mutations due to radiation or chemicals and the elimination of meiosis (the natural production of sex cells). Human cloning would unfortunately make meiosis, along with the critical variation in each generation that it provides, dispensable.

---

[7]The Dolly cloning example is used to emphasize the fact that any attempt to clone humans would probably end in disaster because cloning — at this stage in human biotechnological knowledge — is inefficient.

Genetic engineering, particularly genetic editing, has burst upon the scene, and the accuracy of the techniques involved improves every year. The US National Academies of Sciences, Engineering, and Medicine stated that the science of genome editing has advanced rapidly since 2015 and powerful, precise editing tools are being employed to treat diseases. But, there are two major concerns on governance of human genome editing and use of genome editing on the egg and/or sperm resulting in heritable alterations.[8] Clustered regularly interspaced short palindromic repeats/Cas9 (CRISPR/Cas9) is a technique based on the way that bacteria immunize themselves (and future progeny) against viral infection. Cas9 (which stands for CRISPR-associated nuclease) is an enzyme that cuts DNA of essentially any cell type. Since CRISPR/Cas9 is not tightly restricted research (as in, for example, the gain-of-function H5N1 avian influenza studies commonly known as bird flu) and is available to researchers globally, it could be used to edit the genomes of human reproductive cells or human embryos, resulting in designer babies carrying heritable, CRISPR/Cas9-introduced genetic changes.[9] Genomic edits, such as trying to revert the mutant cystic fibrosis transmembrane conductance regulator gene (*CFTR*) that causes cystic fibrosis back to the normal sequence, might occur in the wrong place (off-target editing), leading to a new disease. This is the general concern. A related concern is the use of CRISPR/Cas9 for genetic enhancement with accompanying negative side effects. For example, parents-to-be who desire smarter babies could choose short-term benefits and reject negative side effects, resulting in some babies developing into children who are autistic or have the debilitating illness of amyotrophic lateral sclerosis (also called Lou Gehrig's disease — a nervous system disorder).

Other concerns are unregulated clinics performing unsafe procedures and the danger of bad actors using the technology for negative purposes,

---

[8]There is a global ban on embryo editing (germline editing) because mistakes or mutations would be passed down for generations. Unfortunately, this ban does not work in countries with little to no regulation of their bioengineering programs.

[9]The gene editing technology CRISPR/Cas9 and PGD together can be employed to design babies (i.e., eye color, etc.)

so-called dual use. One recent example was a biohacker — motivated by the democratic ideal — selling do-it-yourself genome-editing kits, allowing anyone with college-level genetic lab skills to edit or add genes to bacteria or yeast to create glow-in-the-dark bacteria.[10] This is misguided and irresponsible, because someone could randomly create a dangerous microbe. Another danger comes from gene drives, in which wild animals or plants are obtained, have their genomes edited, and are released back into the wild. A bad actor could cause catastrophic harm to a fast-reproducing species, either directly or indirectly, through disease, starvation, or extinction.

The genie is out of the bottle, so to speak; the technology is available. Some of these concerns prompted biochemist and 2020 Nobel laureate Jennifer Doudna, one of the creators of CRISPR/Cas9, to write, "With our mastery over the code of life comes a level of responsibility for which we, as individuals and as a species, are woefully unprepared."[11] Generally, I agree with Doudna's statement. However, I take issue with the suggestion that bioengineers have "mastery over the code of life." The complexity of life and its interaction with the environment makes real mastery of life impossible; therefore, the beginning of Doudna's statement is more hyperbole than hubris.

The goal of scientific advances in general, and those in biotechnology in particular, is to make life better for humankind (e.g., repairing damaged cells and tissues, curing diseases, and improving and protecting crop plants from insects and adverse climate). These goals indicate that scientists in the "trenches" doing the hard work are honorable men and women serving humankind. Of course, there will always be the dangers of dual use, gene drives, and countries with limited laws to regulate their scientists (to prevent, for example, risky gain-of-function virus studies that

[10]The gene editing technology CRISPR/Cas9 is user-friendly and is being used prolifically all over the world; many biological organisms have been edited. The fear is dual use: an individual or group could use CRISPR to modify a plant (i.e., crop plant) or animal species (i.e., livestock) leading to extinction of these species and, subsequent, deprivation in one or more developing countries.

[11]Jennifer Doudna is absolutely correct; we may not be ready to care for, protect, and use — responsibly — the products of scientific advancement. Scientists plough ahead, with hubris, not realizing how their research impacts the natural and social environment.

could spark deadly pandemics through mishandling and faulty containment). And do not forget that market forces — with the indirect help of so-called transhumanists who advocate the use of technology to make humans better — will seek large profits by exploiting our human desire to be mentally and physically fitter than we already are.

Nevertheless, five key questions must be answered:

1. How and where should the line be drawn for the Promethean task of genetic engineering? In other words, how should public safety and scientific advancement be balanced? Or are these entities two sides of the same coin — in other words, a false dichotomy?
2. How should therapy be differentiated from elective genetic enhancement within the social context?
3. What are the microevolutionary (short-term) and macroevolutionary (long-term) effects of genetic engineering on the genome of biological organisms?
4. Do the risks of genome alterations in biological organisms outweigh the benefits, in that inherited deleterious mutations will occur gradually over time, resulting in species extinctions?
5. If only the rich have access to genetic enhancements, will this lead to competition within that social group that further isolates them from the majority of people?

Curiosity is a dangerous trait, but it has allowed the successive lineages of the *Homo* genus, from *Homo erectus* two million years ago to archaic *Homo sapiens* 100,000 years ago, to peer over the next hill and advance into the unknown, including stumbling into agriculture and civilization 10,000 years ago. Will we humans be so lucky in our use of biotechnology, or will we stumble and fall into an abyss?

Conrad B. Quintyn, Ph.D.
Bloomsburg University, Dept. of Anthropology
cquintyn@bloomu.edu

# About the Author

Dr. Conrad B. Quintyn is associate professor of biological anthropology at Bloomsburg University, Bloomsburg, PA.17815, U.S.A. He was educated at Baylor University (B.A., 1991), University of Michigan (M.A., 1993), and University of Michigan (Ph.D., 1999). From 2000 to 2001, he taught at Iowa State University. From 2001 to 2002, he worked as a biological anthropologist for the Joint POW/MIA Accounting Command [JPAC] Laboratory (currently called Defense POW/MIA Accounting Agency [DPAA]) located on Hickam Air Force Base, Honolulu, Hawaii. There he conducted search, recovery, and identification of the remains of U.S. soldiers listed as missing-in-action from World War II and the Vietnam War. Occasionally, he assists the Pennsylvania State Police in the recovery and identification of human remains His interests include worldwide postcranial variation (i.e., osteometric analysis of skeletal collections in various domestic and international museums), evolutionary biology, the problem of species, evolution of human diseases, and forensic anthropology.

Genetic engineering and medical genetics have been an interest (and concern) of Dr. Quintyn since his postgraduate studies in biological anthropology. As a student of evolutionary biology, he praises the scientific breakthroughs in biotechnology leading to cures for some diseases. But, he worries about the impact of genetic engineering on future evolutionary change and the *Anthropocene* — the period of greatest human impact on Earth's biotic and abiotic resources.

In his research, Dr. Quintyn has collected human skeletal data from the Peabody Museum of Archaeology & Ethnology, Harvard University, Cambridge, MA; the Natural History Museum, London, UK; the Museum National D' Histoire Naturelle, Paris, France; and the William S. Webb Museum of Anthropology, University of Kentucky, Lexington, KY. He is the editor of an academic text titled *Readings in Evolutionary Theory, Genetics, and the Origins of Modern Morphology* (San Diego: Cognella Academic Publishing) published in summer 2021.

Dr. Quintyn was born in London, U.K. and he immigrated to the United States in 1978. Two years after graduating high school, Dr. Quintyn enlisted in the U.S. Navy for active duty in January 1983 and was honorably discharged in January 1987. Subsequently, he served two years inactive reserve while attending university.

Webpage: https://www.bloomu.edu/people-directory/conrad-quintyn-phd

Twitter: https://twitter.com/b_quintyn

Facebook: https://www.facebook.com/conrad.quintyn

Email: cquintyn@bloomu.edu

# Acknowledgments

In the summer of 2015, I was compiling materials to teach Anthropology 290: Race & Racism — an elective for Bloomsburg University students that I teach every two years. I read Daniel J. Kevles' book *In the Name of Eugenics: Genetics and the Uses of Human Heredity* as I was compiling materials for the course, and I found myself particularly interested in Chapter 17, titled "A New Eugenics." This was the launching point for this book. I began by taking notes on several tentative titles and chapters, building the framework of the book. I then collected a variety of sources (scientific books; the National Academy of Sciences, Engineering, and Medicine summit reports; historical medical reports; national newspapers; electronic reports; etc.) that I would use during the writing process. I began writing in late May 2017 and completed the first draft in January 2020. I did most of my writing in the summer months and during my free time in the fall, winter, and spring. I enjoyed the journey that led to this book, but the journey that would eventually lead to publication was not so easy. This process was difficult and disappointing; the book was rejected a number of times — the worst of these rejections were the silent ones. I believe all book publishers, big and small, should have the decency to reply to queries and not keep the writer (particularly unknown writers or professors from tier three or four universities) waiting indefinitely. It is always devastating when your creation is rejected. You reflect on the possibility that "your baby" may not be as beautiful as you think, and it compels you to seek professional help to improve the book. In the end, I was forced to go alone in getting this book published.

First, I would like to thank Drs. Kady Bell-Garcia and Sara Cole and their colleagues at Dissertation Editors, LLC, for providing critical review and feedback on the strength and structure of my argument. Their efforts have advanced the writing in this book. I take great pleasure in thanking the numerous scientists, scholars, historians, bioethicists, journalists, and institutions whose sources were indispensable in writing and organizing the chapters. It is an unfortunate fact of all academic disciplines that one cannot possibly acknowledge everyone, so please know that if I have omitted any names, I have done so unintentionally: Daniel J. Kevles, Peter Raven, George B. Johnson, Kenneth A. Mason, Jonathan B. Losos, Susan R. Singer, Jean Baudrillard, Rainer Breitling, Eriko Takano, Timothy S. Gardner Human, Robert G. McKinnell, Paul Knoepfler, Jim Kozubek, Jennifer A. Doudna, Samuel H. Sternberg, Harry F, Tibbals, Ross D. Park, Christine W. Gailey, Scott Coltrane, M. Robin DiMatteo, Henry T. Greely, Michael Sandel, Diane B. Paul, Howard Markel, Jared M. Diamond, Mara Hvistendahl, David Joravsky, Julian Savulescu, Jonathan Pugh, Thomas Douglas, Christopher Gyngell, William Brahms, Bao Xichen, Zhu Xihua, Liao Baojian, Miguel A. Esteban, Maywa Montenegro, Nisar A. Wani, Min J. Kim, Hyun J. Oh, Geon A. Kim, Erif M. N. Setyawan, Yoo B. Choi, Byeong C. Lee, Gordon L. Woods, Kenneth L. White, Dirk K. Vanderwall, Guang-Peng Li, Kenneth I. Aston, Thomas D. Bunch, Lora N. Meerdo, Barry J. Pate, Qi Zhou, Jean-Paul Renard, Gaëlle Le Friec, Vincent Brochard, Nathalie Beaujean, Yacine Cherifi, Alexander Fraichard, Jean Cozzi, Taeyoung Shin, Duane Kraemer, Jane Pryor, Ling Liu, James Rugila, Lisa Howe, Sandra Buck, Keith Murphy, Leslie Lyons, Mark Westhusin, Patrick Chesné, Pierre G. Adenot, Céline Viglietta, Michel Baratte, Laurent Boulanger, Irina A. Polejaeva, Shu-Hung Chen, Todd D. Vaught, Raymond L. Page, June Mullins, Suyapa Ball, Yifan Dai, Jeremy Boone, Shawn Walker, David L. Ayares, Alan Colman, and Keith H. Campbell, Jean-Paul Renard, Zhen Liu, Yijun Cai, Yan Wang, Yanhong Nie, Chenchen Zhang, Yuting Xu, Xiaotong Zhang, Yong Lu, Zhanyang Wang, Muming Poo, Qiang Sun, Kelly Servick, Masayo Takahashi, Shinya Yamanaka, Ryan Lister, Mattia Pelizzola, Y.S. Kidda, David R. Hawkins, Joe Nery, G. Hon, Jessica Antosievicz-Bourget, Ronan O'Malley, Rosa Costanon, S. Klugman, M. Downes, R Yu, Ron Stewart, Bing Ren, James A. Thomas, R. M. Evans, Joseph R. Ecker, Ivan Gutierrez-Aranda,

Veronica Ramos-Mejia, Clara Bueno, Martin Munoz-Lopez, Pedro J. Real, Angela Macia, Laura Sanchez, Gertrudis Ligero, Jose L. Garcia-Parez, Eric R. Scerri, Michael Specter, Milly Dawson, Ronald K. Wetherington, Eric Lander, Ruud Jansen, Jan. D.A. van Embden, Wim. Gaastra, Leo. Schouls, Stan J. Brouns, Matthijs M. Jore, Magnus Lundgren, Edze R. Westra, Rik J. Slijkhuis, Ambrosius P. Snijders, Mark J. Dickman, Kira S. Makarova, Eugene V. Koonin, John van der Oost, David A. Wah, Jurate Bitinaite, Ira Schildkraut, and Aneel K. Aggarwal, Matthew H. Porteus, Hélène Deveau, Rodolphe Barrangou, Josaine E. Garneau, Jessica Labonté, Christophe Fremaux, Patrick Boyaval, Dennis A. Romero, Philippe Horvath, Sylvain Moineau, Luciano A. Marraffini, Erik Sontheimer, Elitza Deltcheva, Krzysztof Chylinski, Cynthia M. Sharma, Karine Gonzales, Yanjie Chao, Zaid A. Pirzada, Maria R. Eckert, Jörg Vogel, Emmanuelle Charpentier, Martin Jinek, Ines Fonfara, Michael Hauer, Giedrius Gasiunas, Prashant Mali, Luhan Yang, Kevin M. Esvelt, John Aach, Marc Guell, James E. DiCarlo, Julie E. Norville, George M. Church, Le Cong, F. Ann Ran, David Cox, Shuailiang Lin, Robert Barretto, Naomi Habib, Patrick D. Hsu, Xuebing Wu, Wenyan Jiang, Feng Zhang, Jon Cohen, Dennis Normile, Puping Liang, Yanwen Xu, Xiya Zhang, Chenhui Ding, Rui Huang, Zhen Zhang, Jie Lv, Heidi Ledford, Fuguo Jiang, Kaihong Zhou, Linlin Ma, Saskia Gressel, Osagie K. Obasagie, Katie Moisse, Jacob Brogan, Judith Daar, Bethany Johnson, Margaret M. Quinlan, Paul Berg, Francisco Juan Martinez Mojica, César Diez-Villaseñor, Jesús Garcia-Martínez, Elena Soria, Alexander Bolotin, Benoit Quinquis, Alexei Sorokin, S. Dusko Ehrlich, Christine Pourcel, Grégory, Gilles Vergnaud, Cesar Diez-Villaseñor, Elena Soria, Guadalupe Juez, Jason Carte, Ruiying Wang, Hong Li, Rebecca M. Terns, Michael P. Terns, Sonia M. Suter, David Robson, Jun Wu, Aida Platero-Lugengo, Masahiro Sakurai, Atsushi Sugawara, Maria Antonia Gil, Takayoshi Yamauchi, Keiichiro Suzuki, Juan Carlos Izpisua Belmonte, Sara Reardon, Philip Cohen, Rick Weiss, Patrick Dixon, Ian Wilmut, D. Eugene Redmond Jr., J.D. Elsworth, R. H. Roth, C. Leranth, T.J. Collier, B. Blanchard, K.B. Bjugstad, R.J. Samulski, P. Aebischer, J.R. Sladek, Dustin Wakeman, Hemraj R. Dodiya, Jeffrey H. Kordower, Susan Scutti, David Robson, Kelly Service, Michael Bess, Hamayoun Vaziri, Samuel Benchimol, Juilius Halaschek-Weiner, Jaswinder S. Khattra, Sheldon

McKay, Anatoli Pouzyrev, Jeff M. Stott, George S. Yang, Robert A. Holt, Steven J. M. Jones, Marco A. Marra, Angela R. Brook-Wilson, Donald L. Riddle, Antonio Regalado, Henk van den Belt, Rainer Breitling, Eriko Takano, Timothy S. Gardner, Jon Turney, Clyde A. Hutchison III, Ray-Yuan Chuang, Vladimir N. Noskov, Nacyra Assad-Garcia, Thomas J. Deerinck, Mark H. Ellisman, John Gill, James R. Clapper, Jeronimo Cello, Aniko V. Paul, Eckard Wimmer, Ariella M. Rosengard, Yu Liu, Zhiping, Robert Jimenez, Martin Enserink, Marc Lipsitch, Joshua B. Plotkin, Lone Simonsen, Barry Bloom, Masaki Imai, Tokiko Watanabe, Masato Hatta, Subash C. Das, Makoto Ozawa, Kyoko Shinya, Gongxun Zhong, Sander Herfst, Eefje J. Schrauwen, Martin Linster, Salin Chutinimitkul, Emmie de Wit, Vincent J. Munster, Erin M. Sorrell, David Malakoff, Jocelyn Kaiser, Ryan S. Noyce, Seth Lederman, David H. Evans, Kai Kupferschmidt, Verena Stolcke, Philomena Essed, Gabrielle Schwad, Scott Gottlieb, Anna Abram, Sarah Taddeo, Jason S. Robert, William H. Velander, Jocelyn Kaiser, Kathleen M. Vogel and Students From PLCY 306, Woo suk Hwang, Young June Ryu, Jong Hyuk Park, Eul Soon Park, Eu Gene Lee, Ja Min Koo, Hyun Yong Jeon, Sung Il Roh, Byeong Chun Lee, Sung Keun Kang, Dae Kee Kwon, Sue Kun, Sun Jong Kim, I. Liebaers, S. Demyttere, W. Verpoest, M. De Rycke, C. Staessen, K. Sermon, P. Devroey, P. Haentjens, M. Bonduelle, Ettore Caroppo, Luis Lugo, Maggie Fox, Yuyu Niu, Bin Shen, Yiaqiang Cui, Yongchang Chen, Jianying Wang, Lei Wang, Yu Kang, Pablo Tebas, David Stein, Winson W. Tang, Ian Frank, Shelley Q. Wang, Gary Lee, S. Kaye Spratt, Pasqualino Loi, Grazyna Ptak, Barbara Barboni, Josef Fulka Jr., Pietro Cappai, Michael Clinton, Deshun Shi, Fenghua Lu, Yingming Wei, Kuiqing Cui, Sufang Yang, Jingwei Wei, Qingyou Liu, Ziyi Li, Xingshen Sun, Juan Chen, Xiaoming Liu, Samantha M. Wisely, Qi Zhou, Jean-Paul Renard, Martha C. Gómez, C. Earle Pope, Angelica Giraldo, Leslie A. Lyons, Rebecca F. Harris, Amy L. King, Alex Cole, Min Kyu Kim, Goo Jang, Hyun Ju Oh, Fibrianto Yuda, Hye Jin Kim, Woo Suk Hwang, Mohammad Shamin Hossein, Gretchen Vogel, Takanori Takebe, Keisuke Sekine, Masahiro Enomura, Hiroyuki Koike, Masaki Kimura, Takumori Ogaeri, Ran-Ran Zhang, Tea S. Park, Imran Bhutto, Ludovic Zimmerlin, Jeffrey S. Huo, Pratik Nagaria, Diana Miller, Abdul Jalil Rufaihah, M. Grskovic, A. Javaherian, B. Strulovici, G. Q. Daley, Junying Yu, Maxim A. Vodyanik,

Kim Smuga-Otto, Jessica Antosiewicz-Bourget, Jennifer L. Frane, Shulan Tian, Jeff Nie, Ivan Guiterrez-Aranda, Veronica Ramos-Mejia, Clara Bueno, Martin Munoz-Lopez, Pedro J. Real, Angela Mácia, Laura Sanchez, I. H. Park, N. Arora, H. Huo, N. Maherali, T. Ahfeldt, A. Shimamura, M. W. Lensch, C. Cowan, K. Hochedlinger, G. Q. Daley, Satoshi Nori, Yohei Okada, Akimasa Yasuda, Osahiko Tsuji, Yuichiro Takahashi, Yoshiomi Kobayashi, Kanehiro Fujiyoshi, Nikhil Swaminathan, R. M. Mario, Katerina Strati, Han Li, Matilde Murga, Raquel Blanco, Sagrario Ortega, Oscar Fernandez-Capetillo, Manuel Serrano, Maria A. Blasco, Andre Nel, Tian Xia, Lutz Mädler, Ning Li, Eugenia Valsami-Jones and Iseult Lynch, Rudy Baum, Julian Savulescu, Jonathan Pugh, Thomas Douglas, Christopher Gyngell, Richard Taite, Matthew Mientka, David Haig, Eva Jablonka, Marion J. Lamb, Eugene V. Koonin, Yuri I. Wolf, Madeleine J. Van Oppen, James K. Oliver, Hollie M. Putnam, Ruth D. Gates, Kevin M. Esvelt, Andrea L. Smidler, Flaminia Catteruccia, John Harris, Ronald M. Dworkin, Lee M. Silver, David Cyranoski, Michael Specter, Andrew Hammond, Roberto Galizi, Kyros Kyrou, Alekos Simoni, Carla Siniscalchi, Dimitris Katsanos, Matthew Gribble, Dean Baker, Eric Marois, Nigel Williams, Colin Russell, Judith M. Fonville, André E. X. Brown, David F. Burke, David L. Smith, Sarah L. James, Katherine Harmon, Jon Cohen, Heidi Ledford, Federal Drug Administration; U.S. National Academy of Sciences, Engineering, and Medicine; The Royal Society & The Royal Academy of Engineering; National Research Council; Fertilisation and Embryology Authority; International Union for Conservation of Nature, and Humanity+.

Second, several books (in addition to Kevles' *In the Name of Eugenics*) guided me in articulating the issues addressed in this book. They are Jennifer Doudna and Samuel Sternberg's *A Crack in Creation: Gene Editing and the Unthinkable Power to Control Evolution*; Jim Kozubek's *Modern Prometheus: Editing the Human Genome with CRISPR-CAS9*; Judith Daar's *The New Eugenics: Selective Breeding in an Era of Reproductive Technologies*; Paul Knoepfler's *GMO Sapiens: The Life-Changing Science of Designer Babies*; Michael Sandel's *The Case Against Perfection: Ethics in the Age of Genetic Engineering*; Michael Bess' *Our Grandchildren Redesigned: Life in the Bioengineered Society of the Near Future*; Howard Jones' *Personhood Revisited: Reproductive*

*Technologies, Bioethics, Religion and the Law*; Greely's *The End of Sex and the Future of Human Reproduction*; Mara Hvistendahl's *Unnatural Selection: Choosing Boys Over Girls, and the Consequences of a World Full of Men*; Nathaniel Comfort's *The Science of Human Perfection: How Genes Became the Heart of American Medicine*; Harry Tibbals' *Medical Nanotechnology and Nanomedicine*; and Jon Turney's *Frankenstein's Footsteps: Science, Genetics, and Popular Culture*.

Finally, I would like to thank World Scientific Publishing for attention to detail in content evaluation, production, and marketing of this book.

Thank you.

# Contents

# List of Tables

# Chapter 1

# Defining Bioengineering: The New Eugenics?

> The horizons of the new eugenics are in principle boundless for we should have the potential to create new genes and new qualities yet undreamed ... Indeed, this concept marks turning point in the whole evolution of life. For the first time in all time, a living creature understands its origin and can undertake to design its future. Even in the ancient myths man was constrained by his essence. He could not rise above his nature to chart his destiny. Today we can envision that chance — and its dark companion of awesome choice and responsibility.
>
> — Robert Sinsheimer, molecular biologist and chancellor of the University of California, Santa Cruz, 1969[1]

In 1985, Kevles quoted the 1969 ruminations of distinguished molecular biologist Robert Sinsheimer (1920–2017) on trends in molecular biology:

> A new eugenics has arisen, based upon the dramatic increase in our understanding of the biochemistry of heredity and our comprehension of the craft and means of evolution. ... The old eugenics would have required a continual selection for breeding of the fit, and a culling of the unfit. The new eugenics would permit in principle the conversion of all of the unfit to the highest genetic level. The old eugenics was limited to a numerical enhancement of the best of our existing gene pool.[2]

Bioengineering — the application of molecular principles to biological organisms — has revolutionized biology and medicine. And the techniques in biotechnology have, for better or for worse, advanced exponentially. Beginning in the mid-1970s, the ability to isolate DNA was made possible by the discovery and use of plasmids — small, circular pieces of extrachromosomal DNA found in bacterial cells.[3] Plasmids enabled the replication of the modified DNA or recombinant DNA when it was introduced into the cell. From the mid-1990s to the start of the 21st century, biologists managed to sequence the complete genomes of first *Haemophilus influenzae*, a bacterial species, and then *Homo sapiens*.[4] Techniques for creating transgenic animals and plants, gene therapy, producing DNA vaccines, cloning mammals, isolating ES cells, and genetic editing all emerged within the last 20 years.

How should bioengineering be defined? In this book, it is defined as applying the following techniques to improve human health, ecology, and safety: manipulation of DNA (including animals and plants) for health and enhancement by means of gene therapy; induced pluripotent stem cells (iPSCs), cloning, xenobiology, in vivo and ex vivo genetic editing to cure genetic illnesses, genetic editing applied to assisted reproductive technologies (ART), such as in vitro fertilization (IVF) and preimplantation genetic diagnosis (PGD); and genetic editing used to increase or drive new genes. All these biotechnologies presently being used to understand DNA and the complexity of human heredity are the unintended consequences of the old eugenics, which was obsessed with preventing certain races, ethnicities, classes, and individuals with visible deformities from producing offspring; the old eugenicists did not fully understand genetics and the complexity of human variation.

Interestingly, another definition for bioengineering could be "the application of science, technology, and engineering to facilitate and accelerate the design, manufacture and/or modification of genetic materials in living organisms."[5] This definition of synthetic biology, which overlaps to some extent to the definition of the bioengineering in this book, came from a joint panel of the European Commission Scientific Committees that was under pressure to issue a draft opinion on whether existing risk-assessment methods were adequate for synthetic biology. Bioengineering also involves modifying nonhumans to benefit *Homo sapiens*. This is why the word *ecology* (which includes nonhuman biological organisms) appears in my definition. Nanobiotechnology and nanomedicine are also relevant because of the application of nanomaterials in drug delivery, tissue regeneration, fighting bacteria, and prosthetics.[6] The danger of the convergence of biotechnology and nanobiotechnology compels discussion of the latter.

As mentioned in the preface, gene therapy, synthetic biology, and genetic editing by either suppression of a mutant gene or addition of a critical gene could not be done if DNA was not practically universal. At a macro level, for example, mammalian embryogenesis has many similarities to embryogenesis in reptiles and birds.[7] One such similarity is the development stage called gastrulation that occurs in mammals and birds, during which the flattened collection of cells in the embryo move and form three germ layers: ectoderm, mesoderm, and endoderm.[8] Another example comes from plants — a respectable number of non-plant genes in plant genomes (genes not involved in photosynthesis and photosynthetic anatomy) are also found in animal and fungal genomes.[9] These genes are involved in general intermediary metabolism repair, genome replication and repair, and transcription and protein synthesis. This is why biopharming (the medicinal use of plants) has become lucrative for pharmaceutical companies. For example, cancer chemotherapeutic agents such as taxol, vinblastine, and vincristine were all isolated from plant sources.[10] Another applied example is that of the human protein serum albumin, which was produced in genetically engineered tobacco and potato plants.

In general, crops are being engineered to resist disease, insects, and herbicides, and domesticated animals are being engineered for economically desirable traits.[11] This is big business for biotechnology (and pharmaceutical) companies. Commercially important crops such as corn, cotton, soybeans, and canola have been genetically modified (had foreign genes added) for resistance to the herbicide glyphosate as well as resistance to insects via expression of the Bt toxin.[12] These crops have also been genetically edited (had existing genes altered) to repair or knock out functions. Soybeans, for example, had genes altered to produce fewer fatty acids. These crops are sold and consumed in nine countries, with a high proportion of one or more of these crops being genetically engineered in each country.[13] Concerns include dietary safety and the risk of the introduced genes merging with genes of wild species, leading to loss of species diversity or other unknown consequences. In one experiment, a cyclin-dependent kinase (Cdk) inhibitor gene was inserted into *Arabidopsis thaliana* (wall cress) plants, resulting in reduced cell division in leaf meristems that led to significant changes in leaf size and shape.[14]

Transgenic animals are designed to be larger or smaller, to have different hair colors, to be virus resistant, to lack horns, or to excrete less phosphate (an environmental pollutant). As with genetically engineered crops, there is concern that introduced genes could merge with the genes of wild species, spread by transformation or introgression, and lead to a reduction of biodiversity.[15] Genetic editing could cause unintended mutations or the introduction of foreign DNA into the genomes of plants and animals.

Free market capitalism is one accelerator of the new eugenics. Pharmaceutical and biotechnology companies motivated by profits exploit the human desire to be mentally and physically better than nature allows. This desire has transformed into a movement known as transhumanism that advocates for human genetic enhancement to improve *Homo sapiens*.[16] This movement indirectly provides free advertising for biotechnology companies offering nontherapeutic genetic enhancements. There is, in all of this, a faint imprint of the old eugenics. Suppose, for example, that genes related to memory or human growth hormones (and thus height) were discovered; biotechnology companies would be working overtime to develop these drugs for patients suffering from memory loss and muscle deficiency, but they would also offer the drugs to the general public — most likely to the people who could afford them. Another example of a medical product that is marketable is IVF, which is very expensive and can be restricted by class, ethnicity, and education. Ross Parke, emeritus professor of psychology at the University of California, Riverside, and colleagues stated that at the end of a successful in vitro fertilization treatment, the overall cost can be US$55,000 for the first cycle to US$73,000 by the sixth cycle.[17] And the price is even higher for twins and triplets. They wrote that "many insurers and public health systems do not cover or only partially cover these procedures, making IVF and related therapies available only to those with the ability to pay."[18] It follows that the Americans who are able to use these services, based on the national history of race and socioeconomic status, would be white, college-educated women with above-average incomes. The downsides of in vitro fertilization that are (obviously) missing from the marketing literature are the possibility of spontaneous abortion, physical or mental defects if the baby survives birth, and more importantly, reduction in diversity.[19]

## 1.1. Genetic Engineering: Treating Illnesses versus Personal Enhancement

During the Pew Research Center and Brookings Institute cosponsored event The Pursuit of Perfection: A Conversation on the Ethics of Genetic Engineering, held in March 2004, three major themes stood out: (1) the ethic of giftedness (natural genetic inheritance or abilities) and social solidarity, (2) therapy versus enhancement in the social context, and (3) how and where to draw the line in genetic engineering.[20] Michael Sandel, one of the guest speakers at this event, made some thought-provoking statements based on the above themes. On the theme of giftedness, Sandel used features such as muscles, height, weight, intelligence, memory, and selected behaviors to support his argument. Muscles, he stated, were needed in a highly physical sport such as American football, and some football players had the natural genetics for greater muscle growth compared to others.[21] This difference, Sandel continued, is not viewed as unfair by the National Collegiate Athletic Association or National Football League. He posed a question that everyone must have thought about from time to time: What if all of the small-bodied players injected a new, "safe," synthetic muscle-growth gene into their muscles?[22] Would this not be leveling the playing field, so to speak?

This is not leveling the playing field. It is simply elective genetic enhancement for recreational purposes rather than to alleviate a medical illness. Additionally, this idea of giftedness is relative in Darwinian evolution because what might be a gift (adaptive) in one environment might not be in another. The quotes around the word *safe* above are to emphasize that, in general, the long-term effects of inserted genes on the human body are unknown; they may result in a desired function but could also trigger multiple unknown detrimental effects, such as cancer due to excessive cell division. George M. Church, professor of genetics at Harvard Medical School and the Massachusetts Institute of Technology, argued that there is a lower risk of off-target mutations editing a single sperm or egg cell, saying that "fewer mutations or risk are actually incurred when editing the germline than in somatic cells, since it involves fewer cells by orders of magnitude."[23]

If a memory-enhancing drug were developed to help Alzheimer's patients, and trials of the drug were a total success, giving participants total memory recall plus additional skills they lacked before, would it be fair to offer this drug to the public? More than likely, the drug would be very expensive, only a small portion of the population could afford it, and they would probably stockpile it. So, if similar memory-enhancing genes could be inserted into human embryos, should it be done?

Sandel would agree that, except for the original development of the memory-enhancing drug to treat Alzheimer's, both the muscle-growth and memory-enhancement scenarios are examples of genetic enhancement reminiscent of eugenics. What if, posed Sandel, everyone had equal access to the memory-enhancing drug? Would the scenarios be objectionable? Using the latter scenario above, one could replace the memory-enhancing drug with a human growth hormone for height enhancement. Human growth hormone has in fact been approved by the Food and Drug Administration (FDA) for children of very short stature since the 1980s.[24] Again, would offering the hormone to children of all statures be objectionable? Both scenarios are objectionable because, again, they are reminiscent of eugenics. Sandel believed that as the genetically enhanced people became taller, the unenhanced people would fall greatly below the new mean for stature.[25] Not wanting to be left behind, the unenhanced would seek hormonal treatment. The long-term evolutionary effects within and between populations for inserting memory-enhancing or height-enhancing genes in germ line cells are unknown, but it is possible to speculate that the long-term evolutionary effects for height enhancement would be an artificial directional selection toward tall stature that would remove normal variation. In most human populations, excluding environments of starvation or high altitude, there is a higher frequency of the average height because it is adaptive. In other words, there is a higher survival rate among people having average stature (in the United States, the average height is 5 feet 9 inches for males and 5 feet 4 inches for females).[26] The people at the extremes of the normal distribution of height — very tall or very short — make up the smallest percentage, because these heights are nonadaptive. In short, science and technology may not always be a good thing.

Moreover, these examples are instructive because they present two uses for genetic engineering: (1) genetic enhancements for health purposes and (2) elective genetic enhancements to satisfy personal desires. This leads back to the question posed earlier in this chapter: where should the line be drawn? How far do we go before we see shadows of the old eugenics? Sandel stated that "the deeper danger is that genetic enhancement and bioengineering represent a kind of hyper-agency, a kind of Promethean aspiration to remake nature, including human nature, to serve our purposes and to satisfy our desires. ... The problem is drive to mastery."[27]

In essence, Sandel believed that the ethic of giftedness is under siege because human beings are not satisfied with their natural genetic inheritance; they want to be taller or resemble their favorite Hollywood star.[28] This desire is fueled by scientists (some transhumanists) in collaboration with pharmaceutical companies, biotechnology companies, universities, and private research organizations whose research into biological modification, although initiated to seek helpful new knowledge, is mongrelized by free market genetic engineering, prestige, and lucrative patents. These benefits might be just as dangerous as the scientists' experiments in the use of biotechnology. These personal desires are further motivated by the fantasies embedded into reality in films and mass media. A perfect example is Captain America, because this fictional comic book character is an enhanced (via growth hormones) human who is physically strong, honorable, and fighting the bad guys. We accept, albeit fictionally, the enhanced human Captain America as a testament to scientific advancement. But we reject muscle-enhanced football players. It is indeed within the murkiness of social context that drawing lines on how to use genetic engineering becomes difficult. Related to this is the eternal choice between public safety and scientific advancement in the modern age of science. Or, do these choices form a false dichotomy? The federal government would answer yes! The case of nanotechnology is instructive here. In 2008, the goals of the US National Nanotechnology Initiative (NNI), which is managed by the executive branch's Office of Science and Technology Policy, were noted in a review for National Research Council (NRC) of the National Academy of Sciences:

> Because the NNI is responsible for ensuring US competitiveness through the development of a robust research and development program and ensuring the safe development of nanotechnology, it may be perceived as having a conflict of interest. . . . *The committee concludes that the conflict constitutes a false dichotomy* and that strategic research on potential risks posed by nanotechnology can be an integral and fundamental part of the sustainable development of nanotechnology.[29] (italics added)

If the reader scoured federal government policy documents on most research, she would find that this review expressed the general sentiment for most emerging technologies in the United States.

Furthermore, the randomness of meiosis, Mendelian inheritance, and complex traits motivates parents to use genetic engineering to eliminate randomness and "design" their children's height, eye color, hair color, athletic prowess, musical ability, and intelligence.[30] Sandel had deep moral objections to these designing parents:

> This helps us to see that the deepest moral objection to enhancement or to the pursuit of designer children lies not in the perfection that it seeks, but in the human disposition that it expresses and promotes. The problem is not that parents usurp the autonomy of the child whose sex they choose or whose traits they design, because the child wouldn't otherwise choose her genetic traits for herself. Autonomy is not what's at issue. The problem lies in the hubris of the designing parents in their drive to master the mystery of birth.[31]

While some of the qualities parents want in their children are still fantasy, such as intelligence or musical ability, it seems that the specter of the old eugenics has merged with the new eugenics.

Some may ask what is wrong with parents trying to do the best for their children. Most parents want their children to be healthy and intelligent; they want their children to surpass them in health, intelligence, and socioeconomic status. Consequently, they spend much time and money getting their children into the best schools and on preparatory courses before their children take standardized tests for college admission. If other parents, who also wanted the best for their children, used genetic technologies to enhance their children's intelligence or athletic ability, why would this be objectionable? It is objectionable because the former is normal social-cultural enhancement based on socioeconomic status. The latter is genetic engineering (new eugenics), of which a small percentage of the population will reap the benefits (old eugenics). Sandel stated that "the shadow of eugenics hangs over today's debates about genetic engineering and enhancement. Critics say what we're witnessing with enhancement is nothing more than privatized or free market eugenics. Defenders of enhancement say, No, no, it's not eugenics, provided there is no state imposition; provided there is no coercion."[32] But in the United Kingdom, which historically did not have coerced or state-imposed sterilization laws, voluntary sterilization still fell under the banner of eugenics.[33] In the early 20th century, for instance, British eugenicists (radicals and mainliners) wanted their race-improvement programs — such as sterilization to prevent the reproduction of the "unfit" — to be voluntary and not imposed compulsorily by the state.[34] They did not want to be viewed as tyrannical; instead, they wanted to emphasize education, self-control, contraception, civil duty, social inducements, and responsibility to the race. Additionally, volunteerism was deemed important because geneticists in the movement were prescient in realizing the harmful evolutionary changes that could occur if they advanced too aggressively in ignorance of the laws of heredity in humans. This caution appears to be lacking at present in many fields of science worldwide. Nevertheless, free market genetic engineering is dangerous for the simple reason that the procedures will not really be free, because most people will not be able to afford them, and, more importantly, ignorance of the long-term evolutionary effects of genetic enhancement is an ongoing problem. In the short-term, there may be social conflict between the genetically modified humans (the designer babies) and "natural" humans. The group having no economic or technological advantage will be viewed as "the other."

Parents make difficult choices every day. Lee Silver, a Princeton University molecular biologist and advocate for designer babies attending the Pursuit of Perfection forum discussed a 2002 *Harper's* magazine index of questions concerning genes. He highlighted one of the questions that emphasized choices parents have to make in the age of biotechnology: "If you could give your child enhanced genes that would increase his resistance to disease, would you do it?"[35] Seventy-five percent of the respondents said yes. Silver stated that nature is mean and nasty, using an example implying the wastefulness of a turtle laying 1,000 eggs when only a single survivor would likely reach the water.[36] Many similar examples are found in nature. In evolutionary biology, words like *mean* and *nasty* or implications of wastefulness are subjective; the environment is always changing, and what is mean and nasty for one organism might be good and nice for another. As Silver intimated, leveling the playing field upsets the balance in nature.[37] Genetic engineering, although used with good intentions in many cases, upsets the balance of nature in unknown ways.

Sandel presented another hypothetical case where a gene for homosexuality is discovered and a pregnant woman finds out that her fetus carries the gene.[38] Should the woman be free to abort the fetus? (While the abortion question is beyond the scope of this book, the woman should be advised to discuss her options with her immediate family.) What if the homosexuality gene could be blocked or suppressed through gene therapy (see Chapter 2)? Should the woman (if she can afford it) have the procedure? Although the woman has the freedom to have this procedure, she should not, because of the dangers of turning on other genes with negative consequences. Also, as humans, we are all connected; we live in a social environment in which genome manipulation impacts the entire population for better and for worse. Sandel wrote:

> But what happens when genetic engineering enables us to override the results of the natural lottery, to replace chance with choice? The gifted character of human powers and achievements would recede and with it, perhaps, our capacity to see ourselves as sharing a common fate. The successful, thinking themselves wholly self-made men and women, would become even more likely than now to view themselves as self-sufficient and, hence, wholly responsible for their success.[39]

The knowledge and techniques used today in the new eugenics have been accumulated over time through hard work by many brilliant

**Table 1.1.** Selected timeline of genetic engineering

| Year | Biotechnology |
|---|---|
| 1971 | Ananda M. Chakrabarty GMO microbe patent filing |
| 1972 | Paul Berg, Stanley N. Cohen, and Herbert Boyer Recombinant DNA |
| 1974 | Rudolf Jaenisch makes first GMO animal |
| 1975 | Asilomar Conference on Recombinant DNA |
| 1976 | Genentech |
| 1977 | DNA sequencing and first human IVF |
| 1980 | Ananda M. Chakrabarty patent granted |
| 1981 | Calgene is born — Mouse embryonic stem cells (ESC) |
| 1982 | FDA approves insulin from microbes |
| 1987 | 'Ice-minus' microbe; first GMO released in the wild |
| 1989 | First human preimplantation genetic diagnosis (PGD) |
| 1990 | Human genome project begins; first gene therapy in the U.S. occurs |
| 1996 | First cloned mammal: Dolly the sheep |
| 1997 | Monsanto buys Calgene |
| 1998 | Human ESC generated |
| 1999 | Jesse Gelsinger dies from gene therapy |
| 2001 | Stanley N. Cohen first germline human GM paper; FDA stops ooplasm transfer |
| 2003 | First GMO pet |
| 2006 | Shinya Yamanaka induced pluripotent stem cells (iPSCs) |
| 2010 | First synthetic bacterial genome |
| 2012 | Clustered Regularly Interspaced Short Palindromic Repeats (CRISPR)-Cas9 works as gene editing tool |
| 2013 | Shoukhrat Mitalipov human Somatic Cell Nuclear Transfer (SCNT) |
| 2014 | GM monkeys via CRISPR; UK approves three-parent IVF |
| 2015 | First International Summit on Genome Editing: Calls for a moratorium on germline GM; first CRISPR GM human embryos |
| 2019 | Second International Summit on Genome Editing: Calls for a moratorium on germline GM; first CRISPR GM human embryos |

*Source*: Adapted from Knoepfler 2016.[44]

scientists (working generally without malice) motivated by the possibility of scientific discovery and prestige (e.g., scientific prizes and patents) while simultaneously helping humankind (see Table 1.1). Helping humankind or, put another way, saving humankind — public safety — is quite honorable. For example, CRISPR/Cas9 — discussed in the preface and in great detail in Chapters 5 and 6 — can cause new genes and their related traits to spread or drive through wild populations faster than normal sexual reproduction. This has been effective in stopping the spread of the lethal malaria parasite *Plasmodium falciparum*, which infects and kills millions of humans in the tropical regions of Africa, the Middle East, Near East, and South Asia.[40] The new sequence inserted by CRISPR/Cas9 after the DNA is cut has the genetic information that encodes CRISPR itself. Consequently, the CRISPR gene drive will copy itself into new chromosomes via meiosis, "propelling a respective suite of genes throughout a fast-reproducing species over many generations."[41] The drive was very successful in spreading a gene that gave mosquito descendants resistance to the malaria parasite, hence preventing the spread of this parasite and saving lives. Other gene drives were created to spread genes for sterility in the mosquito population. One scientist remarked that "if sustained in wild-mosquito populations, it could eventually lead to outright extermination of an entire mosquito species."[42] If one removed *mosquito* and *species* from that sentence and replaced them with *humans*, then the quote would read like old eugenics. Nevertheless, these bioengineers demonstrate little or no interest in causing the extinction of a biological life-form, though there are many bad actors out there who could weaponize gene drives to cause a pandemic or target a country's food crops.

New scientific discoveries advance civilization for the better and for the worse, because scientific advancement and public safety are inextricably linked. In this dangerous era of light-speed scientific advancement, especially in biotechnology, can humankind endure the long-term cost in detrimental changes to human and nonhuman life-forms? Fortunately, international summits on genetic engineering in general and gene editing in particular sponsored by the US National Academy of Sciences, the US National Academy of Medicine, the Royal Society, the Chinese Academy of Sciences, and the Academy of Sciences of Hong Kong are addressing this concern and several others discussed in this book (see Chapter 10).[43]

## 1.2. Emergence of the New Eugenics

When did the old eugenics transform into the new eugenics? It seems that there was a subtle shift in meaning from promoting the genetics of one social or racial group to using biotechnology to promote the genetics and health of all people in the population regardless of race, social status, or health (see Table 1.1).[44] According to Kevles, the beginnings of the new eugenics after World War II can be traced to two highly personal events in the lives of American and British families: the rise of genetic counseling and advances in reproductive medicine, such as artificial insemination, in vitro fertilization, amniocentesis, and early ultrasound fetal scans.[45] This is reasonable, because these two large medical phenomena and their accompanying technologies carried the faint stench of the old eugenics. In addition, however, the new eugenics began with the successful cloning of frogs by Briggs and King in 1952 (which Kevles mentioned in his brief survey of the biotechnological advances from the 1960s to the late 1970s), because the definition used in this book focuses heavily on the biotechnical aspects of genetics.[46] Collectively, there was an obsession among North American and western European scientists (and the lay public in these regions) during the first half of the 20th century to eliminate hereditary disease from the human genome.

Kevles wrote that the growth in genetic counseling clinics stemmed from people worried about health and family history or, more importantly, children being born with a deficiency or suspected heritable disease.[47] He also noted that by the 1950s, simple biochemical tests were available, so that counselors could test a couple for Mendelian disorders (simple inheritance of discrete traits only) and then inform the couple regarding which of them carried a deleterious recessive gene.[48] Additionally, counselors could give information on genotype frequencies for the next generation. In other words, counselors could tell couples the degree of likelihood that their future offspring would inherit this deleterious recessive gene. The only way a recessive trait is expressed, according to Mendelian inheritance, is that the gene coding for the respective trait must be passed to the offspring by both parents. It follows that both parents would be carriers of the recessive gene; in the terms of population genetics, they would be heterozygote dominants. In this situation, there is a 25 percent chance that offspring will be homozygous recessive, meaning that they inherit the disease (in the case of sickle cell anemia, the universal abbreviation is $Hb^s$ $Hb^s$ –"$Hb^s$" denoting a mutation in hemoglobin chain). With both parents as heterozygote dominants, there is also a 50 percent chance that the offspring will be heterozygote dominant and a 25 percent chance that the offspring will be homozygous dominant and carry no recessive genes.

The preoccupation with heredity in the first half of the 20th century thrust several genetic disorders to the forefront. Some of them had a racial component, which stoked the dying embers of mainline eugenics. Three of the Mendelian diseases that Kevles discussed were PKU (Phenylketonuria), Tay-Sachs disease, and sickle cell anemia.[49] Kevles wrote that by the late 1960s to early 1970s, pediatricians had come to realize that human genetics was a useful tool in testing for a disease or deficiency so that treatment could be started immediately if the results were positive. First, PKU was aggressively studied in the 1960s because it accounted for one out of eight infant deaths and, more importantly, because it was indirectly linked to one of the top defective conditions mainline eugenicists wanted to remove from the gene pool, mental retardation.[50] For instance, Diane B. Paul, professor of political science at the University of Massachusetts in Boston, writing in 1997, quoted an unnamed respondent writing in 1961 for the magazine *Family Weekly*: "PKU strikes only one child in 20,000. But circumventing this disease has opened a way toward eradicating the blight of mental retardation which, in the United States alone, afflicts 5,500,000 persons."[51] According to Paul, in the past, PKU resulted in severe mental retardation, and approximately 90 percent of those afflicted had IQs of less than 50. PKU is caused by a deficiency in a liver enzyme that normally catalyzes the conversion of phenylalanine to tyrosine; the deficiency results in the accumulation of toxic levels of phenylalanine in the blood.[52] Today, PKU can be prevented by screening newborns and special diets from which most phenylalanine has been removed.

As early as the 1930s, noted Paul, American biochemists George Jervis and Richard Block, along with British geneticists and statisticians, proposed treating affected infants with low-phenylalanine diet, but the high cost of producing synthetic food prevented states from pursuing these proposals.[53] Nonetheless, a phenylalanine-restricted diet was tested on small children in Britain and the United States (a retrospective statistical study assessing the benefits of dietary therapy would not occur until 1960) with promising results coupled with questionable claims of markedly increased IQs of severely retarded children.[54] Paul noted that these claims were wrong, and it became increasingly clear to serious researchers that any mental defects were irreversible. Researchers in the 1950s believed that cognitive improvement would be more successful if they could start diet therapy within the first few months of life. This led to identification of newborns with PKU via population-wide screening.

In 1960, microbiologist Robert Guthrie in the School of Medicine at the State University of New York at Buffalo, New York, (whose son and niece were both mentally ill and allegedly diagnosed with PKU) developed an inexpensive, sensitive, and simple bacterial inhibition assay using a strain of *Bacillus subtilis* that would grow in a cultured sample of blood only if the sample was abnormally rich in phenylalanine.[55] The test could be administered a few days after birth. Guthrie publicized his unpublished findings with encouragement from the executive director of the National Association of Retarded Children but without vetting from the scientific community (although a peer-reviewed report appeared in 1963).

In late 1961, field trials began involving 400,000 infants in 29 states to assess the assay's suitability for a national screening program.[56] Between 1966 and 1974, according to Kevles, a screening program in New York City identified 51 PKU infants.[57] Subsequently, the Guthrie test was found to be cost-effective (approximately $1 million spent on the city's screening program compared to the projected cost of $13 million to keep the identified children in institutions).[58] The effect of all of this was rethinking of the traditional methods used to help mentally ill patients (i.e., educational, social, and rehabilitative). The focus at this time was on preventive measures by screening all American infants — the new eugenics. By 1975, according to Paul, 43 states had enacted mandatory screening, and 90 percent of all newborns were tested.[59] Today, every American state mandates screening of newborns for PKU, congenital hypothyroidism, galactosemia, and other metabolic disorders.[60]

The motivation for the new screening program was to prevent and reduce the scourge of the old eugenics and help the afflicted (as opposed to preventing the reproduction of the afflicted). But the Guthrie test and the intentions of the people conducting the screening took a slightly ominous turn (ominous turns in science are a past and present danger and a great concern in this era of biotechnology). Scientists and health workers began screening groups that were suspected to have high risk for particular genetic disorders. In their enthusiasm, they did not realize the impact they were having on the greater culture. What should have given them a clue, especially in the shadow of the old eugenics, was that individuals in these groups most shared socioeconomic status, ethnicity, or race. Publicly acknowledging high-risk groups for a disease might leave a stamp of inferiority on the groups, resulting in discrimination in the social environment. But this is the nature of science's unintended consequences: enthusiastic, well-intentioned scientists narrowly focused on trying to be the first to discover a technique or cure to save the world are often totally oblivious to how their science impacts the social and natural environment. This is a worrying but subtle danger of scientific advancement.

As the offensive on PKU was well underway, attention turned to sickle cell anemia and Tay-Sachs disease. Sickle cell anemia occurs when the DNA nucleotide triplet (codon) coding for the amino acid glutamic acid (C*T*T) randomly changes to one that codes for valine (C*A*T).[61] This change manifests as a point mutation in the sixth position of the beta chain in the hemoglobin molecule and results in sickle-shaped red blood cells. These cells are very different from the normal red blood cells in that they restrict normal blood flow by clogging the arteries, preventing oxygenated blood from supplying vital regions of the body. These cells have a very short life span, resulting in the reduction of red blood cells — hence the term sickle cell *anemia*. As mentioned above, implementation in the 1960s and 1970s of mandatory genetic disease screening laws, particularly those for sickle cell anemia, brought back memories of mainline eugenics. It is difficult to understand why the federal government, public health agencies, and African American activists (particularly) chose to focus on screening for sickle cell anemia — a disease with no curative treatment (unlike PKU) that affected one out of every 600 (or 0.2 percent) African Americans at the time — over practical issues such as better jobs, equal pay, better housing, and others.[62] They could have chosen treatable medical issues such as type 2 diabetes or hypertension.[63] Perhaps African American activists saw an opportunity to exploit the attention around sickle cell anemia and use it as a platform to expedite other pressing medical and socioeconomic needs.

Social forces in the African American community in the early 1970s (the civil rights movement, wider voter registration, and an increased number of African American elected officials) prompted President Richard Nixon to increase federal support for sickle cell research.[64] And the US Senate followed by establishing a national sickle cell anemia program. According to Howard Markel, professor of pediatrics and communicable diseases at the University of Michigan, Ann Arbor, 12 states and the District of Columbia enacted mandatory sickle cell laws for African Americans between 1970 and 1972, and the majority of those screened for sickle cell were school-aged children and young adults.[65] These laws, however, failed to consider that a group already stigmatized by skin color would be also stigmatized by sickle cell anemia (homozygotes and, primarily, heterozygotes, because they survive as carriers). In essence, sickle cell anemia became known among the majority of Americans as a "black" disease. This label fueled some misguided eugenic-type laws. For example, a New York state law ordered that all persons "not of the Caucasian, Indian, or Oriental races" be tested for the sickle cell gene before they were allowed to obtain a marriage license.[66] To mask the specter of race, a subsequent New York state law required all urban (not suburban or rural) schoolchildren to be screened, fully aware that most African American children in the state lived in the urban regions of New York City.[67]

The late American physical anthropologist Frank Livingstone suspected since 1958 that the sickle cell gene (Hemoglobin *S* or $Hb^S$) occurred with high frequency not only in sub-Saharan Africa (and inherited by African American descendants) but also in other tropical regions such as Greece, Turkey, South India, and Southeast Asia, because of natural selection.[68] Thousands of years of slash-and-burn agriculture have created an environment optimal for the *Anopheles* mosquitoes, which carry the malaria parasite. The malaria parasite cannot survive in a sickled red blood cell (the parasite needs a normal red blood cell to obtain specific metabolites); consequently, heterozygotes survive — contracting a mild case of malaria — and increase the frequency of the sickle cell gene. Unfortunately, the lay public, politicians, and some physicians would never have grasped this complexity. And the fact the sickle cell gene was found in other ancestral and descendant populations came as a major revelation. Sickle cell anemia being an alleged "black" disease was neat and simple for the new eugenicists. This association of race and disease was easily remembered by the American public not only because of its brevity — two words — but also because of its link to pigmentation. But the damage was already done, and the screening laws provided no privacy rights.[69] As a result, many African Americans were denied life and health insurance, employment opportunities, and acceptance into the US Air Force Academy — the rationale being in the 1960s that low oxygen saturation at high altitudes precludes African Americans with the sickle cell disease from becoming pilots.[70] The situation was not helped when, in February 1968, chemist and Nobel laureate Linus C. Pauling reflected on the new biology in a foreword for the *UCLA Law Review*: "There should be tattooed on the forehead of every young person, a symbol showing possession of the sickle cell gene [so as to prevent] two young people carrying the same seriously defective gene in single dose from falling in love with one another."[71] Unfortunately for Pauling, his writings about the new biology certainly helped to usher in the new eugenics.

Race or ethnicity was also a key factor for Tay-Sachs disease, but this time African Americans were not the focus of attention. Tay-Sachs disease occurs with high frequency (one in every 2,500 births) among Ashkenazi (eastern European) Jews.[72] It is caused by the accumulation of lipid-sugar molecules called gangliosides, which are important constituents of cell membranes.[73] Normally, gangliosides are broken down by the enzyme hexosaminidase A. Due to a random point mutation, this enzyme is not produced in individuals with Tay-Sachs disease (homozygous dominants), resulting in the accumulation of excess ganglioside in neuronal cells. While survival to adulthood is possible for individuals with sickle cell anemia, an individual afflicted with Tay-Sachs disease will not survive infancy.

Like the natural selection hypothesis for the high frequency of the sickle cell gene in regions where there is malaria, one could also speculate that the high frequency of the Tay-Sachs gene in Ashkenazi Jews may be related to this population's history of adapting in environments of deprivation and disease. According to Jared Diamond of the Department of Geography at the University of California, Los Angeles, Jews who originally settled in France and Germany during the eighth and ninth centuries were later persecuted during the period of the Crusades.[74] Fleeing eastward into Lithuania and western Russia, they remained there until the

19th century, when anti-Semitic attacks drove millions of them out of eastern Europe. European Jews were confined to crowded towns for more than 1,000 years because they were forbidden to own land or be employed in certain professions, such as farming. Tuberculosis and typhoid thrived in those crowded conditions.[75] These diseases were also prevalent in the crowded ghettos of Poland during the forced internment of European Jews by the Nazi regime during World War II. Heterozygote individuals had lower mortality rates in these environments, survived longer, and had higher fertility rates that balanced out the loss of infants to Tay-Sachs disease over time.[76]

Genetic screening for Tay-Sachs disease began, like sickle cell screening, in the late 1960s and early 1970s. However, Markel explained that the experiences with genetic screening programs for Ashkenazi Jews and African Americans were different:

> In the first place, the reproductive choices for Tay-Sachs disease were less ambiguous when compared to sickle cell anemia; prenatal diagnosis for Tay-Sachs was possible. Then, too, there were striking social differences between them. The Jewish-American community was no stranger to discrimination particularly in its relation to the application of genetic theory to social policy. Many of those enrolling in the Tay-Sachs screening programs of the 1970s were literally, the grandchildren of the East European Jewish immigrants who were stigmatized during the 1920s and accused of importing inferior genes and "protoplasm" into the United States. Yet with the passage of time and acquisition of the confidence of assimilation, Jewish-Americans of the 1970s generally expressed fewer fears of discrimination than their African-American counterparts when confronted with these new screening technologies.[77]

Nevertheless, the plan of public health officials was similar to the philosophy of mainline eugenicists of the 1920s and 1930s and was to *avoid* (the word used in the past had been *prevent*) the birth of a Tay-Sachs-afflicted (heterozygous or homozygous) fetus.

In 1971, Michael Kaback at the Johns Hopkins University School of Medicine and Robert Zeiger at the National Cancer Institute, both located in Maryland, began a pilot program of voluntary screening for Tay-Sachs gene carriers "among a total of 8,000 Ashkenazic Jews of childbearing age in the Greater Baltimore area."[78] This led to thousands of young Jewish Americans volunteering to be tested and a more accurate measurement of the frequency of the Tay-Sachs gene in the Jewish community.

In advance of implementing screening programs, meetings and discussion forums were held between physicians, ethicists, rabbis, and other members of the Jewish communities involved.[79] Markel noted that not all Jewish communities accepted the provisions of the screening programs. For example, Orthodox Hasidic Jews, with their practice of arranged marriages, strongly opposed abortion and contraception (although the matchmaker — armed with the screening test results of the prospective bride and groom — would declare a "bad match" if both were positive for the Tay-Sachs gene).[80] Despite this, the primary concern of public health officials was development of screening programs that focused on young Ashkenazi Jewish married couples who wanted to have children and then all unmarried individuals in the community who were 18 years old or older.[81]

Those individuals who tested positive for the Tay-Sachs gene were informed immediately. Unlike early sickle cell screenings, where there were no provisions for confidentiality, steps were taken to ensure it for Tay-Sachs screening. Those who tested positive for the Tay-Sachs gene, especially those couples who both tested positive, were offered genetic counseling.[82] Generally, the bride in such couples underwent prenatal testing of amniotic fluid when she became pregnant. Kevles wrote that the procedure used, known as amniocentesis, was popular in the late 1960s.[83] Prior to that time, it was used to detect the Rh-induced hemolytic disease erythroblastosis fetalis (the mother's immune system creates antibodies against fetal red blood cells when mother is Rh-negative and fetus is Rh-positive) and determine the sex of the fetus prenatally.[84] Amniocentesis involves inserting a long needle into the uterus and withdrawing fluid containing fetal cells from the amniotic sac. Using biochemical means, fetal cells are cultured to diagnose chromosomal and genetic disorders. By the mid-1970s, "virtually all of the hundred or so known chromosomal disorders could be detected in utero, and so could twenty-three inborn errors of metabolism, including the error that produced Tay-Sachs disease; almost forty more seemed potentially detectable."[85] One of these inborn errors was the high-profile chromosomal disease hemophilia, which is carried on the X chromosome and expressed in males; females are always heterozygotes.[86] If the sex of the fetus is determined to be male based on amniocentesis, the mother can abort to avoid giving birth to a hemophilic child.

Returning to the discussion of Tay-Sachs disease, Markel noted that the majority of couples who found out that the wife carried a fetus that was heterozygous or homozygous for the Tay-Sachs gene elected to terminate the pregnancy.[87] The right to an abortion was secured in 1967 in the United Kingdom and 1973 in the United States, and aspects of the old eugenics — preventing the birth of children who would have severe physical or mental abnormalities — might have had some influence on this.[88] It is ironic that the new eugenic factor in all of this was that modern technology was used to accomplish the lingering goals of the old eugenics.

Amniocentesis and the genetic-disease-screening technologies would have been a blessing for mainline eugenicists of the early 20th century. But in the late 1960s and early 1970s, these technologies would help to usher in the new eugenics and have a psychological impact on the greater culture by creating a market for prenatal genetic diagnosis. According to Kevles, approximately 5,000 prenatal diagnoses of genetic disorders occurred in the United States, and approximately 7,500 in the United Kingdom, before 1976.[89] After that date, 20,000 prenatal diagnoses had occurred in the United States, and 7,000 had occurred in the United Kingdom. The number of genetic counseling clinics increased from 40 in 1960 to approximately 400 in 1974, and this increase was accompanied by a change in philosophy in the advice they gave to patients.[90] Counselors no longer gave eugenically oriented advice, such as that Americans must reduce the frequency of "bad" genes in the US gene pool or that the individual must think of the "welfare of the gene pool rather than that of the family."[91] Instead, counselors hoped that couples who solicited advice would act responsibly and avoid the risk of giving birth to a child with a probably deleterious gene.

Another technology that transformed the old eugenics was the application of fetal ultrasound scans. Ultrasound waves are like normal sound waves, but humans cannot hear them. Animals such as bats and porpoises use ultrasound to locate prey and obstacles.[92] French and British submarines used ultrasound to navigate underwater and detect enemy submarines during World War I.[93] Later, ultrasound was applied to medicine in general and reproductive medicine in particular, because doctors could not perform x-ray examinations of pregnant women because of risk to the fetus.[94] The large, intimidating (to the patient) machines with very low resolution simultaneously fascinated and scared the 1960s American public. Writing in 2011, Mara Hvistendahl, Beijing-based correspondent for *Science,* quoted the headline on the cover of the September 10, 1965, issue of *Life* magazine, which foreshadowed some of the concerns applicable to the new eugenics: "Control of Life: Audacious Experiments Promise Decades of Added Life, Superbabies with Improved Minds and Bodies, and Even a Kind of Immortality."[95]

To safeguard precious American "protoplasm," Congress passed the National Genetic Diseases Act in early 1976.[96] This act provided for research, screening, counseling, and education in sickle cell anemia, thalassemia, Tay-Sachs disease, Down syndrome, cystic fibrosis, muscular dystrophy, galactosemia, hemophilia, and Huntington's disease among others. At this time, close to one million people had been screened for Tay-Sachs disease and hemoglobinopathies such as sickle cell anemia and thalassemia.[97] Despite the new standards of genetic counseling discussed above, the goal of health officials and scientists under the new eugenics was to use the new medical technologies to "improve" the gene pool (as opposed to the philosophy of the old eugenics, which was to prevent people with genetic disorders from having children).

One of these scientists was the late Hermann J. Muller, American geneticist, 1946 Nobel laureate in physiology or medicine, and son of German-Jewish immigrants. In the eugenics movement of the early 20th century, Muller could be considered a social radical in the tradition of the British eugenics movement, because some in this wing of the movement were convinced "that in almost all cases sterilization must be voluntary rather than coercive."[98] According to Kevles, while some American social radicals dabbled in eugenics, many were not drawn to it because the protocols of the movement were antifeminist, anti–birth control, and, most of all, anti-immigrant.[99] Muller was keen to protect the gene pool, as demonstrated by his long and untiring efforts — after his discovery of the mutagenic effects of ionizing radiation — to make the genetic hazards of radiation understood and to limit these hazards. His interest in the evolutionary future of humans and how to control it, humankind's "genetic load," as he called it, prompted him to propose voluntary seminal choice (artificial insemination by voluntary choice).[100] The late geneticist Guido Pontecorvo, writing Muller's biographical memoir for the Royal Society in 1968, stated: "The advent of the atomic age and the feeling that the highest recognitions imposes a tremendous responsibility towards society, made him intensify his public and scientific efforts to limit the load of genetic effects of radiations on future generations."[101]

Muller's presidential address, read before the American Society of Human Genetics in New York on December 28, 1949, expanded on an earlier paper read before the Eighth International Congress on Genetics in Stockholm, Sweden, on July 7, 1948, and proposed the theoretical quantitative consequences of what he assumed at the time to be the relaxation of both natural selection and inbreeding.[102] Therefore, he advocated "purposive *voluntary* control of human reproduction as the long-term counter measure, with the alternative of letting future generations pay for our lack of humanity."[103] Kevles speculated that Muller viewed advanced medical technology used in extending the lives of diseased individuals at the expense of healthy ones as counterproductive to the health of the gene pool, but it could mitigate any dysgenic cost imposed on the human gene pool.[104] Soon, the antievolutionary Lysenkoism (Soviet-era pseudoscientist Trofim Lysenko) firmly established in Soviet Russia and the descent of the iron curtain on central-eastern Europe caused Muller to modify his socialism.[105] Kevles stated, however, that "his theory differed from the mainline creed in that it did not identify dysgenic trends with race or class — mutations occurred in all sectors of society — and was couched in socially antiseptic, genetic language."[106]

Muller knew back then what evolutionary geneticists know today: the genomes of all biological life-forms on this planet are constantly subject to mutational change; some mutations are advantageous, but most are deleterious (and deadly in the homozygous recessive form). He also knew that although mutations can occur at random, they can also be caused by chemicals or radiation, leading to a greater susceptibility to cancer and many infectious or mental disorders. In Muller's estimation, the gradual accumulation of these mutations over time spread through breeding constituted the "genetic load" of the human species or "the total number of potentially lethal genes in the human gene pool."[107]

The question for geneticists during 1950s and 1960s was how to use advanced technology to offset the effects of increased genetic load, not just tolerate it. Muller, whose reformed eugenics rested on the framework of evolution, proposed reproduction control. According to Kevles, Muller explained to a physician in 1954 that "the fact that the so-called eugenics of the past was so mistaken … is no more argument against eugenics as a general proposition than, say, the failure of democracy in ancient Greece is a valid argument against democracy in general."[108] This statement is an instance of twisted or false logic from a brilliant scientist. With the advent of genetic engineering, this type of logic becomes very dangerous. By modifying Muller's statement using a different example, the false logic is further highlighted: the fact that using retroviruses in the first gene therapy trials was so mistaken (several patients died during the trials) is no more argument against using retroviruses in gene therapy as a general proposition than, say, the low efficiency of cloning is a valid argument against cloning in general. Nevertheless, Muller was, unwittingly, aiding in the birth and development of the new eugenics by applying modern scientific advancements. For example, recent developments in the field of artificial insemination were key to Muller's overall positive eugenic purpose, which was to reduce the genetic load over time. Kevles wrote:

> The demonstrated success of freezing and storing sperm, then thawing them for vaginal injection — enhanced the plan's prospects for success. The preservation techniques — dry ice had been used first, then the much colder liquid nitrogen — allowed for the accumulation of sperm from a given donor, and as the number of sperm increased so did the chance of producing a pregnancy … the frozen sperm might well be stored until, say, twenty years after the death of the donor. By that time, it could be better judged whether the donor, outstanding in life, seemed truly outstanding in calm retrospection. Thus, the effort to guide man's evolution could be kept to the highest standard.[109]

In essence, Muller hoped that responsible couples with deleterious genes would sacrifice their reproductive potential, and, over time, the absence of deleterious genes would increase the genetic quality of *Homo sapiens*.

In the early 1960s, Muller began the process of establishing the Foundation for Germinal Choice (a fancy name for sperm bank) and garnered much interest and many donations from scientists and businessmen.[110] He dissociated himself from the project when high intelligence (i.e., Nobel laureates and altruism) were the primary criteria for donors. It seemed that these prominent scientists and businessmen, who were probably middle-class or upper-class young men during the 1930s and 1940s, were enculturated in the social-cultural environment of mainline eugenics; some of them still clung to the remnants of the old eugenics — tradition dies hard. In summary, Kevles quoted science reporter Albert Rosenfeld, who in 1969 made a prescient statement in *Life* magazine that could be applied to cloning or any other genetic engineering technology today:

> We are now entering an era when, as a result of new scientific discoveries, some mind-boggling things are likely to happen. Children may routinely be born of geographically separated or even long dead parents, virgin births may become relatively common, women may give birth to other women's children, romance and genetics may finally be separated, and a few favored men may be called upon to father thousands of babies.[111]

The advances in medical science during the 1950s and 1960s were critical to the mechanical and philosophical (as far as primary goals) transformation of the old eugenics into the new eugenics. Biotechnology cannot be divorced from the definition of the new eugenics, because technological advancement has fueled the new eugenics. For instance, the mixed success in cloning frogs during the 1950s and the 1960s was a eugenicist's dream, especially if cloning could be applied to humans. A Nobel laureate, for example, could clone himself rather than subject himself to the highly personal procedure of donating sperm. However, even the future outcome of a child born via voluntary artificial insemination is by no means certain.

In the 21st century, geneticists have sequenced the three billion base pairs of the human genome. And, with better machines, they can sequence 20 million base pairs in months rather than years (it took 15 years to sequence the human genome). This exponential increase in scientific advancement has armed well-meaning scientists in the never-ending battle to prevent or repair the defects of humankind (see Table 1.1). Specifically, zinc finger nucleases (ZFNs) and CRISPR/Cas9 gene-editing systems have been used to correct the mutation in the breast cancer suppressor gene 1 (*BRCA1*); to snip out a mutated copy of the gene that encodes a protein called dystrophin, which is essential for proper muscle function (malfunction of the dystrophin gene causes the X-linked recessive genetic disease Duchenne muscular dystrophy that results in progressive systemic muscle degeneration); to disrupt the erythroid-specific *BCL11A* fetal globin suppressor to increase fetal hemoglobin, replace the mutated beta hemoglobin, and create a possible cure for sickle cell anemia by isolating stem cells from a patient's bone marrow and repairing the mutated beta-globin gene; to edit the chemokine receptor 5 (CCR5) site on the T cell (an important cell in the immune system) to prevent HIV from binding to it and causing infection; and to effect other therapies that will be discussed in detail in Chapters 5 and 6.[112]

It is interesting that Muller's lamentations about the genetic load of the human species can be found in writings today. For example, Australian bioethicist Julian Savulescu and colleagues at Oxford University's Uehiro Center for Practical Ethics stated the following in their paper advocating for no moratorium on gene editing research on embryos: "Advanced gene editing techniques could reduce the *global burden of genetic disease* and potentially benefit millions worldwide" (italics added).[113] The authors were probably unaware that their tone echoed one aspect of the new eugenics. Moreover, scientists are presently fully engaged in the fight to repair humans, and some have become so lost or absorbed in this battle that they are not aware (or are unsure) of the variable effects of their therapy (i.e., genetic modification resulting in lethal mutations and death or long-term, deleterious, heritable effects if germ line modifications are made). These effects of biotechnology exemplify the very meaning of the new eugenics.

## Notes

1. Daniel Kevles, *In the Name of Eugenics: Genetics and the Uses of Human Heredity* (Cambridge, MA: Harvard University Press, 1985), 268; Robert L. Sinsheimer, "The Prospect of Designed Genetic Change," *Engineering and Science* 32, no. 7 (April 1969): 8, 13.
2. Kevles, *Name of Eugenics*, 267–268; Sinsheimer, "Designed Genetic Change," 8, 13.

3. Peter Raven, George B. Johnson, Kenneth A. Mason, Jonathan B. Losos, Susan R. Singer, *Biology*, 9th ed. (New York: McGraw-Hill, 2008), 330.
4. Ibid., 352.
5. Rainer Breitling, Eriko Takano, Timothy S. Gardner, "Judging Synthetic Biology Risks," *Science* 347, no. 6218 (January 9, 2015): 107.
6. Harry F, Tibbals, *Medical Nanotechnology and Nanomedicine* (Boca Raton: CRC Press, 2011), 20–24.
7. Raven *et al.*, *Biology*, 1112.
8. Ibid., 1113.
9. Ibid., 476.
10. Ibid., 348–349.
11. Ibid., 347–349.
12. Ibid., 347.
13. Ibid., 347.
14. Ibid., 375.
15. Ibid., 348.
16. Paul Knoepfler, *GMO Sapiens: The Life-Changing Science of Designer Babies* (Hackensack, NJ: World Scientific Publishing Co. Ptc. Ltd., 2016), 181.
17. Ross D. Park, Christine W. Gailey, Scott Coltrane, M. Robin DiMatteo, "The Pursuit of Perfection," in Philomena Essed and Gabrielle Schwad, eds., *Clones, Fakes, and Posthumans* (Thamyris/Intersecting: Place, Sex, and Race) (Amsterdam: Rodopi Press, 2012), 115.
18. Ibid., 115.
19. Henry T. Greely, *The End of Sex and the Future of Human Reproduction* (Cambridge, MA: Harvard University Press, 2016), 57.
20. Michael Sandel, "The Pursuit of Perfection: A Conversation on the Ethics of Genetic Engineering," March 31, 2004, http://www.pewforum.org/2004/03/31/the-pursuit-of-perfection-a-conversation-on-the-ethics-of-genetic-engineering, 7–11.
21. Ibid., 6–8.
22. Ibid., 6–8.
23. Paraphrased in Jim Kozubek, *Modern Prometheus: Editing the Human Genome with CRISPR-CAS9* (New York, NY: Cambridge University Press, 2016), 310.
24. Sandel, "The Pursuit of Perfection: A Conversation on the Ethics of Genetic Engineering," 5. The pharmaceutical company Eli Lilly persuaded the FDA to approve its human growth hormone for healthy children whose projected adult height would be in the bottom one percentile.

25. Sandel, "The Pursuit of Perfection: A Conversation on the Ethics of Genetic Engineering," 5–6.
26. Ibid., 6.; US Department of Health and Human Services, "Anthropometric Reference Data for Children and Adults," *U.S. National Health Statistics Reports* 11 (2011–2014), 11.
27. Sandel, "The Pursuit of Perfection: A Conversation on the Ethics of Genetic Engineering," 6–7.
28. Ibid., 8.
29. The National Academies of Sciences, Engineering, and Medicine, *Review of the Federal Strategy for Nanotechnology-Related Environmental, Health, and Safety Research* (Washington, DC: The National Academies Press, 2009), 50.
30. Sandel, "The Pursuit of Perfection: A Conversation on the Ethics of Genetic Engineering," 8; Raven *et al.*, *Biology*, 1110, 199. After fertilization, the diploid zygote goes through a period of rapid mitotic divisions. In animal embryos, the timing and number of these divisions are controlled by a set of molecules that "drive" the cell cycle called the cyclins and cyclin-dependent kinases (*Cdks*).
31. Sandel, "The Pursuit of Perfection: A Conversation on the Ethics of Genetic Engineering," 8.
32. Ibid., 10.
33. Kevles, *Name of Eugenics*, 90–91, 106.
34. Kevles, *Name of Eugenics*, 60–61, 96–97, 106–107. In the early 20th century, there were two key reasons why the British government could not enact legislation on eugenics, specifically sterilization. First, the American and British government systems were different. Specifically, Parliament was/is the entire government in the United Kingdom, as opposed to state and federal components in the United States government. While federal laws supersede state laws, states had/have the autonomy to enact laws governing the respective states. Consequently, American eugenics (i.e., forced incarceration and forced sterilization of anyone deemed "unfit" — mentally ill, epileptics, criminals, alcoholics, homeless, prostitutes, handicapped, etc.) was successful because it was in the province of state legislatures. In fact, the first state sterilization law was passed in 1907. Between 1907 and 1917, sterilization laws were enacted by 15 more states in every region of the country except the South. Some states had wide-ranging laws making individuals involved in white slavery eligible for sterilization. Second, the Offence Against Persons Act of 1861, which made abortion illegal, blunted any support for sterilization.

35. Sandel, "The Pursuit of Perfection: A Conversation on the Ethics of Genetic Engineering," 15.
36. Ibid. Dr. Lee Silver basically believes that nature is unfair and technology provides a leveled "playing field." He states: "The problem I have is that I don't see a secular reason to stop parents from giving a child a gene that other parents give their children naturally, not in a society where we know that the playing field is not fair to begin with. If a playing field was fair, I think it would be different, and I think that's why the mythology of everybody starts at the same place,' of 'the playing field is level' that liberal mythology is put upon us to try to convince people that anybody can get anywhere they want. It's not true we know, and that causes a severe problem for which I have no answer" (p. 16).
37. Sandel, "The Pursuit of Perfection: A Conversation on the Ethics of Genetic Engineering," 15–16.
38. Ibid., 10.
39. Ibid., 11.
40. Jennifer A. Doudna and Samuel H. Sternberg, *A Crack in Creation: Gene Editing and the Unthinkable Power to Control Evolution* (New York: Houghton Mifflin Harcourt, 2017), xiv, xv.
41. Doudna and Sternberg, *A Crack in Creation*, 148–153; Ewen Callaway, "U.S. Defence Agencies Grapple with Gene Drives: National Security Community Studies Risks of Method to Quickly Spread DNA Modifications," *Nature* 547, no. 7664 (July 27, 2017): 388–389.
42. Ibid., 150.
43. The National Academies of Sciences, Engineering, and Medicine, *International Summit on Human Gene Editing: A Global Discussion*, December 1–3, 2015 (Washington, DC: The National Academies Press), 4.
44. Adapted timeline from Paul Knoepfler, *GMO Sapiens: The Life-Changing Science of Designer Babies* (Hackensack, NJ: World Scientific Publishing Co. Ptc. Ltd., 2016), 134.
45. Kevles, *Name of Eugenics*, 253–255.
46. Ibid., 264.
47. Ibid., 267–268.
48. Ibid., 253.
49. Ibid., 254–257.
50. Ibid., 255.
51. Diane B. Paul, "The History of New Born Phenylketonuria Screening in the U.S." in Neil Holtzman and Michael S. Watson, eds., "Promoting Safe and Effective Genetic Testing in the U.S.: Final Report of the Task Force on

Genetic Testing" (National Institutes of Health — Department of Energy Working Group on Ethical, Legal, and Social Implications of Human Genome Research, September 1997), appendix 5, 137–149. https://biotech.law.lsu.edu/research/fed/tfgt/appendix5.htm.

52. Raven *et al.*, *Biology*, 255.
53. Paul, "The History of New Born Phenylketonuria Screening in the U.S.," 138.
54. Ibid.
55. Ibid.
56. Kevles, *Name of Eugenics*, 267–268; Diane B. Paul, "The History of New Born Phenylketonuria Screening in the U.S.," 139. https://biotech.law.lsu.edu/research/fed/tfgt/appendix5.htm.
57. Ibid.
58. Kevles, *Name of Eugenics*, 255.
59. Ibid.; Paul, "The History of New Born Phenylketonuria Screening in the U.S.," 139.
60. Centers for Disease Control and Prevention: Newborn Screening Portal. https://www.cdc.gov/newbornscreening/index.html. Accessed 9/12/2019.
61. Raven *et al.*, *Biology*, 249–250.
62. Howard Markel, "Scientific Advances and Social Risks: Historical Perspectives of Genetic Screening Programs for Sickle Cell Diseases, Tay-Sachs Disease, Neural Tube Defects, and Down Syndrome 1970–1997" in Neil Holtzman and Michael S. Watson, eds., "Promoting Safe and Effective Genetic Testing in the U.S.: Final Report of the Task Force on Genetic Testing" (National Institutes of Health — Department of Energy Working Group on Ethical, Legal, and Social Implications of Human Genome Research, September 1997), appendix 6, 2. https://biotech.law.lsu.edu/research/fed/tfgt/appendix6.htm.
63. Ibid.
64. Ibid., 2.
65. Ibid., 3.
66. Ibid.
67. Ibid.
68. Frank B. Livingstone, "Anthropological Implications of Sickle Cell Gene Distribution in West Africa," *American Anthropologist* 60, no. 3 (June 1958): 533–562.
69. Markel, "Scientific Advances and Social Risks," 3.
70. Ibid., 4.

71. Linus C. Pauling, "Reflections on the New Biology: Foreword," *UCLA Law Review* 15 (February 1968): 267–272.
72. Stephen Molnar, *Human Variation: Races, Types, and Ethnic Groups*, 4th ed. (Englewood Cliffs, NJ: Prentice Hall, 1998), 128–130.
73. Ibid.
74. Jared M. Diamond, "Curse and Blessing of the Ghetto," *Discover* 12, no. 3 (March 1991): 60–65.
75. Ibid.
76. Ibid.
77. Markel, "Scientific Advances and Social Risks," 5.
78. Kevles, *Name of Eugenics*, 256.
79. Markel, "Scientific Advances and Social Risks," 6.
80. Ibid.
81. Ibid.
82. Ibid.
83. Kevles, *Name of Eugenics*, 257.
84. Ibid. Rh-induced hemolytic disease — Erythoblastosis fetalis: The Rhesus factor is a protein on surface on blood cells. There is a positive reaction with serum. If mother is Rh negative (dd) and father is Rh positive (DD or Dd), the offspring will always be Rh positive (Dd, heterozygote). This will cause mother-fetal incompatibility. The mother's immune system will create anti D antibodies (traveling via umbilical cord) to destroy fetal cells. Occurs after the first pregnancy.
85. Kevles, *Name of Eugenics*, 257.
86. Ibid.
87. Markel, "Scientific Advances and Social Risks," 5.
88. Kevles, *Name of Eugenics*, 257.
89. Ibid.
90. Ibid.
91. Ibid., 258.
92. Raven *et al.*, *Biology*, 717.
93. Mara Hvistendahl, *Unnatural Selection: Choosing Boys Over Girls, and the Consequences of a World Full of Men* (New York: PublicAffairs, 2011), 116.
94. Ibid.
95. Ibid., 117.
96. Kevles, *Name of Eugenics*, 256.
97. Ibid.

98. Ibid., 106.
99. Ibid.
100. Ibid., 259–262.
101. Guido Pontecorvo, "Hermann J. Muller. 1890–1967," *Biographical Memoirs of Fellows of the Royal Society* 14 (November 1968): 360.
102. Hermann J. Muller, "Our Load of Mutations," *American Journal of Human Genetics* 2, no. 2 (June 1950): 111–176; Herman J. Muller, "Genetics in the Scheme of Things," *Proceedings of the 8th International Congress on Genetics; Hereditas*, supplement, no. 1 (1949): 96–127.
103. Pontecorvo, "Hermann J. Muller. 1890–1967," 350.
104. Kevles, *Name of Eugenics*, 260.
105. David Joravsky, *The Lysenko Affair*, (Chicago: University of Chicago Press, 2010); B. M. Cohen, "The Descent of Lysenko," *The Journal of Heredity* 56 (1965): 229–233. Lysenkoism is a term derived from the name of Soviet-era agronomist Trofim Lysenko who rejected the established tenets of early 20th-century plant biology and genetics (i.e., Mendelian inheritance, DNA, etc.) in favor of Jean Baptiste Lamarck's disproven eighteenth-century proposal called the "inheritance of acquired characteristics." Lysenko worked on converting winter wheat into spring wheat. By treating wheat seeds with moisture and cold, he induced them to bear a crop when planted in the spring — a process he called vernalization. He believed this condition could be inherited by the progeny and they would be able to survive in variable conditions. Lysenko's ideas were favored by Stalin, and he (Lysenko) soon became the director of Genetics for the Academy of Sciences — essentially controlling genetics in the Soviet Union. Soviet scientists who accepted mainline genetics or refused support Lysenko's ideas were dismissed from their posts, imprisoned, or executed as enemies of the state. Some examples of Lysenko's ideas are as follows: 1) He claimed that plants do not die due to lack of water and sunlight but sacrifice themselves for the healthy plants. 2) Obtaining more milk from cows did not depend on biology but on how you treat them. With the rise of Soviet science after the Stalin era, mainstream scientists emerged. One of them, physicist Andrei Sakharov, spoke out against Lysenko in the General Assembly of Russian Academy of Sciences in 1964: "He is responsible for the shameful backwardness of Soviet biology and of genetics in particular, for the dissemination of pseudo-scientific views, for adventurism, for the degradation of learning, and for the defamation, firing, arrest, even death, of many genuine scientists" (p. 233).

106. Kevles, *Name of Eugenics*, 260.
107. Ibid., 259; Muller, "Our Load of Mutations,"142–143.
108. Kevles, *Name of Eugenics*, 261.
109. Ibid.
110. Ibid., 262.
111. Ibid., 263–264; Albert Rosenfeld, "The Second Genesis," *Life* 23 (June 13, 1969): 40.
112. Jim Kozubek, *Modern Prometheus: Editing the Human Genome with CRISPR-CAS9* (New York, NY: Cambridge University Press, 2016), 293–296, 350–351.
113. Julian Savulescu, Jonathan Pugh, Thomas Douglas, and Christopher Gyngell, "The Moral Imperative to Continue Gene Editing Research on Human Embryos," *Protein Cell* 6, no. 7 (May 2015): 478.

# Chapter 2

# Genetic Engineering in the Twenty-First Century: Genetically Modified Organisms

> I don't know what I may seem to the world. But as to myself I seem to have been only like a boy playing on the seashore and diverting myself now and then in finding a smoother pebble or a prettier shell than the ordinary, whilst the great ocean of truth lay all undiscovered before me.
>
> — Isaac Newton, 1727[1]

In early plans of this book, this chapter was titled "Genetic Engineering in the Twenty-First Century: Humans." But it was impossible to write about genetic engineering solely in humans because the success, advances, techniques, and science of genetic engineering could not have occurred without using microorganisms, insects, animals, and plants as model organisms. (An additional advantage to using nonhuman organisms is that scientists are free from legal liability and government regulation when using nonhuman genomes.) All life on this planet is linked by the universality of the genetic code made up of the four nitrogenous bases that code for amino acids and proteins: adenine (A), thymine (T), guanine (G), and cytosine (C). This is why nonhuman biological life-forms are included in discussion of genetic engineering and why the title of this chapter encompasses all biological life. In the mid-1970s, the construction of recombinant DNA molecules (in which segments of DNA from two different organisms are joined to produce a single DNA molecule) made genetic engineering possible (see Table 1.1). But the story begins with the unsung heroes called nucleases, which are enzymes that degrade DNA. They were not commonly known until the first restriction enzyme (*Hin*dII) was isolated in 1970.[2] Restriction enzymes or, more specifically, restriction endonucleases cleave (cut) DNA internally at specific sites.[3] Molecular biologists had labored for years to learn the mechanism responsible for DNA cleavage at specific sites, but the mechanism was eventually discovered serendipitously — like many scientific discoveries are — while a researcher was studying why bacterial viruses infected some cells but not others. Bacteria create enzymes to cut out the infection (the invading viral DNA) and protect noninfected DNA from cleavage by modifying the DNA at the cleavage sites.[4]

There are three types of restriction enzymes, designated Type I, Type II, and Type III. Type II enzymes cut with precision compared to Type I and Type III enzymes and are used in cloning and DNA manipulation: "Type II enzymes recognize a specific DNA sequence, ranging from 4 bases to 12 bases, and cleave the DNA at a specific base within this sequence."[5] Type II enzymes are palindromic, meaning that the same DNA sequence from 5′ to 3′ (5′-A/GCGCT-3′) and 3′ to 5′ (3′-TCGCG/A-5′) occurs on both complementary strands.[6] This palindromic nature allows creation of recombinant DNA. For instance, the restriction endonuclease will cleave the sequence between A and G on both strands because they are the same but running in opposite directions. Cutting the DNA at the same base on either strand results in single-stranded tails that are complementary to each other. These complementary ends can then be joined (by DNA ligase) to a fragment from another DNA source (that has been cut with the same enzyme) to produce a recombinant molecule. Scientists later exploited this knowledge to create the CRISPR/Cas9 gene-editing technology (see Chapters 5 and 6).

One might argue that without the discovery of restriction endonuclease cleavage, in which DNA molecules are identified and recombinant molecules are created, DNA manipulation and cloning would not be possible. In short, there would be no reason to write this book. But this is not the case; the human propensity for wanting to know what's over the next hill, for better or for worse, leads to further discoveries such as transgenic techniques.

The next discovery was *transformation*, by which some bacterial species exchange genetic material, and this enabled geneticists to reintroduce recombinant DNA into the original cells from which it had been isolated to form adult cells or adult tissue.[7] In bacterial species that did not naturally transfer genetic material, such as *Escherichia coli* (*E. coli*), artificial transformation techniques (using electric charge) were developed to make the *E. coli* membrane permeable for the introduction of foreign DNA.[8] Bacteria that incorporate recombinant genes can synthesize large amounts of the proteins those genes specify, such as human insulin used to treat diabetes and human growth hormone used to treat muscle-wasting diseases. However, a specialized vector is needed to drive the expression of inserted DNA in a specific cell type to permit protein synthesis.[9] This type of genetic engineering led to the creation of transgenic organisms that bring to the layperson's mind Frankenstein's monster as seen in numerous films throughout the 20th century. Chapters 7 and 8 address this subject in greater detail. Some geneticists, however, believe that transgenic organisms are beneficial in medicine and agriculture. But do the benefits of genetic engineering and DNA manipulation outweigh the risks? This persistent dichotomy is a critical factor in all human decisions and warrants further exploration.

To understand how genes work or how genes are expressed, geneticists — in most cases — choose organisms that are easy to manipulate in the laboratory, have short generation times, have male and female sex organs, and have well-delineated genomes to perform effect-on-function or gain-of-function experiments. In the study of the development of the eye in both vertebrates and insects, for example, the two prominent organisms of choice were the overused fruit fly (*Drosophila melanogaster*) and mouse (*Mus musculus*). In each case, a gene was discovered in the 1990s that codes for lens formation: the fly gene was called *eyeless* because a mutation in this gene resulted in no eye development, and the mouse gene was given the name *Pax6*.[10] It soon became apparent that the *Pax6* gene could trigger lens formation in both insects and vertebrates. Essentially, *Pax6* was homologous to the *eyeless* gene. To go beyond the sequencing of these genes and find support for homology, Swiss developmental biologist Walter Gehring introduced the mouse *Pax6* into the genome of the fruit fly.[11] In this transgenic fly, an eye was formed on the fly's leg. Why? The *Pax6* gene was "turned on" by regulatory factors in the fly's leg and organized the formation of eyes. In short, "the *Pax6* master regulator gene can initiate compound eye development in a fruit fly or simple eye development in a mouse."[12] In this and other amazing discoveries due to genetic engineering, simple long-standing observations are overshadowed or forgotten. In cave-fish, for example, eyes begin to develop, and then degeneration occurs because the *Pax6* gene expression is greatly reduced or "turned off."[13] The environment is the moderator in that, senses other than sight are required in a dark cave. Genes have potential but are mediated by the environment, and there are many genes for eye expression. This means that eye expression is a complex trait (i.e., genetic with a strong environmental component). Complex traits are found in many biological species, but as humans, we encounter these traits more frequently in our own species. For example, the inactive form of the enzyme tyrosinase in people of Asian and European ancestry originating in regions of low solar radiation (this enzyme is critical in the physiological pathway leading to melanin formation necessary for humans originating in regions of high solar radiation) or lactose intolerance in populations with no tradition of using dairy products occur because the gene coding for the enzyme lactase is "turned off." The multiple effects of genes have tremendous implications for the short-term and long-term effects of genetic engineering.

One of the primary motivations for effect-on-function experiments is cancer research, specifically of genes that promote or suppress cancerous tumors. For instance, the gene for the p53 tumor suppressor — which stops cell division in damaged cells, preventing proliferation of mutated cells — is also important in cell regulation.[14] This gene is usually found mutated in many human cancers. In one DNA manipulation experiment (in vitro mutagenesis), p53 is inactivated by introducing an antibiotic resistant gene that allows mouse cells to survive and recombine when grown in a medium containing the antibiotic. The altered p53 gene is introduced into ES cells, which are cells derived from early embryos that can develop into any adult cells or tissues.[15] In these cells, the altered p53 gene combines with the chromosomal copy of the gene. Subsequently, the ES cells containing the inactivated p53 gene are injected into a blastocyst-stage embryo and then implanted into a female with a receptive uterus. The transgenic $F_1$ (first generation) offspring are heterozygous (meaning that the offspring has the dominant inactivated p53 gene but still carries a recessive gene for the normal p53 gene) for inactivated p53, and the $F_2$ (second generation) offspring can be homozygous.

When this gene is inactivated in mice, the mice experience normal growth and development. As they age, however, tumors develop in several tissues.[16] This indicates that later or long-term effects of genetic engineering are the real danger, and this danger increases tenfold if the alterations are heritable. Furthermore, biologists believed that the inactive p53 gene (loss of function) would be detrimental to mice early in development, but they were wrong. The same or identical genes are expressed in different tissues, at different times, and in different combinations in different organisms. This result serves as a cautionary tale of the numerous known and unknown functions affected by genetic engineering.

Science writer and computational biologist Jim Kozubek discussed another effect-on-function experiment conducted by Luhan Yang (cofounder and chief scientific officer of eGenesis, a life science company in Cambridge, Massachusetts, focusing on gene-editing technology) and David Sinclair (Harvard Medical School geneticist) in which genetic variants contributing to a risk for breast cancer were edited.[17] First, the researchers obtained the ovaries of a woman undergoing surgery for ovarian cancer caused by a mutation in the *BRCA1* gene. Subsequently, they tried to create a viable egg without the cancer-causing mutation by extracting immature egg cells and using CRISPR to edit the *BRCA1* gene ex vivo.

As mentioned earlier in this chapter, the use of bacteria that can incorporate recombinant genes to synthesize large amounts of the proteins specified by those genes is the key driving force behind genetic engineering. Several substances are produced this way: forms of human insulin and human growth factor hormone; tissue plasminogen activator, which causes blood clots to dissolve preventing disability or death; erythropoietin, which is important in red blood cell production; atrial peptides, which are new substances for treating hypertension and kidney failure; and interferon, an immune system protein.[18]

Vaccines are difficult to produce and expensive in the age of communicable diseases, but recombinant DNA has the potential to simplify vaccine production. For instance, genes encoding a small segment of the polysaccharide coat of a virus, such as the herpes simplex virus, are spliced into a fragment of the harmless cowpox (vaccinia) virus that is used as a vector to carry the viral coat genes into cultured mammalian cells.[19] Once infected, "these cells produce many copies of the recombinant vaccinia virus, which has the outside coat of a herpes or hepatitis virus."[20] Injected as a vaccine, the recombinant virus should cause an immune response in the animal that directs antibodies against the antigens of the viral coat. The vaccinated animal now enjoys immunity to herpes or hepatitis.

Another more controversial and dangerous approach to genetically engineered vaccines are DNA vaccines. These vaccines are designed to stimulate the cell-mediated response, specifically cytotoxic T cells (or killer T cells, as they are commonly known), which identify and kill "altered-self" cells such as tumor cells or cells infected by viruses.[21] In the first DNA vaccine, an influenza virus encoding a nucleoprotein was spliced into a plasmid and subsequently injected into mice.[22] While the mice developed a strong cell-mediated response to the virus, there are clear dangers to introducing engineered viruses, particularly influenza viruses, into the body: the engineered virus can mutate, creating new surface proteins or new strains leading to new disease in the host and loss of immunity. Viruses in general, and influenza viruses in particular, have RNA genomes replicated by RNA polymerase. RNA polymerase lacks "proofreading" ability (in contrast to DNA polymerase), which results in antigenic drift or antigenic shift — accumulated mistakes or mutations in the genes for the influenza strains (i.e., H1N1, H5N1, H7N3, etc.): hemagglutinin (HA) and neuraminidase (NA) that over time increase its resistance to vaccines.[23] The long-term effects of genetic engineering are still unknown, but (although it may be a provocative concept) engineered pathogens, like many other natural pathogens, might be able to shift their surface antigens to survive attack from the immune system. In another dangerous move, scientists are using nanoparticles against certain strains of bacteria and viruses. What if there were a synthesis between a bacterial or viral species and these nanoparticles? What new plague would be unleashed against the world?

In recent years, geneticists have worked tirelessly to "[weaponize] a patient's own immune cells into a precision-guided strike force."[24] Bruce Levine at the Perelman School of Medicine of the University of

Pennsylvania and his collaborators in cell therapy and cancer immunotherapy, David Porter, Stephan Grupp, and Carl June, have engineered T cells to fight cancer by adding a gene that did not exist in nature. The gene was designed on a computer and then constructed using fragments of genetic code from mice, cows, and woodchucks, essentially resulting in a chimera, a single organism composed of cells with distinct genotypes.[25] This chimera was packaged in a deactivated HIV virus, which would travel to and bind with a T cell. The chimeric genetic code would build a receptor on the surface of the T cell that would guide it to a unique marker on a cancer cell (i.e., on solid tumor of mesothelioma or breast, brain, or pancreatic cancer). The researchers called their synthetic creation chimeric antigen receptor (CAR) T cells.[26] The initial patient trials in the 1990s were promising; doctors saw significant reduction in the weight and size of cancer in a matter of days in three patients. But the immune cells did not turn off, resulting in nonstop cytokine release (excessive and uncontrolled inflammation), fevers, and collapse of blood pressure.

The innocuous and almost soothing sound of the words *gene therapy*, compared to the more mechanical sound of the words *genetic engineering*, is deceptive because gene therapy involves the complexity of finding the appropriate viral vector (i.e., retrovirus or adenovirus — see Chapter 11) and adding controls to prevent expression of oncogenes (before gene transfer), which could lead to cancer. If an individual inherits a defective gene, the therapy is to replace the defective gene with a copy that works: a cloned copy of the normal gene. In the 1990s, diseases such as hemophilia, Lesch-Nyhan syndrome, sickle cell anemia, rheumatoid arthritis, and several cancers (breast, colon, leukemia, ovarian, etc.) were being treated in clinical trials of gene therapy (Table 2.1).[27]

Another strategy of gene therapy was to block or silence a "bad" gene by using small RNAs. While studying development and regulation of gene expression in the worm *Caenorhabditis elegans* (*C. elegans*), scientists discovered that a particular small RNA (now called micro RNA or miRNA) affected gene expression.[28] In 1992, for instance, Rosalind C. Lee, Rhonda L. Feinbaum, and Victor Ambros of Harvard University's Department of Cellular and Developmental Biology isolated a gene (*lin-4*) that they believed was a mutant that affected heterochrony (developmental timing) and, interestingly, did not encode a protein product.[29] This mutant gene,

**Table 2.1.** Gene therapy clinical trials to treat diseases

| **Diseases** |
|---|
| Acquired immunodeficiency syndrome (AIDS) |
| $\alpha_1$-Antitrypsin deficiency |
| Batten disease (neurological disorder) |
| Cancer (brain, breast, colon, head/neck, leukemia, liver, lung, lymphoma, melanoma, mesothelioma, multiple myeloma, neuroblastoma, ovarian, prostate, renal cell) |
| Chronic granulomatous disease |
| Cystic fibrosis |
| Duchenne muscular dystrophy |
| Familial hypercholesterolemia |
| Fanconi anemia |
| Gaucher disease |
| Hemophilia |
| Hunter syndrome |
| Lesch-Nyhan syndrome |
| Macular degeneration (wet variety) |
| Peripheral vascular disease |
| Purine nucleoside phosphorylase deficiency |
| Rheumatoid arthritis |
| Severe combined immunodeficiency (SCID) |
| Sickle cell disease |
| Thalassemia |

*Source*: Adapted from Raven *et al.* (2008)[27]

they learned, encoded two small RNA molecules that are derived from each other and complementary to a region in another heterochronic gene (*lin-14*). To test gene expression, the scientists developed a model where the *lin-4* RNA acted as a translational repressor of the *lin-14* mRNA. In 1999, scientists discovered a second gene (*let-7*) in the same pathway in *C. elegans* that encodes similar RNA molecules.[30] Soon, homologous genes for *let-7* were found in both *Drosophila* and in humans.[31] These discoveries add support for including all biological life-forms in the discussion of genetic engineering.

Furthermore, miRNAs are pertinent in the generation of iPSCs. Pluripotent stem cells can give rise to any cell type in the adult organism, such as neurons or cells of the heart, skin, liver, or kidneys (see Chapters 3 and 11). The iPSCs begin as adult tissues and are then induced to become pluripotent by introducing genes encoding transcription factors (genes that code for proteins). The differentiation potential of iPSCs can be predicted by measuring variations in miRNAs, and several mechanisms have been proposed for using miRNA to enhance the potential of iPSCs.[32] For example, ES-cell-specific miRNA molecules enhance the efficiency of iPSCs by acting downstream of gene *c-Myc* (which encodes transcription factors). Additionally, miRNAs can block expression of repressors of transcription factors.

Another small RNA or small interfering RNA has been used to turn off gene expression. These small interfering RNAs or double-stranded vectors have been used to interfere with mRNA before translation into the corresponding protein needed for proliferation of the defective gene.[33] In macular degeneration, for example, in which individuals lose their sight because of the uncontrolled proliferation of blood vessels under the retina, the interfering RNA blocks protein production needed for blood vessel development. Therefore, the potential to help many people suffering from this devastating disease is great. But the dangers of suppressing a protein that may have more than one function — such as a regulatory protein — or suppressing a protein that may automatically activate a disease-causing gene are also great. For instance, severe combined immunodeficiency (SCID) is a disease in which the patient has no lymphocytes: T cells or B cells (immunoglobulins or antibodies).[34] The added complication is that there are multiple forms of this disease, such as a form where the enzyme adenosine deaminase is absent (ADA-SCID) and an X-linked form (X-SCID).[35] A danger of gene therapy, as mentioned above, is that the wrong gene might be turned on, such as the proto-oncogene *LMO2*, which can cause childhood leukemia. This occurred in an X-SCID trial in which a few patients developed the same rare form of leukemia and the gene therapy was implicated (see Chapter 11). Apparently, "the vector used to introduce the X-SCID gene integrated into the genome next to oncogene *LMO2* in all three cases."[36] While a small percentage of children survived for more than four years after successful treatment, insertion of a gene can trigger unknown events in an individual based on a myriad of factors that include genetics, population history, health, diet and nutrition, and environment.

## 2.1. Genetically Modified Or Genetically Edited: Are the Resultant Products of these Technologies Dangerous?

In agriculture, animals and plants are engineered to produce pharmaceuticals, and crops are modified to be nutritious, tolerant of herbicides and the natural environment, and resistant to disease. While these goals may seem practical, is it wise to modify plant systems without a clear idea of the long-term effects of genetic engineering in animals and humans? Are pharmaceutical companies in a race to increase their profits in the relatively new market of biopharming? If so, these companies are just responding to the pressures of a capitalist system, but what is the long-term risk? Before considering these questions, first consider some examples of current technology.

Tomato, tobacco, and soybean plants are infected by the bacterium *Agrobacterium tumefaciens* using its (the bacterium) tumor-inducing plasmid.[37] How can biologists alter the characteristics of these plants? They isolate a tumor-preventing gene from the DNA of another organism and insert it into the plasmid.[38] The plasmid is then returned to the *Agrobacterium*. When *Agrobacterium* infects a plant, the new tumor-preventing gene is transferred into a chromosome of the plant cell.

Glyphosate is a powerful herbicide that kills most actively growing plants by inhibiting an enzyme called 5-enolpyruvyl-shikimate-3-phosphate (EPSP) synthetase, which plants need to produce aromatic amino acids.[39] Extra copies of the EPSP synthetase gene are inserted into plants via tumor-inducing plasmid. Subsequently, these new transgenic plants produce more than the normal level of EPSP synthetase and are glyphosate resistant.

At present, corn, cotton, soybeans, and canola have been genetically engineered to be glyphosate resistant.[40] These modified crop plants are popularly known as genetically modified (GM) plants, and glyphosate-resistant soybeans are grown in nine countries.[41] In fact, 90 percent of soybeans grown in the United States are GM soy. Historically, Western countries have taken steps to protect the public from scientific research. However, in this new frontier of genetic engineering, and GM crops in particular, US biologists are moving quickly and somewhat recklessly.

Interestingly, their speed in creating transgenic plants is similar to that of biologists in Asia who have generated the largest growth of GM crops in an environment of limited safety regulations.[42] In contrast, European governments are being more cautious about GM crops.

Biologists have also engineered staple crops to express key vitamins and minerals. For example, they genetically modified rice to produce β-carotene (provitamin A), which gives the engineered rice a golden color in the outer layer or endosperm.[43] This GM rice is possible because rice generates a precursor, geranylgeranyl diphosphate, in the endosperm tissue, which can be converted by three enzymes (phytoene synthase, phytoene desaturase, and lycopene β-cyclase) to β-carotene.[44] Genes for these three enzymes were introduced into the rice genome to allow expression of the pathway producing β-carotene in the endosperm. Using the gene for phytoene synthase from maize (instead of the original daffodil gene) leads to even higher levels of β-carotene. In essence, this so-called Golden Rice provides provitamin A in the diet, which enzymes in the human body can convert to vitamin A.[45] Deficiency in this vitamin affects preschool children worldwide.

Insects feast on many commercially important plants, which costs billions of dollars each year. The usual defense against such feasts is to apply insecticides. However, within several generations, insects become resistant. In addition, chemical and pharmaceutical companies do not produce large stocks of insecticides because they are expensive to make, and the demand is small.[46] At present, the targets of more than half of all insecticides used are boll weevils, bollworms, and other insects that eat cotton. As a result, biologists have engineered plants that are resistant to insects.

Obviously, any engineered gene must be harmful only to the targeted organisms — boll weevils and bollworms, for example — and not other organisms. This is the source of long-term uncertainty. Nevertheless, genes for the Bt toxin produced by the soil bacterium *Bacillus thuringiensis* have been introduced in crop plants.[47] When insects ingest the engineered crop plants, Bt toxin enters their systems. For the insects, irreversible cascading events of paralysis and death occurs as their endogenous enzymes convert Bt toxin into an insect-specific toxin. It is absolutely critical that these enzymes are not found in humans or animals; otherwise, consumption of these GM crops would be catastrophic. Fortunately, the global distribution of Bt crops is similar to the distribution of herbicide-resistant crops. Bt corn is the second most common GM crop, representing 14 percent of the global area of GM crops in nine countries.[48]

At this point, CRISPR/Cas9 enters the conversation, because biochemist and 2020 Nobel laureate Jennifer Doudna, professor in the molecular and cell biology departments at the University of California, Berkeley, and one of the discoverers of the CRISPR/Cas9 technology, makes a distinction between GMOs and gene-edited organisms.[49] European governments have made a similar distinction as they struggle to fit genetic editing into their current regulatory framework, which was designed for GMOs (see Chapter 10). Doudna stated:

> Conventional GMOs contain foreign genes randomly inserted into the genome; these genes produce novel proteins that give the organism a beneficial trait it did not previously possess. Gene-edited organisms, by contrast, contain tiny alterations to existing genes that give the organism a beneficial trait by tweaking the levels of proteins that were already there to begin with — without adding any foreign DNA. In this respect, gene-edited organisms are often no different than those organisms produced by mutation-inducing chemicals and radiation. Furthermore, scientists have used methods to avoid leaving any traces of CRISPR in plant genome once the gene-editing task is complete.[50]

But members of the public are clearly not educated enough in molecular genetics nor have the patience for the fine distinctions of scientists; they feel that "the gene-edited crops are nothing but hidden GMOs and that scientists are trying to sneak them into grocery stores through the back door."[51] Nevertheless, CRISPR has been used to edit genes in rice and oranges to make them resistant to different bacterial species.[52] For potatoes, corn, and soybeans, genes have been edited so that they have resistance to herbicides. Additionally, Doudna noted that potatoes have been edited to inactivate a gene that produces sugars such as glucose and fructose (produced during cold storage preservation and converted to acrylamide, a neurotoxin, during cooking); soybeans have been edited to generate seeds with less fatty acids; and mushrooms have been edited to prevent premature spoiling.[53]

The distinction between GMOs and gene-edited organisms seems specious; there is one common fact: humans are genetically modifying nonhuman species. To date, GM and gene-edited crops have not caused any major medical issues.[54] But what are the potential long-term evolutionary biological changes for humans, animals, and plants? Environmental scientist Maywa Montenegro in the Department of Environmental Science at the University of California, Berkeley, noted the emphasis on genes and their effects as separate from the environmental context. She wrote:

> In what scholar Donna Haraway calls the "god-trick," we thought of genetics as the key to scientific mastery of nature, as if there was no context, no agency in the object, no imperfection in human knowledge. *Molecular science somehow licensed us to treat genes as separate from ecology and bodies*. Now we fathoming intricate interactions between genes and environments, and ecosystems whose changes aren't smooth or predictable, but that bristle with threshold effects and emergent properties. We've come to appreciate the inseparability of nature and culture in complex systems.[55] (italics added)

In short, molecular scientists consumed by "correcting" nature forget about the complex bioenvironmental interactions. There is now fear that after a particular genetic modification, modified genes might spread to wild relatives through a process called introgression, leading to biodiversity loss and extinction (gene drive mechanisms in the wrong hands might do the same — see Chapters 1 and 14).[56] Also, there is a good chance that the engineered gene could trigger an allergic reaction based on the genetics of the individual consuming a product derived from it. Despite these concerns, some beneficial human proteins have been produced in genetically engineered plants. In fact, serum albumin was produced in 1990 by genetically engineered tobacco and potato plants.[57]

For multinational pharmaceutical companies, using transgenic plants for the production of useful compounds is cost-effective. For instance, a recombinant subunit vaccine against Norwalk virus is being produced in genetically modified potatoes, and a vaccine against rabies is being produced in spinach.[58] What is amazing is that "250 acres of greenhouse space could produce enough transgenic potato plants to supply Southeast

Asia's need for hepatitis B vaccine."[59] These vaccines are presently in clinical trials.

However, the controversy increases when considering the research into DNA modification of animals. Studies have concentrated on (1) experiments with gene expression, an example of which was the insertion of a gene from a jellyfish that encoded green fluorescent protein into a blastocyst-stage pig embryo, resulting in a piglet with a yellow rather than pink snout and hooves; (2) effect-on-function experiments, in which, for example, knocking out the gene for the p53 tumor suppressor resulted in an increased incidence of tumors in mice; and (3) desirable animal products, such as increased growth, improved flavor in food, pharmaceuticals in milk, texture specificity of hair or fur, and other features.[60] The temptation within the field of biotechnology, despite overwhelming public opinion against it, is to clone a human. If this ability were discovered today in the United States, there would be strict laws against it. Since the cloning of Dolly the sheep in 1997, however, geneticists have successfully cloned cattle, sheep, goats, calves, mice, pigs, rabbits, cats, rats, mules, horse, ferrets, wolves, buffalos, camel, dogs, and old-world monkeys.[61] Chapter 3 addresses cloning in more detail.

One could argue that GMOs in general, and the ability to clone domesticated animals in particular, could produce "improved" animals with economically desirable traits. The word *improved* is in quotes because this gets at the heart of the very meaning of eugenics, which is "improving the stock." (In this particular instance, *stock* does mean animals.) But the human impact is through medical, dietary (agriculture), and environmental effects. While the so-called EnviroPig excretes up to 70 percent less phosphorus because it is engineered with the phytase gene that breaks down phosphorus in the feed, the pig engineered to overproduce growth hormone had lower fat levels and other deleterious effects, including reduced flavor.[62]

Before the curious EnviroPig discussed earlier, Genie was the world's first pig to produce appreciable amounts of human protein C (which acts to control blood clotting) in her milk.[63] Biomolecular engineer William H. Velander at the University of Nebraska–Lincoln and colleagues at the National Institutes of Health and Jerome H. Holland Laboratory chose to work with pigs for no specific reason other than short gestation and generation times and large litter sizes.[64] As for milk production, Velander and colleagues were keenly interested in the production of useable quantities of human protein in whatever milk was produced. They created a segment of DNA that contained the human gene for the target protein C and the mouse promoter gene for mouse milk protein. They injected this DNA fragment into several pig embryos (rather than mice embryos, as Hennighausen and colleagues did) and then implanted the embryos into a surrogate mother pig.

A female piglet, Genie, was born that carried the foreign DNA in all its tissues. In adulthood, Genie produced protein C in substantial amounts in her milk, "with about one gram of protein C in each liter of milk — 200 times the concentration at which this protein is found in normal human blood plasma."[65] The unknown factor in the resultant pig-made human protein was biological activity. There is a posttranslational process after protein synthesis in which the newly generated protein is modified in shape and chemical composition to function properly.[66] Genetic engineering has great capacity for causing malfunctions in this process with deleterious effects. In Genie's case, the researchers got lucky; they isolated protein C from pig milk and found it to be biologically active.[67] Genie's mammary tissues made the posttranslational modifications. Additional experiments were performed to increase the concentration of active human protein C in pig's milk (some of the protein C remained in an immature inactive form) by introducing another foreign gene that would allow more of the needed processing enzyme, furin, to be made.

Closer to home, in terms of taxonomy, and more problematic, particularly to Kathleen Conlee, vice president of animal research issues at the Humane Society of the United States, is genetically designing a monkey (like humans, a primate) that will become ill in order to study the progression of a disease.[68] Among nonhuman primates, *Macaca* (rhesus macaques) are the most popular genus of research monkeys and have been used in research for a long time. However, enthusiasm for the new-world monkeys commonly known as marmosets (*Callithrix jacchus*) peaked in 2009, mainly because they were the first primates to pass a genetic modification to offspring in their sperm and eggs.[69] A research team headed by geneticist Erika Sasaki of the Central Institute for Experimental Animals and neuroscientist Hideyuki Okano of Keio University, Tokyo, injected embryos with the gene for a fluorescent protein, resulting in offspring with skin and hair that glowed green under ultraviolet light.[70] In addition to the ease of genetic modification, marmosets are smaller, easier to house than rhesus macaques or larger monkeys, and have a faster generation time (giving birth twice a year), which helps researchers studying long-term heritable characteristics.[71] Data on diseases that affect development and aging can be obtained faster with marmosets due their faster maturation.

Marmosets have now been genetically engineered to make their brains serve as models for neurological disorders such as autism and Parkinson's disease. In fact, Kelly Servick, writing for *Science*, reported on a Society for Neuroscience meeting held on November 5, 2018, in San Diego, at which Sasaki and Okano discussed their progress with transgenic marmosets with Parkinson's disease and Ret syndrome (a neurodevelopmental disorder).[72] Sasaki and Okano had a unique opportunity to watch the progression of the disease — a marmoset's brain is less convoluted than a macaque's and thus is easier to image — and simultaneously analyze the brain to find a cause for the illness and hopefully translate the results into a cure.

Now, the race for cures and patents for different therapies continues as we humans begin to cross the Rubicon of violating our own order: primates. But scientists see no substitute for primates in some studies. Servick quoted Joshua Gordon, director of the National Institute of Mental Health in Bethesda, Maryland, who expressed his views on genetically engineered nonhuman primates at an October 4, 2018, meeting at the Institute for Laboratory Animal Research, part of the National Academies of Sciences, Engineering, and Medicine: "When it comes to [studying] cognitive processes and other complex behaviors, some things you just need to do in a primate model. … The study of mental illness requires an understanding of brain structures that don't exist in rodents."[73] Gordon did acknowledge that the public needs to be involved in decisions on primate research. Nevertheless, Servick reported that Japanese researchers received 40 billion yen ($350 million) from a government initiative to map the marmoset brain, and the first marmoset with a mutation in the gene *SHANK3* (implicated in some cases of autism) was born at the Massachusetts Institute of Technology in April 2018.[74]

The motivation to help humankind fuels scientific advancement. This fuel is unending, and it propels scientists faster and faster into unknown and hostile territory, such as pathogens being transmitted to patients, adverse effects of drugs made from transgenic livestock, and the long-term changes to the human genome. These issues recur throughout this book, as scientists cross the threshold into the murkiness that is genome modification.

## 2.2. Induced Pluripotent Stem Cells: The Complexities of Reprogramming Human Adult Cells

Earlier in this chapter, iPSCs appear in relation to miRNAs. Now, it is important to discuss iPSCs as they relate to regenerative medicine and changes — for better or worse — in body tissues. This story begins with biologist and 2012 Nobel laureate Shinya Yamanaka and other researchers at the Department of Stem Cell Biology in Kyoto University's Institute for Frontier Medical Sciences who showed in 2006 that the introduction of four specific genes encoding transcription factors (*Oct4*, *Sox2*, *c-Myc*, and *Klf4*) could convert adult cells into pluripotent stem cells.[75] This specific set of pluripotent-associated genes, or *reprogramming factors*, is the most conventional combination used in producing iPSCs, but the genes can be replaced by related transcription factors (e.g., miRNAs or plasmids carrying transcription factors).

Generating iPSCs is a complex and time-consuming process. For instance, it takes two weeks for mouse cells and four weeks for human cells, both with low efficiency.[76] Before deriving the first-generation mouse iPSCs, Yamanaka and colleagues believed that genes (or transcription factors) important to ES-cell function might be able to return adult cells to their embryonic state.[77] These ES-cell genes were introduced into mouse fibroblasts using retroviruses. There was an important ES-cell gene (*Fbx15*) in the fibroblasts that the scientists wanted to reactivate and isolate, and they hoped that these incorporated genes would stimulate reprogramming.[78] ES-cell-like colonies subsequently formed in the petri dish, indicating *Fbx15* was reactivated by the four specific factors.

Interestingly, these iPSCs could give rise to any cell type in the body (as discussed earlier in the chapter) and contribute to all three germ layers: ectoderm, endoderm, and mesoderm. But, on close examination of these cells' gene expression, epigenetics, and so on, the scientists found them to have structures of both ES cells and fibroblasts.[79] And these first-generation iPSCs failed to produce viable chimeras — tissues of diverse genetic makeup.

In 2007, biologist Marius Wernig and colleagues at the Massachusetts Institute Technology, Harvard, and the University of California, Los Angeles (in addition to Yamanaka and colleagues working separately), generated iPSCs indistinguishable from ES cells.[80] According to the scientists, they succeeded because they reactivated and isolated a different ES-cell gene, *Nanog* (using the same four factors), resulting in viable chimeric mice. Soon, Yamanaka and colleagues reported reprogramming of human adult cells as iPSCs in a paper titled "Induction of Pluripotent Stem Cells from Adult Human Fibroblasts by Defined Factors" (2007), and biologist Junying Yu and colleagues at the University of Wisconsin–Madison reported the same achievement.[81] Both teams used the same principles to transform human fibroblasts into iPSCs. However, Yu and his colleagues used genes for two extra factors (*Nanog* and *Lin28*) in addition to the factors previously used (*Sox2* and *Oct4*).[82] Furthermore, they used a lentivirus instead of a retrovirus to deliver the genes.

In 2011, however, Ryan Lister and colleagues at the Salk Institute for Biological Studies in La Jolla, California, found some hot spots of aberrant epigenomic reprogramming in human iPSCs.[83] Human iPSCs (produced using the Yamanaka factors) are indistinguishable from ES cells, as already mentioned. However, nearly 1,000 DNA regions showed differences in iPSC lines, with half resembling the somatic cell line the iPSCs were derived from — not reprogrammed to the ES cell state — and the remainder remaining iPSC specific.[84]

As far as applications go, iPSCs eliminate the need to produce an embryo and eliminate the need for immune-suppressant drugs after transplant, because the transplanted cells come from the patient. But is this DNA manipulation technique safe, particularly using retroviruses? Additionally, will the iPSCs change over time? Genes have multiple effects. One major concern is the propensity of iPSCs to form or eventually form tumors. In fact, *c-Myc* and *Oct4* are already implicated in several cancers.[85] While Chapter 11 addresses details of the benefits and risks in more detail, it is worth reiterating here the dichotomy of public safety and scientific advancement. How are benefits and risks to be apportioned to what might be two sides of the same coin, or a false dichotomy?

## Notes

1. William Brahms, *Last Words of Notable People: Final Words of More Than 3500* (Haddonfield, ST: Reference Desk, 2012), 451.
2. Peter Raven, George B. Johnson, Kenneth A. Mason, Jonathan B. Losos, and Susan R. Singer, *Biology*, 9th ed. (New York: McGraw-Hill, 2008), 327.
3. Ibid., 327–328.
4. Ibid., 328.
5. Ibid.
6. Ibid.
7. Ibid., 329–330.
8. Ibid., 328.
9. Ibid., 330.
10. Ibid.
11. Ibid., 502.
12. Ibid.
13. Ibid.
14. Ibid., 202–203.
15. Ibid. The p53 protein's actual function is to check for and repair damage to DNA. If cell damage is found, p53 will stop cell division and begin the process to repair the damage to cell. If the damage could not be repaired, p53 will direct the cell to destroy itself.
16. Ibid., 202–203.
17. Jim Kozubek, *Modern Prometheus: Editing the Human Genome with CRISPR-CAS9* (New York, NY: Cambridge University Press, 2016), 99.
18. Raven *et al.*, *Biology*, 343–344.
19. Ibid.
20. Ibid., 344.
21. Ibid., 345.
22. Ibid.
23. Ibid., 1079.
24. Jim Kozubek, *Modern Prometheus: Editing the Human Genome with CRISPR-CAS9* (New York, NY: Cambridge University Press, 2016), 244.

25. Ibid., 243–245.
26. Ibid.
27. Raven *et al.*, *Biology*, 345.
28. Ibid., 317–319.
29. Ibid., 317–319; Ronda C. Lee, Rhonda L. Feinbaum, and Victor Ambros, "The *C. elegans* Heterochronic gene lin-4 encodes small RNAs with Antisense Complementarity to lin-14, *Cell* 75, no. 5 (December 1993): 843.
30. Ibid.
31. Ibid.
32. Bao Xichen, Zhu Xihua, Liao Baojian, Miguel A. Esteban …, "MicroRNAs in somatic cell programming," *Current Opinion in Cell Biology* 25, no. 2 (January 2013): 201–214. ES cell-specific microRNA molecules, miR-291, miR-294, miR-295, are short (20 to 24 nucleotides) noncoding RNAs that are involved in posttranscriptional regulation of gene expression in multicellular organisms by affecting both the stability and translation of mRNAs; *Let-7* (lethal-7) gene was first discovered in the translucent worm (*C. elegans*) as a key developmental regulator and became one of the first two miRNAs known. Soon *let-7* was found in the fruit fly and vertebrates. Most animals with loss of function *let-7* mutation die, which means the mutation is lethal (*let*); *Lin-4* is a miRNA that was discovered during study of developmental timing in *C. elegans*. *Lin-4* controls the onset of larval stages in *C. elegans*. *Lin-4* loss of function mutations lead to changes in heterochronic developmental patterns (i.e., loss of structures, etc.). *Lin-4* is a negative regulator of *lin-14* repressing *lin-14* protein.
33. Raven *et al.*, *Biology*, 319–320, 345.
34. Ibid., 345.
35. Ibid.
36. Ibid., 345. Oncogene is a mutant form of a growth-regulating gene that should not be "on," causing "runaway mitosis" or unrestrained cell growth and division, G-15.
37. Raven *et al.*, *Biology*, 346.
38. Ibid., 345.
39. Ibid., 346–347.
40. Ibid., 347.
41. Ibid.
42. Ibid.
43. Ibid., 348.
44. Ibid.

45. Ibid., 348.
46. Ibid., 347.
47. Ibid.
48. Ibid., 347.
49. Jennifer A. Doudna and Samuel H. Sternberg, *A Crack in Creation: Gene Editing and the Unthinkable Power to Control Evolution* (New York: Houghton Mifflin Harcourt, 2017), 125–126.
50. Ibid.
51. Ibid., 127.
52. Ibid., 124.
53. Ibid., 122–123.
54. Ibid., 124; Raven *et al.*, *Biology*, 348.
55. Maywa Montenegro, "CRISPR Is Coming to Agriculture — With Big Implications for Food, Farmers, Consumers and Nature," *Ensia*, January 28, 2016. Accessed May 24, 2019. https://ensia.com/voices/crispr-is-coming-to-agriculture-with-big-implications-for-food-far.
56. Raven *et al.*, *Biology*, 348.
57. Ibid., 345.
58. Ibid., 349.
59. Ibid.
60. Ibid., 284, 343, 349.
61. Nisar A. Wani, U. Wernery, F. A. Hassan, R. Wernery, and J. A. Skidmore, "Production of the First Cloned Camel by Somatic Cell Nuclear Transfer," *Biology of Reproduction* 82, no. 2 (October 2009): 373–379; Min J. Kim, Hyun J. Oh, Geon A. Kim, Erif M. N. Setyawan, Yoo B. Choi … Byeong C. Lee, "Birth of clones of the World's First Cloned Dog," *Scientific Reports* 7, no. 15235 (November 10, 2017): 1–4; Gordon L. Woods, Kenneth L. White, Dirk K. Vanderwall, Guang-Peng Li, Kenneth I. Aston, Thomas D. Bunch, Lora N. Meerdo, and Barry J. Pate, "A Mule Cloned from Fetal Cells by Nuclear Transfer," *Science* 301, no. 5636 (August 2003): 1063; Qi Zhou, Jean-Paul Renard, Gaëlle Le Friec, Vincent Brochard, Nathalie Beaujean, Yacine Cherifi, Alexander Fraichard, and Jean Cozzi, "Generation of Fertile Cloned Rats by Regulating Oocyte Activation," *Science* 302, no. 5648 (November 2003): 1179; Taeyoung Shin, Duane Kraemer, Jane Pryor, Ling Liu, James Rugila, Lisa Howe, Sandra Buck, Keith Murphy, Leslie Lyons, and Mark Westhusin, "A Cat Cloned by Nuclear Transplantation," *Nature* 415, no. 6874 (February 2002): 859; Patrick Chesné, Pierre G. Adenot, Céline Viglietta, Michel Baratte, Laurent Boulanger, and Jean-Paul Renard,

"Cloned Rabbits Produced by Nuclear Transfer from Adult Somatic Cells," *Nature Biotechnology* 20, no. 4 (April 2002): 366–369; Irina A. Polejaeva, Shu-Hung Chen, Todd D. Vaught, Raymond L. Page, June Mullins, Suyapa Ball, Yifan Dai, Jeremy Boone, Shawn Walker, David L. Ayares, Alan Colman, and Keith H. Campbell, "Cloned Pigs Produced by Nuclear Transfer from Adult Somatic Cells," *Nature* 407, no. 6800 (September 2000): 86–90; Zhen Liu, Yijun Cai, Yan Wang, Yanhong Nie, Chenchen Zhang, Yuting Xu, Xiaotong Zhang, Yong Lu, Zhanyang Wang, Muming Poo, and Qiang Sun, "Cloning of Macaque Monkeys by Somatic Cell Nuclear Transfer," *Cell* 172, no. 4 (February 2018): 881–887.

62. Raven *et al.*, *Biology*, 349.
63. Ibid.
64. Ibid., 71; Gestation periods for pigs is four months, and one generation is 12 months. Litter size is 10 to 12 piglets, and a lactating pig generates 300 liters of milk in a year.
65. Velander, "Transgenic Livestock as Drug Factories," 72.
66. Ibid., 71.
67. Ibid.
68. Kelly Servick, "U.S. Labs Clamor for Marmosets: Shortage Develops as New Transgenic Models for Neurological Diseases Stoke Interest," *Science* 362, no. 6413 (October 25, 2018): 383–384.
69. Ibid., 383.
70. Ibid.
71. Ibid.
72. Ibid.
73. Ibid.
74. Ibid.
75. Masayo Takahashi and Shinya Yamanaka, "Induction of Pluripotent Stem Cells from Mouse Embryonic and Adult Fibroblast Cultures by Defined Factors," *Cell* 126, no. 4 (July 2006): 663–676. Transcription factors: *c-Myc* family are proto-oncogenes causing cancer. Twenty-five percent of mice transplanted with *c-myc*-induced iPSCs developed lethal teratomas (a type of germ cell tumor that may contain several different types of tissues, i.e., hair, bone, etc.) (Yu *et al.* 2007; Guiterrez-Aranda *et al.* 2010; Okita *et al.* 2007); *Oct-3/4* family of octamer transcription factors give rise to the pluripotency and differentiation potential of ESCs; *Sox* family of transcription factors are associated with pluripotent, multipotent, and unipotent stem cells; *Nanog* transcription factor works with *Sox2* and *Oct-3/4* to generate iPSCs; *Klf*

family of transcription factors was identified by Takahashi and Yamanaka (2006) and Wernig *et al.* 2007 as a factor for the generation of mouse iPSCs and by Takahashi and Yamanaka (2006) as a factor for generation of human iPSCs; *Lin-28* is an mRNA binding protein resulting in differentiation and proliferation in ESCs and embryonic carcinoma cells; *Fbx15* is a protein expressed in undifferentiated ESCs (simultaneously expressed with *Oct3/4*, *c-Myc*, *Klf4*, and *Sox2*).

76. Ibid., 663.
77. Ibid.
78. Ibid.
79. Ibid., 669–670.
80. M. Wernig, A. Meissner, R. Foreman, T. Brambrink, M. Ku, K. Hochedlinger, B. E. Bernstein, R. Jaenisch, "In Vitro Reprogramming of Fibroblasts into a Pluripotent ES-cell-like State," *Nature* 448, no. 7151 (July 2007): 318.
81. J. Yu, M. A. Vodyanik, K. Smuga-Otta, J. Antosiewicz-Bourget, J. L. France, S. Tian, J. Nie, G. A. Jonsdottir, V. Ruotti, R. Stewart, I. I. Slukvin, J. A. Thomson, "Induced Pluripotent Stem Cell Lines Derived from Human Somatic Cells," *Science* 318, no. 5858 (November 20, 2007): 1917.
82. Human IPSCs is generally dubbed "Yamanaka factors" even though scientists at Harvard University and MIT were involved in the creation of a second-generation iPSCs because Yamanaka pioneered the iPSC technology.
83. Ryan Lister, Mattia Pelizzola, Y.S. Kidda, David R. Hawkins, Joe Nery, G. Hon, Jessica Antosievicz-Bourget, Ronan O'Malley, Rosa Costanon, S. Klugman, M. Downes, R Yu, Ron Stewart, Bing Ren, James A. Thomas, R. M. Evans, Joseph R. Ecker, "Hotspots of Aberrant Epigenomic Reprogramming in Human Induced Pluripotent Stem Cells," *Nature* 471, no. 7336 (March 3, 2011): 68.
84. Ivan Gutierrez-Aranda, Veronica Ramos-Mejia, Clara Bueno, Martin Munoz-Lopez, Pedro J. Real, Angela Macia, Laura Sanchez, Gertrudis Ligero, Jose L. Garcia-Parez, "Human Induced Pluripotent Stem Cells Develop Teratoma More Efficiently and Faster Than Human Embryonic Stem Cells Regardless the Site of Injection," *Stem Cells* 28, no. 9 (September 2010): 1568–1570; Satoshi Nori, Yohei Okada, Soraya Nishimura, Takashi Sasaki, Go Itakura, Yoshiomi Kobayashi, Francois Renault-Mihara, Atsushi Shimizu, Ikuko Koya, "Long-Term Safety Issues of iPSC-Based Cell Therapy in Spinal Cord Injury Model: Oncogenic Transformation with Epithelial-Mesenchymal," *Stem Cells Reports* 4, no. 3 (March 10, 2010): 360–373; Keisuke Okita, Tomoko Ichisaka, and Shinya Yamanaka,

"Generation of Germline-Competent Induced Pluripotent Stem Cells," *Nature* 448, no. 7151 (July 19, 2007): 313–317.

85. Gutierrez-Aranda *et al.*, "Human Induced Pluripotent Stem Cells Develop Teratoma More Efficiently and Faster Than Human Embryonic Stem Cells Regardless the Site of Injection," 1568.

# Chapter 3

# Cloning and In Vitro Fertilization

> Our announcement of Dolly's birth in February 1997 attracted enormous press interest, perhaps because Dolly drew attention to the theoretical possibility of cloning humans. This is an outcome I hope never comes to pass.
>
> — Ian Wilmut, 1998[1]

> Cloning technology presents humankind with the very real possibility that it may one day control not only its destiny but also its origin. Human cloning allows man to fashion his own essential nature and turn chance into choice . . . human cloning and genetic manipulation intrude on upon the profound nature of the inherently unknowable; they represent the bottleless depths of human arrogance and irresponsibility.
>
> — Patrick Stephens, 2001[2], Objectivist Center manager

In 1985, Kevles quoted science reporter Albert Rosenfeld, who wrote in *Life* magazine in 1969:

> We are now entering an era when, as a result of new scientific discoveries, some mind-boggling things are likely to happen. Children may routinely be born of geographically separated or even long dead parents, virgin births may become relatively common, women may give birth to other women's children, romance and genetics may finally be separated, and a few favored men may be called upon to father thousands of babies.[3]

While Rosenfeld does not use the word *clone* (a genetically identical copy), it is certainly implied along with in vitro fertilization.[4] There are actually two types of cloning: *gene (molecular) cloning* and *reproductive (organismal) cloning*. Chapter 2 discusses some aspects of molecular cloning, in which a specific sequence of DNA encoding a protein is isolated. If the DNA sequences are short, they can be synthesized in the laboratory (in vitro). But it is difficult to clone large, unknown sequences that require the replication of recombinant DNA molecules in a cell (in vivo). Then a vector — plasmid or artificial chromosome — is required to carry the recombinant DNA molecule into the host cell so that it can be replicated when introduced.[5] The most practical host used for molecular cloning is the bacterium *E. coli*, although geneticists frequently reintroduce cloned eukaryotic DNA using other hosts, such as mammalian tissue culture cells, yeast cells, and insect cells. As discussed in the preface and Chapter 1, the universality of the genetic code makes these experiments possible. It follows that any discussion of genetic engineering today must include all living organisms.

Plasmid vectors are indispensable for the replication of recombinant DNA molecules in a cell. Plasmid vectors are small, circular extrachromosomal pieces of DNA that are used to clone small segments of DNA that do not exceed a maximum size of 10 kilobases.[6] While plasmids are dispensable to the bacterial cell, they must have three components in order to be effective: (1) an origin of replication allowing them to be replicated in, for example, yeast cells independent of the host chromosomes; (2) a selectable marker allowing their presence to be easily identified through genetic selection; and (3) one or more unique restriction sites where foreign DNA can be added.[7] To identify the plasmid, an antibiotic-resistant gene that would select for plasmid-containing cells and not kill them is inserted in a region of the plasmid called a multiple cloning site. This is the same strategy used in examining the effect on function in inactivated genes discussed in Chapter 2. The plasmid is then introduced into cells by transformation.

What if foreign DNA is inserted into the plasmid region? What are the changes in corresponding physiological processes? β-galactosidase, an enzyme that cleaves galactoside sugars such as lactose, is a perfect test case to answer these questions.[8] Foreign DNA inserted into the β-galactosidase gene interrupts the coding sequence, thus inactivating the gene and preventing a functional enzyme from being produced. The evidence for activity and inactivity of the gene comes from plating cells on a medium containing both antibiotic (selecting for plasmid-containing cells) and an artificial substrate.[9] When the β-galactosidase gene is active, the expressed enzyme cleaves to the substrate in the medium, producing blue colonies. When the β-galactosidase gene is inactive, the substrate is not cleaved, and there is no reaction. This is one of the concerns of DNA manipulation: the unintentional consequences of alterations in the genome that result in detrimental effects to the individual and the individual's descendants.

Since plasmid vectors can only clone small segments of DNA molecules, geneticists have created artificial chromosomes to tackle large-scale analysis of genomes.[10] The development of yeast, bacterial, and mammalian artificial chromosomes have all been accomplished due to the rapid (albeit frightening) pace of biotechnology.

Understanding the cellular mechanism of development is key to successful experiments in genetic engineering (cloning, in vitro fertilization, and ES-cell development and isolation). Due to the complexity and sensitivity of cellular development — most multicellular organisms develop according to molecular mechanisms — genetic engineering may unintentionally cause deleterious alterations to a human embryo's genome. If such an embryo develops and the resulting baby survives childbirth, these genetic alterations could be passed on to the next generation. This is one of the main concerns of the US National Academy of Sciences, Engineering, and Medicine, the Royal Society, the Chinese Academy of Sciences, and the Academy of Sciences of Hong Kong.[11]

Biological organisms have many differentiated, or specialized, cells, which can be distinguished from one another by their morphologies, their specific functions, and the respective proteins they synthesize. For brain cells to develop into brain structures, a molecular decision or cell determination must occur prior to any changes in the cell or differentiation.[12] After the molecular determination is made, the cell is committed to a specific pathway. Interestingly, mouse ES cells have been used to demonstrate in culture media the principal stages of development. These ES cells direct differentiation into the ectoderm, endoderm, and mesoderm, and then these three cell types give rise to different cells: cells from the three germ layers form epidermis and nervous system; pancreas, lungs, and liver; and notochord, bones, blood vessels, connective tissue, muscles, kidneys, and gonads.[13] These experiments enable scientists to understand the molecular cues that are involved in cell determination and the subsequent differentiation of cell types. Nevertheless, ES cells isolated from the blastula or early gastrula stage of amphibian embryos and the blastocyst stage of mammalian embryos are not yet determined. If, for example, undetermined cells from the brain region are transplanted to another region in the embryo, these cells will develop according to the new region.[14] If these cells are already determined, however, brain cells will develop in the new transplanted region. There is additional complexity because cell determination is not a uniform process; it takes place in stages. First, the cell becomes partially committed, in that only its position and location are established. For example, suppose that cells from a chicken embryo's tissue at the base of the leg bud (that normally gives rise to the thigh) are transplanted to the tip of the identical-looking wing bud (that normally gives rise to the wing tip).[15] Based on the earlier discussion on cell determination, one might assume that the transplanted leg bud tissue would develop into a thigh. But this assumption is wrong; the transplanted leg bud tissue will develop into a toe rather than a thigh. The location (the leg) and the position (positional signaling to form a tip of the leg, or toe) have all been established, but there is no commitment to forming a thigh.[16] In summary, once the cell determination switch is thrown, the cell is committed to its future developmental pathway. In contrast, ES cells are not committed to a single developmental pathway, which makes them indispensable (controversially so) to genetic engineering.[17]

Furthermore, stem cells can be classified into several categories according to the cells and tissues they generate. But, before classifying them, it is important to dispel a fantasy about cloning (in this case, *reproductive* cloning: a copy of the entire organism) derived from books and Hollywood films, in particular *The Boys from Brazil* (1978), which was based on the life of the notorious Nazi doctor Josef Mengele (portrayed by Gregory Peck), who conducted studies on Jewish prisoners in the Auschwitz concentration camp during World War II.[18] Mengele was interested in various abnormal traits, but he was particularly intrigued by identical twins due to their rarity and, for practical purposes, the Nazi regime's desire to improve the reproductive rate of racially desirable Germans by increasing the probability of desirable parents having twins. In fact,

Mengele intentionally infected one twin with a disease or transfused blood from one twin to another to investigate the effect on function. Mengele was also curious about individuals with eyes of different colors (heterochromia iridum).[19] His experiments with eyes included attempts to change eye color by injecting chemicals into the eyes of living subjects and killing prisoners with heterochromatic eyes so that their eyes could be removed for study. In the waning months of the Nazi regime, the story goes, Mengele escapes to Brazil and subsequently continues his research into identical twins, injecting chemicals into the eyes of selected Brazilian children to make them blue. His darker purpose is to create a clone of Adolf Hitler from cells donated by the dictator.[20]

The purpose of summarizing a few of Mengele's documented experiments is to show the early attempts by Nazi doctors to modify human biology against the backdrop of genetics, hereditary, and eugenics. The experiments of Mengele and other Nazi doctors on defenseless human beings was a barbaric, unscientific, and criminal way to try to study and modify human biology. It is now well-known in biology that cells taken from animal A, whether sex cells or somatic cells, cannot by themselves produce an identical copy of animal A. Even ES cells can only give rise to cells or tissues in an adult organism: "the only cells that can give rise to both the embryo and the extraembryonic membranes in mammals are the zygote and early blastomeres from the first few cell divisions."[21]

Stem cells can be totipotent, meaning that they can give rise to any tissue in an organism.[22] As mentioned in Chapter 2, stem cells that give rise to all the cells in the organism's body are pluripotent. Furthermore, stem cells that are limited and only give rise to blood cell types are multipotent.[23] Finally, stem cells that give rise to only a single cell type (e.g., sperm cells) are unipotent.[24] Pluripotent ES cells have been derived in the laboratory from mammalian embryos that have undergone the cleavage stage of development to produce the ball of cells called a blastocyst.[25] The blastocyst consists of an outer cell mass, the trophectoderm, that will become the placenta, and an inner cell mass — where ES cells can be isolated — that will become the embryo proper. Based on ES-cell studies in mice, it was not long before human ES cells were isolated and grown in the laboratory.[26] In summary, human ES cells can be used in regenerative medicine because of their potential to produce any cell type, such as hematopoietic stem cells to replace blood cells in patients with diseases that originate from those blood cells or cardiomyocytes to replace damaged heart tissue.[27] And the revolutionary iPSC technology that can convert adult cells into pluripotent cells could eliminate the controversial situation in which parents have to create an embryo just to isolate ES cells in order to help their sick child. But are the safety and challenges of this biotechnological advancement worth the benefits? Chapter 11 addresses this question in greater detail.

***

Biologists and some members of the general public (including a respectable percentage of scholars in the arts, humanities, and social sciences) have a different perception and understanding of cloning. In their discussions of the possibility of human cloning, the motivations of scientists to attempt it stems from improving health. This is regarded as therapeutic cloning, which is more palatable to most scientists than reproductive cloning. Nevertheless, the objection to human cloning by some in the general public is based on present and historical evidence of the propensity of humans to abuse one another based on misguided notions of race, ideal types, and perfection. Other members of the general public, who are not scientists or scholars, tend to merge fact with fantasy regarding cloning. David Berreby, in his book *Us and Them*, captures this involuntary mental activity very well: "We live in a virtual world, where belief matters more than fact, and perception shapes reality."[28]

First, consider the perceptions of some in the general public who are allied against cloning in general and human cloning in particular (if it becomes a reality in the near future). Their fear of human cloning is magnified by films like *The Boys from Brazil*, mentioned above. This film provoked nightmares about inadvertently cloning psychopathic mass murderers like Adolf Hitler. Furthermore, their perception of cloning originates from a larger cultural context where humans, in general, feel comfortable with sameness, repetition, predictability, norms, and traditional values instead of difference or diversity.[29] They fear that the preferred types that would be cloned are male, white, healthy, heterosexual, Protestant, highly educated, and rich. It is no surprise that 90 percent of Americans would choose white, blonde, and blue-eyed babies if babies could be designed, according to a documentary on ART technologies called *Frozen Angels*.[30] Additionally, American culture is saturated with replication. For example, phones are cloned, documents are copied or their facsimile (commonly known as a fax) is sent, artifacts are copied, and the biological mechanism of twins fascinates (coupled with the complexities of the nature-nurture debate). Philomena Essed, professor of critical race, gender, and leadership studies at Antioch University's Graduate School, New Hampshire and Gabrielle Schwab, distinguished professor of comparative literature, School of Humanities, University of California, Irvine articulated these views very well:

> Cloning we argue, is not an isolated biotechnological phenomenon. There are other cloning-like phenomena, such as, for example, copies and fakes of famous art works or designer products. Many cultural forms of cloning have often not really been recognized as such, because they have not been considered part of an encompassing new epistemological configuration inspired by new technologies of replication. We are therefore looking at cloning from a larger cultural and political perspective by asking in what ways technologies of cloning have affected the cultural imaginary more generally. We are posing the question of whether the availability of cloning technologies and the feasibility of the cloning of humans has both been preceded by and then reproduced by a form of "mental cloning" that deeply affects the ways we think about sameness and difference.[31]

As human beings, we are immersed in cultures of replication and dichotomy (e.g., common versus unique, original versus copy, duplication versus replication, and tradition versus innovation), which may be at the root of "mental cloning." Could it be, asked Essed and Schwab, that a "'cloning culture' has made biotechnological cloning desirable in the first place ...?"[32]

Biotechnological advances, economic power, and control of nature (or the place of humans in nature) factor into some people's fears of human cloning. Such people see shadows of colonialism as powerful multinational corporations promote genetic engineering in capitalist and emerging capitalist countries. What will follow, they believe, is the same historical dynamic: the dominant race and culture is seen as the ideal that should be replicated; the same profitable crops are replicated year after year, leading to environmental degradation and subjugation of indigenous people as a cheap labor force; and ultimately the life and death of indigenous peoples are controlled by these corporations.[33] Reversing the question posed by Essed and Schwab above, Does biotechnology reinforce mechanisms of "cultural cloning"? The reality cable television series *Extreme Makeover*, for example, represents an industry geared to replicating the cultural ideal. Moreover, the fear is that the most desirable characteristics — as defined by the majority culture — of people, objects, practices, and cognition become the ideal and are reproduced. In the commercial market of buying eggs and sperm, these desirable characteristics are advertised in respectable newspapers with promise of a monetary reward attached.[34]

There is also a fear of loss of diversity among humans and animals: "Cloning technology … engenders an unforeseen gender equalizer: human embryo cloning makes sperm dispensable, while female 'golden eggs' remain highly priced items in the market economy."[35] In agriculture, genetically engineered crop plants in general, and cloned plants in particular, could cause mutations that would spread to wild types, resulting in loss of biodiversity and other unknown consequences to the plants, insects, animals, and humans.

Another fear or, more accurately, an ambivalence stems from cloning a loved one who died. But would the clone have the same memories and behaviors as the original individual? The fictionalized Mengele, in *The Boys from Brazil*, raises his boys — clones of Hitler — in a Brazilian environment of economic deprivation mimicking post–War World I Germany.[36] How would such a clone be treated — as something less than human (abnormal) or disposable, with a batch number? Dolly, a fertile — albeit cloned — sheep was catalogued as 6LL3, and Cumulina, a fertile cloned mouse, was B6D2F1-derived.[37] The mass production of nonwhite, healthy, good-natured, docile, caring, mildly intelligent, loyal, and obedient clones to perform labor and fight wars, as depicted in the Hollywood film *Star Wars: Attack of the Clones*, is a dark example of how clones could be put to use.[38] Again, fantasy intrudes — for better or for worse — on the consciousness of the anticloners.

Some people have no qualms about human cloning, if it were possible, because they have been raised to believe that cultural homogeneity is the norm. As Essed wrote with David Golberg, director of the Humanities Research Institute, University of California, Irvine, in 2012, "It is not that the interest in cloning has belatedly given rise to the preference for likeliness. Rather, the longstanding drives and demands for more of the same have made possible the very conceivability of cloning as a supposed ideal worthy of pursuit."[39] Three worthy applications of cloning that capture the imagination of nonscientists and nonscholars are replacing a dead loved one (child, wife, husband, pet, etc.); remedying infertility; and having a spare self from which to obtain cells, tissues, or organs after disease or trauma.[40] These fantasies are the result of wishful thinking that ignores the harsh realities of the society's current level of science and technology. For example, human cloning might become a reality in a few years, but personality and experiences, which are influenced largely by culture, may not be subject to cloning.

Fantasy regarding cloning technology is not something one would find among scientists who practice and support cloning in the United States. Robert McKinnell, professor emeritus of genetic and cell biology, and the late Marie Di Berardino, professor emerita of biochemistry, argued in 1999 that cloning, like other discoveries in science, was initiated to find new knowledge. As such, "cloning was never intended as a procedure for the simple multiplication of animals. Frogs are cheap in the United States, as are sheep in Scotland. The reasons for cloning are more complex than simply producing identical animals."[41] They defined human cloning as "the attempt to produce a human organism by any cloning procedure: blastomere isolation, bisection (splitting) of preimplantation embryos, and nuclear transplantation."[42] They stated in great detail (1) why human cloning is scientifically and ethically unsound and (2) why they opposed human cloning.

First, nuclear transplantation of embryonic, fetal, or adult cells resulted in animals with serious defects, and these defects correlated positively with age of the donor cell. These abnormalities, which could be inherited — one important argument against genetic engineering, resulted from incomplete nuclear programming and failed cell-cycle matching. Lifespan of the offspring, on average, was short. And there would be birth defects. Furthermore, defective nuclear transplants from all donor stages stopped developing at various stages, such as "nuclear transfer and activation, cleavage, organogenesis, tadpole, and juvenile."[43] In viviparous species, faulty transplanted embryos could fail to implant or, if implant was successful, spontaneously abort at various embryonic and fetal stages. Based on cellular analysis, abnormal nuclear transplant embryos from mice, cattle, rabbits, and pigs revealed chromosomal and molecular changes that might have caused the morphological abnormalities.[44] Furthermore, an important aspect of nuclear transfer is that even in some successful nuclear transplants of frog blastulae derived from embryonic or adult nuclei, cells of the same blastulae had normal and abnormal chromosomes.

Second, the possibility of cloning humans is becoming less of a fantasy as the cloning of adult nuclei becomes efficient. However, as detailed above, the level of difficulty increases exponentially. For example, random mutations might occur due to chemicals in culture, radiation, or aging over time of the donor.[45] Another critical problem is the possibility of telomere shortening in the donor cell, resulting in a shortened life span for the clone. Telomeres are specialized structures found at the ends of eukaryotic chromosomes that consist of simple repetitive DNA; they protect the genome from nucleolytic degradation, unnecessary recombination, repair, and interchromosomal fusion.[46] During growth and development, a small portion of telomeric DNA is lost with each cell division. When telomere length reaches a critical limit — it shortens with age — the cell undergoes senescence (aging) and apoptosis (self-destruction). Additional problems include possible incompatibility between RNA and proteins in one donor's eggs with another donor's nucleus and the probable loss of variation and DNA repair when nuclear transplantation or any other cloning procedure is substituted for meiosis.[47] These problems go to the heart of the new eugenics when scientists, with good intentions, try to genetically engineer biological life to *repair individuals to ultimately make humankind better.*

## 3.1. From Frogs to Dolly the Sheep and Other Animals: A Short History of Cloning Research — 1952 to the Present

The science known as cloning today began as research into the cellular mechanisms of development. Some of the most important discoveries in science begin with a specific research path that randomly follows other serendipitous paths until a scientist accidentally discovers something. During the research into cellular development, for instance, scientists discovered that a donor adult somatic cell, when fused with an enucleated (nucleus-free) egg by an electric discharge, reverts back to its pluripotent or totipotent state and begins directing development of an exact genetic copy of the donor.[48] However, the path to cloning began as an attempt to understand the mechanism of cellular differentiation.

By the first decade of the 20th century, careful embryological research had generated much information about the cell, cell division, and the threadlike structures called chromosomes. The different combinations of pea-plant characteristics obtained by the Augustinian monk Gregor Mendel during his cross-fertilization experiments in the mid-19th century were confirmed by meiosis in the early 20th century. Of course, the greatest problem was finding the genetic material. Biologists knew that chromosomes were composed primarily of both protein and DNA. But they did not know which of these was the genetic material. Experiments performed by British bacteriologist Frederick Griffith, molecular biologist Oswald Avery and coworkers, and bacteriologist Alfred Hershey (1969 Nobel laureate) and geneticist Martha Chase confirmed that DNA — not protein — was the genetic material.[49] Another important, albeit underrated problem in the early 1950s was the mechanism of cell specialization during development. McKinnell and Di Berardino detailed this history, which began in the late 19th century with German evolutionary biologist August Weismann's erroneous belief that cell differentiation resulted from the differential and sequential partitioning of the genome.[50] Subsequent experiments on the blastomere produced interesting information. For instance, killing one blastomere of a two-cell amphibian embryo resulted in a half embryo, and if the blastomeres of two-cell sea urchin embryos were physically separated (or isolated with a portion of cytoplasm), entire embryos formed from each blastomere.

McKinnell and Di Berardino discussed German embryologist and 1935 Nobel laureate Hans Spemann's 1914 experiment that finally set the foundation for modern cloning:

> He constricted a zygote with a noose made of baby hair, causing the egg to assume the shape of a dumbbell. When the cleaving (nucleated)

> portion reached the 8- or 16-cell stage, he loosened the constriction and permitted a nucleus to move to the non-nucleated cytoplasmic portion. Here, too, the non-nucleated portion cleaved and formed a clone of its nuclear donor.[51]

While these experiments confirmed that the complete genome was replicated during cell division, there was no known technique by which to transfer a nucleus from an older embryo into an enucleated cytoplasm.[52]

By 1952, embryologists Robert Briggs and Thomas King of the Institute for Cancer Research in Philadelphia knew that DNA was the genetic material from the work of Griffith and Avery with *Streptococcus pneumoniae*.[53] They also knew that the genetic material, DNA, was in the nucleus of the cell. Regarding cleavage — the second major event in animal development, when the zygote becomes subdivided into many smaller cells called blastomeres — they knew that the nuclei in various blastomeres were equivalent. Knowledge concerning embryo development in the early 1950s was limited, which prompted Briggs to state that "whether they [blastomeres] remain equivalent or become differentiated as the various parts of the embryo differentiate has never been tested."[54] Their interest was to develop a method to test directly whether or not the nuclei of differentiating embryonic cells were themselves differentiated. The method they devised involved manually removing nuclei from irreversibly differentiated embryonic cells (committed developmental pathway) and undifferentiated embryonic cells (undetermined developmental pathway) of one strain of frog, injecting these nuclei into enucleated eggs of another strain, and then observing the resulting development.[55] In the 10 years before their groundbreaking work, Briggs had honed his microsurgical techniques while analyzing the development of haploid and triploid frog embryos. This skill became critical in attempting to transfer a nucleus from a donor cell to a donor egg.

In the class Amphibia, Briggs and King chose the North American leopard frog *Rana pipiens* as their animal model and used embryonic cells from the blastula developmental stage of this species.[56] First, they activated an egg at metaphase II of meiosis (sex cell division) by pricking it with a clean glass needle. *Nuclear reprogramming*, in which a fully differentiated cell is reprogrammed to be totipotent, then occurred.[57] This term *nuclear reprogramming* would not appear in the literature until the

1970s. Writing in 1962, British developmental biologist and 2012 Nobel laureate John Gurdon used the phrase "a reversal of the properties of nuclei" when discussing his results after transplanting a nucleus from one species of the frog genus *Xenopus* to the enucleated egg cytoplasm of another *Xenopus* species.[58]

Analyzing their overall results, Briggs and King were curious about differentiated and undifferentiated cells and their nuclear potential. They did not clone frogs, although eight years later, they did allow embryos to develop into juvenile frogs to study cell determination using donor cells from blastula and early gastrula stages.[59] While the totipotency of the blastula nuclei was suspected in the first half of the 20th century, it was not demonstrated until the early 1960s.[60]

In 1961, Gurdon's transplantation of nuclei between two *Xenopus* species led to more or less the same conclusions that Briggs and King reached using the *Rana* genus.[61] Gurdon was primarily interested in the changes that nuclei of one species undergo, and the resulting development, as a result of transplantation in enucleated cytoplasm of a different species. Additionally, Gurdon used nuclei (extracted from the vegetal pole) from any stage up to the late gastrula because he found "no appreciable decrease in developmental capacity of endoderm nuclei between the late blastula and late gastrula stages."[62] In his results, different combinations of *Xenopus tropicalis* (*X. tropicalis*) and *Xenopus laevis* (*X. laevis laevis*) nucleocytoplasmic transfers all led to arrested development and abnormalities.[63] Gurdon observed that "the *X. tropicalis* egg substances have a more severe effect on the replication of foreign nuclei than do *X. laevis laevis* egg substances. This seems to be the only explanation of the remarkable finding that *X. tropicalis* nuclei can differentiate further with *X. laevis laevis* cytoplasm (up to postneurulae)."[64] Reaching the blastula stage in the *X. tropicalis* cytoplasm, the *X. laevis laevis* nuclei was transplanted back to the *X. laevis* eggs. No matter how many of divisions *X. laevis* nuclei were allowed to complete in *X. tropicalis* egg cytoplasm, the resulting back-transfer embryos died in the late gastrula or early neurula stages.[65] The abnormalities consisted of protruding yolk plugs and subepidermal edema. Collectively, Gurdon believed that the egg cytoplasm contains substances that can induce heritable changes in some parts of a foreign nucleus (donated from a different species) but allow normal replication in other parts of the same nucleus.[66]

A critical aspect of this study and all other nucleocytoplasmic transfer studies was the phases of the nucleus and enucleated egg cytoplasm because these phases had been implicated in completion or arrested development in amphibian and mammalian cloning.[67] This knowledge helped to elucidate the mechanisms behind nuclear reprogramming.

Beginning in the 1990s, mammal embryologists were interested in nuclear transfer procedures that would require serial or multiple generation cloning. This certainly had implications for future mass production of a specific type of human clone. In 1993, biologists Steven Stice, from the Regenerative Bioscience Center at the University of Georgia, and Carol Keeper, from the Department of Animal and Avian Sciences at the University of Maryland, College Park, detailed the efficiencies of recloning bovine embryos.[68] In amphibians, development was enhanced by the transfer of donor nuclei into metaphase I eggs followed by recloning procedures. Stice and Keeper harvested metaphase II eggs from the reproductive tract of ovulated Holstein heifers and then removed the nuclei.[69] To obtain parent embryos, ovulated Holstein heifers were artificially inseminated, and embryos were recovered from the reproductive tract four to six days later. Nuclei from the parent embryos (morula stage — a solid mass of blastomeres) were subsequently fused with enucleated eggs by electrofusion.

Furthermore, the fused embryos were surgically transferred into the oviducts of sheep — a form of embryonic culture — and then recovered from the oviducts after four or five days. Out of the clones that developed to the morula stage, some were used as donor embryos to produce the next generation of embryos, and others were transferred into the reproductive tracts of recipient heifers.[70] The results of this experiment were interesting for several reasons: (1) clones not used as donor embryos and transferred to recipient heifers produced first- and third-generation calves; (2) first-, second-, and third-generation clones resulted in 10 percent, 2 percent, and 3 percent calving rates, respectively; and (3) blastomere-egg fusion rates were higher for day four clones used as donors to produce the next generation than for day five donor clones.[71] This low fusion rate may be due to the fact that blastomeres from day four clones were larger than blastomeres from day five clones (in sheep and rabbits cell-stage 16 to 32 blastomere donors have greater fusion rate than donors from later blastula stages, with inner cell mass and blastocoel).[72] This means that the dreams of a future society where a government can mass produce an army of clones is, for now, fantasy and might not be sustainable. The detailed findings of this experiment showed that in multiple generational cloning, cell-stage 32 blastomeres from donor clones have reduced amounts of cytoplasm — up to one-third less — as a result of the enucleation process compared to blastomeres from stage thirty-two parental embryos.[73] This loss in cytoplasm, accompanied by loss of "specific substances," might have lowered developmental rates.

Japanese developmental biologist Teruhiko Wakayama of the John A. Burns School of Medicine in the University of Hawai'i at Mãnoa, in collaboration with colleagues from Japan, Italy, and UK, successfully cloned the first mouse — Cumulina — and then used clones in subsequent rounds of recloning. They wrote: "These results indicate that clones (series B and C) and cloned clones (series D) are produced with comparable efficiency. This argues that successive generations of clones do not undergo changes (either positive or negative) that influence the outcome of the cloning process."[74] In 1997, Wakayama and colleagues injected enucleated mouse embryos with cumulus cell nuclei. In the first series of experiments, a total of 142 developing embryos were transferred to 16 recipient females. On examination at day 11, all fetuses (five) were dead.[75] In the second series of experiments, a total of 800 embryos were transferred into 54 recipient females. On examination at day 19, 17 fetuses were alive. Seventeen mice were born alive, but seven died within seven days after delivery.[76] The 10 remaining mice (females) — including the firstborn, Cumulina — survived for more than 60 days.[77] In additional series of experiments, Wakayama and colleagues collected cumulus cells from clones and injected their nuclei into enucleated eggs to generate embryos that were transferred as in the earlier experimental series.[78] A total of 287 embryos derived from cloned-cumulus cell nuclei were transferred to 18 recipient females. After 19 days, seven females were born alive and healthy. While the efforts of Wakayama and colleagues should be commended, their cloning efficiency — as in other experiments past and present — was very low (first series: 16 / 142 = 11 percent, 0 / 16 = 0; second series:

54 / 800 = 6 percent, 11 / 54 = 20 percent; additional series: 18 / 287 = 6 percent, 7 / 18 = 39 percent).[79] These figures make cloning in general, and the possibility of human cloning in particular, impractical and dangerous.

On the subject of sex ratios, Wakayama and Japanese development biologist Ryuzo Yanagimachi, also at the University of Hawai'i, conducted experiments to find out if males could be cloned using male-derived cells.[80] By the late 1990s, all reported clones derived from adult animals had been produced using cells related to the female reproductive system (i.e., mammary gland cells, ovarian cumulus/granulosa cells, and oviduct cells).[81] In contrast, Wakayama and Yanagimachi isolated donor cells from the tail-tips of adult male mice and cultured (serum-containing medium versus serum-free medium) the cells for approximately 12 days before transplantation. They subsequently removed metaphase II chromosomes from eggs of ovulating adult female mice. The nucleus was then injected into the enucleated egg. Three hours after transfer in vitro, embryos of two to eight cells morulae stage or blastulae stage developed and were transferred into the oviducts or uteri of recipient females. After 19 days, live fetuses were delivered by cesarean section and raised by lactating foster mothers.[82]

According to Wakayama and Yanagimachi, approximately 60 percent of the nucleocytoplasmic transfer nuclei developed to morulae or blastocyst. Interestingly, only three of 274 transferred embryos reached full-term.[83] Nevertheless, the three mice from tail-tip cells were all males and born alive. In their 1999 correspondence in *Nature*, the experimenters proudly remarked that "cloning using adult somatic cells is not restricted to female or reproductive cells."[84] Furthermore, the birthweights of the male clones were within the normal range of noncloned mice. Unfortunately, two of these males died within one hour of birth of respiratory failure.[85] The remaining male pup developed into a fertile adult male with the same phenotypical traits as the tail-tip cell donor.

Wakayama and Yanagimachi also compared the size and weight of the cloned male and female fetal placentas nineteen days after cloning; there were no great differences (tail-tip cell nucleus male was 0.30 grams, and cumulus cell nuclei female was 0.33 grams). However, these weights were twice those of fetal placentas of noncloned mice (male was

0.12 grams, and female was 0.15 grams).[86] As discussed earlier in this chapter, these problems preclude human cloning. Wakayama and Yanagimachi remarked:

> A high rate of abortion throughout pregnancy, high birth weight and frequent perinatal death are suspected to stem from incomplete nuclear reprogramming of donor cell nuclei after transfer. It is conceivable that imprinted genes involved in the regulation of placentation escape complete "reprogramming" following nuclear transfer and are aberrantly expressed during placental development. Thus incomplete communication between the placenta and embryo/fetus may be the major cause of the problems we have observed although further investigation is needed.[87]

These pre- and perinatal problems in cloning are not unique to mice; they have been seen in other cloning experiments. Understanding these problematic cellular mechanisms became urgent in the wake of the famous sheep Dolly (and subsequent mammalian cloning successes with endangered sheep species, goats, cattle, mice, pigs, rabbits, cats, rats, mules, African wild cats, ferrets, wolves, buffalo, camels, dogs, and old-world monkeys).[88] Dolly was the first mammal to be cloned from adult cells by a technique that is widely referred to in the twenty-first century as somatic cell nuclear transfer (SCNT). And Dolly — more than the results of any of the numerous cloning experiments that came before — refocused attention on the fantasy or theoretical possibility of human cloning.

Dolly's birth was announced to great press interest in February 1997 (she was actually born on July 5, 1996, one year and three months before Cumulina, overshadowing Cumulina, whose birth was equally remarkable).[89] What the public could not appreciate was the difficult journey that led to Dolly. For instance, in the early 1990s, geneticist and developmental biologist Ian Wilmut and his team at the Roslin Institute of the University of Edinburgh were keen to investigate "whether normal development to term is possible when donor cells derived from fetal or adult tissue are induced to exit the growth cycle and enter the $G_0$ phase of the cell cycle before nuclear transfer."[90] They derived three new populations of cells from a day-9 embryo, a day-26 fetus, and the mammary gland of a six-year-old ewe in the last trimester of pregnancy. They recovered eggs and enucleated them as soon as possible. Quiescent donor cells were produced by reducing the concentration of serum nutrients for five days, causing cells to exit the growth phases ($G_1$, S, and $G_2$) and enter the $G_0$ phase.[91] Chinese biologist Zhen Liu and a research team from the Institute of Neuroscience's Chinese Academy of Sciences, who successfully cloned macaque monkeys, first coaxed somatic cells into quiescence at one cell stage after transplantation to improve reprogramming after nuclei transfer.[92] Continuing with Dolly, fusion of donor cell to enucleated egg and activation were induced by electrical pulses. Most of the subsequent embryos (regardless of initial derived cell) were cultured in the oviducts of sheep, but some embryos derived from embryo or fetal cells were cultured in a chemical medium. Embryos that developed to the morulae and blastulae after six days of culture were transferred to recipient ewes and allowed to develop to term.

At about day 110 of pregnancy, the scientists counted four dead fetuses derived from embryo cells; two of them had abnormal liver development.[93] Eight ewes gave birth to live lambs (all cell lines represented), although one lamb died a few minutes after birth. The lambs displayed the phenotypical traits of the nuclei donors. In total, the scientists calculated that 62 percent of the fetuses were lost, "significantly greater than proportion than the estimate of 6% after natural mating."[94] Dolly had reached the respectable age of 6.5 years when she was euthanized because she was suffering from progressive lung disease and arthritis.[95]

Notably, the two macaque monkeys successfully cloned in 2017 by Liu and colleagues showed no enlarged umbilical cord, placenta, or head size (both 40 to 50 days after birth and in January 2018).[96] Pregnancy was confirmed in six surrogates by ultrasound a month after blastulae derived from fetal monkey fibroblasts were transferred to 21 female recipients. Four of them carried five fetuses, and the other two carried only gestational sacs. Two females lost their fetuses in early gestation, and the remaining two females gave birth to live baby monkeys.[97] In Wakayama and Yanagimachi's experiment, pregnancy was confirmed in 22 monkey surrogates after blastulae derived from adult monkey cumulus cells were transferred to 42 surrogates. Twelve of them carried live fetuses but eight aborted within 60 days, and two aborted within 90 days. The last two pregnancies yielded live baby monkeys, but they both died within 48 hours of birth due to postcranial abnormalities and respiratory failure.[98] That respiratory complications resulted in the deaths of Dolly and the macaque monkeys, both derived from adult cells, has not gone unnoticed by researchers. Other pathologies found in clones that died after a few minutes, hours, or days include cardiopulmonary pathology and juvenile-onset diabetes (found specifically in a clone derived from a 21-year-old male buffalo); mastitis-induced sepsis; septic and suppurative pneumonia; congenital megaesophagus pneumonia; dysentery; diarrhea; and incomplete closure of ventral body wall musculature with exteriorization of abdominal organs.[99]

In another incident that might be related, Snuppy, the world's first cloned dog, died at age 10 years while undergoing anticancer treatment.[100] Tai, a male Afghan hound and somatic cell donor for Snuppy, was diagnosed with hemangiosarcoma and euthanized at age 12 years (median life span of Afghan hounds is 11.9 years).[101] Was this cancer a result of the SCNT process (i.e., donor cells cultured prior to SCNT)? Were Tai's somatic cells malignant prior to SCNT, or were his cells benign but had a propensity for change? There is no doubt that population history and pre-existing conditions of donors would be serious complications in human cloning. Incidentally, Min Jung Kim in the College of Veterinary Medicine at Seoul National University, in collaboration with colleagues from Veterinary Medicine at Michigan State University and the University of Illinois at Urbana-Champaign, took adipose-derived mesenchymal stem cells from Snuppy when he was five years old and used them for recloning.[102] A total of 94 SCNT embryos were transferred to the oviducts of seven recipient dogs. Fifty-nine days later, four clones were born. Pregnancy and delivery rates for the reclones were 42.9 percent (three dogs from seven recipients) and 4.3 percent (four clones from 94 embryos), respectively.[103] None of the reclones had abnormal pathology when delivered; however, one clone died due to severe diarrhea.

Overall, it seems that proper growth and development of embryos produced by nuclear transfer depends on the cell cycle stage of donor nuclei and recipient egg and understanding epigenetic changes such as optimal conditions for regulation of gene expression, imprinting, chromatin structure, DNA methylation, and telomere shortening, which may affect the health and life span of the clone. All of these factors increase the difficulty and danger of human cloning tenfold.[104]

As mentioned earlier, rapid aging and shortened telomeres (two ends of a chromosome) are documented in clones. To investigate this, geneticist Paul Shields and colleagues at the Roslin Institute and PPL Therapeutics investigated whether telomere erosion is repaired after nuclear transfer.[105] In their 1999 report, they compared the telomere lengths of an animal derived from nuclear transfer to those of cells taken from the same population as the donor nucleus (using age-matched controls). Initially, the scientists found no difference between the telomere restriction fragment (TRF) of mammary cells derived from six-year-old ewes and age-matched controls.[106] This was a check for any telomere erosion — due to environmental influences — before nuclear transfer. In Dolly's case, one year after nuclear transfer, she showed reduced TRF average mean size compared to age-matched controls (19.14 kilobases versus 24.00 kilobases).[107] Dolly's TRF length was consistent with the TRF length of mammary cells. In the second year, there was a further reduction in Dolly's TRF to 18.24 kilobases.

Do cloned animals inherit the shortened telomeres of their adult predecessors, leading to premature aging? Does TRF length accurately reflect the actual physiological age of animals derived by transfer of adult cell nuclei? These important questions cannot be answered effectively by scientists at this time, which adds to the danger of human cloning. In fact, the evidence collectively indicates that success would be the exception rather than the rule. Nevertheless, Dolly was a healthy two-year-old in 1999 during her veterinary examination, despite having shortened telomeres. At this time, she had already given birth to two live lambs with mean TRF lengths of 17.9 kilobases that were consistent with noncloned lambs.[108] These TRF lengths increased within the first year and then decreased due to normal aging. Shields and colleagues noted this: "One surprising observation from these analyses was that there was an increase in ovine mean TRF size during the first year of life, from average of 18.6 to 24 +/– 0.2 kb [kilobases]."[109] It seems that telomere length is restored by the cloning process and reduced in lambs born by natural fertilization.

Environmental influences, such as frozen storage, laboratory techniques in nuclear transfer (repetition and multiple generation cloning), and length of in vitro cultivation all could lead to cumulative dysregulation of several imprinted genes, resulting in the abnormalities seen in clones.[110] These problems, coupled with the unstable epigenetic state of the ES-cell genome, have tremendous implications for use of human ES-cell technologies in clinics. In summary, the unknown variables in the application of cloning or any biotechnology are the long-term, heritable effects. Rideout and colleagues ended their report with an important statement that could be applied to any area of biotechnology:

> Embryos developing to birth obviously have appropriate expression of genes crucial for early development but may still have epigenetic defects affecting expression of genes activated later in development or adulthood. These considerations raise the possibility that even apparently healthy cloned animals may have subtle gene expression abnormalities that were not severe enough to cause lethality or an obvious postnatal phenotype. *On the basis of the available data, it seems almost certain that clones of all species, including humans, would be subject to these epigenetic abnormalities and their associated phenotypes.*[111] (italics added)

These abnormalities might even, unknowingly, be duplicated in the advent of genome editing. For instance, the rationale for genome editing is to produce better humans. So why not duplicate these better humans via cloning? But suppose an off-target gene edit turns on or off a different gene than intended (as occurred with attempts to correct a mutation in the *CFTR* gene for cystic fibrosis).[112] Such a mistake might be difficult to detect until adulthood, by which time the GM human could have been cloned several times, duplicating the mistake each time — a mistake that results in a series of abnormal GM clones.

## Notes

1. Ian Wilmut, "Cloning for Medicine," *Scientific American* 279, no. 6 (December 1998): 58.
2. Patrick Stephens, "Cloning: Towards a New Conception of Humanity," The Reproductive Cloning Network, January 2001. http://www.reproductivecloning.net/open/objectivist.html.

3. Daniel Kevles, *In the Name of Eugenics: Genetics and the Uses of Human Heredity* (Cambridge, MA: Harvard University Press, 1985), 263–264; Albert Rosenfeld, "The Second Genesis," *Life* 23 (June 13, 1969): 40.
4. Michael J. Sandel, *The Case against Perfection: Ethics in the Age of Genetic Engineering* (Cambridge, MA: Harvard University Press, 2007), 2–5.
5. Peter Raven, George B. Johnson, Kenneth A. Mason, Jonathan B. Losos, and Susan R. Singer, *Biology*, 9th ed. (New York: McGraw-Hill, 2008), 327, 330–331.
6. Ibid., 330–331.
7. Ibid.
8. Ibid.
9. Ibid.
10. Ibid.
11. The National Academies of Science, Engineering, and Medicine International Summit on Human Gene Editing: A Global Discussion, December 1–3 (Washington, DC: The National Academy Press, 2015), 1.
12. Raven *et al.*, *Biology*, 375–377.
13. Ibid., 377.
14. Ibid., 375.
15. Ibid., 376.
16. Ibid.
17. Ibid., 378–379.
18. Ira Levin, *The Boys from Brazil* (New York: Random House, 1976), 212–220, 286.
19. Ibid.
20. Josef Mengele's numerous "scientific" experiments and crimes against humanity are too numerous to discuss and beyond the scope of this book. Nonetheless, there are numerous books and films on Auschwitz, Nazi medicine, the Nuremberg Trials, and Simon Wiesenthal (the Nazi hunter) to do justice to the topic of Josef Mengele.
21. Raven *et al.*, *Biology*, 379.
22. Ibid.
23. Ibid.
24. Ibid.
25. Ibid.
26. Ibid.
27. Ibid.

28. David Berreby, *Us and Them: The Science of Identity* (Chicago: University of Chicago Press, 2005), xix. Berreby explains very clearly how the human brain creates experience and sometimes sacrifices truth/facts. This indirectly relates to the topic of cloning because some people believe human cloning is possible. "Nevermind movie special effects, online avatars, and vlogs — the brain itself is a medium, creating experience more than it perceives. It is, as more than one scientist has remarked, the ultimate virtual-reality chamber. When you feel the warm sun on your skin or taste an all-natural organic apple, your experience is not direct, but an interpretation created by your brain. Indeed, your brain is furnishing experience. It is the mind that makes world, scientists find, as they study the brain's function in all aspects of life" (p. xix).
29. Philomena Essed and David T. Goldberg, "Introduction: Cloning, Cultures, and the Social Injustices of Homogeneities" in Philomena Essed and Gabrielle Schwad, eds., *Clones, Fakes, and Posthumans* (Thamyris/Intersecting: Place, Sex, and Race) (Amsterdam: Rodopi Press, 2012), 106.
30. PBS documentary http://www.pbs.org/independentiens/frozenangels/.
31. Philomena Essed and Gabrielle Schwad, "Introduction: Cloning and Cultures of Replication," in Philomena Essed and Gabrielle Schwad, eds., *Clones, Fakes, and Posthumans* (Thamyris/Intersecting: Place, Sex, and Race) (Amsterdam: Rodopi Press, 2012), 9–10.
32. Ibid., 11.
33. Gabrielle Schwad, "Replacement Humans," in Philomena Essed and Gabrielle Schwad, eds., *Clones, Fakes, and Posthumans* (Thamyris/Intersecting: Place, Sex, and Race) (Amsterdam: Rodopi Press, 2012), 90.
34. Ross D. Parke, Christine W. Gailey, Scott Coltrane, M. Robin DiMatteo, "The Pursuit of Perfection," in Philomena Essed and Gabrielle Schwad, eds., *Clones, Fakes, and Posthumans* (Thamyris/Intersecting: Place, Sex, and Race) (Amsterdam: Rodopi Press, 2012), 112–116.
35. Philomena Essed and Gabrielle Schwad," Introduction: Cloning and Cultures of Replication," in Philomena Essed and Gabrielle Schwad, eds., *Clones, Fakes, and Posthumans* (Thamyris/Intersecting: Place, Sex, and Race) (Amsterdam: Rodopi Press, 2012), 14.
36. Levin, *The Boys from Brazil*, 212–220, 286.
37. Paul Shields, Alexander J. Kind, Keith H. Campbell, Ian Wilmut, David Waddington, Alan Colman, and Angelika E. Schnieke, "Analysis of Telomere Length in Dolly, a Sheep Derived by Nuclear Transfer," *Cloning* 1, no. 2 (June 1999): 121–122; Teruhiko Wakayama, Anthony C. Perry,

Maurizio Zuccoti, Kenneth R. Johnson, and Ryuzo Yanagimachi, "Full-term Development of Mice from Enucleated Oocytes Injected with Cumulus Cell Nuclei," *Nature* 394, no. 6691 (July 23, 1998): 370–371.
38. Philomena Essed and David T. Goldberg, "Introduction: Cloning, Cultures, and the Social Injustices of Homogeneities" in Philomena Essed, ed., *Clones, Fakes, and Posthumans* (Thamyris/Intersecting: Place, Sex, and Race) (Amsterdam: Rodopi Press, 2012), 102.
39. Ibid.
40. Gabrielle Schwad, "Replacement Humans," in Philomena Essed and Gabrielle Schwad, eds., *Clones, Fakes, and Posthumans* (Thamyris/Intersecting: Place, Sex, and Race) (Amsterdam: Rodopi Press, 2012), 80–86; Verena Stolcke, "Homo Clonicus" in Philomena Essed and Gabrielle Schwad, eds., *Clones, Fakes, and Posthumans* (Thamyris/Intersecting: Place, Sex, and Race) (Amsterdam: Rodopi Press, 2012), 29–32.
41. McKinnell and DI Berardino, *The Biology of Cloning: History and Rationale*, 875.
42. Ibid., 883.
43. Ibid.
44. Ibid.
45. Ibid.
46. Masood A. Shammas, "Telomeres, Lifestyle, Cancer, and Aging," *Current Opinion Clinic Nutrition Metabolic Care* 14, no. 1 (January 2011): 28–34.
47. McKinnell and DI Berardino, *The Biology of Cloning: History and Rationale*, 883.
48. Ibid., 881–882; Raven *et al.*, *Biology*, 380.
49. Raven *et al.*, *Biology*, 257–259. In 1928, British microbiologist Frederick Griffith was trying to make a vaccine that would protect the population against influenza, which was believed to be caused by *Streptococcus pneumoniae*. In the course of culturing this bacterium, he learned that there were two strains. He designated one strain the S form because it formed smooth colonies on the petri dish and the other R form because it formed rough colonies. Griffith also found that the S form had a polysaccharide coat. To test which strain was virulent, he injected each strain into mice. The mice injected with the S strain died of pneumonia. The mice injected with the R strain survived. At this point, Griffith knew that the S strain was virulent, and the polysaccharide coat was necessary for virulence. Subsequently, he injected one mouse with a mixture of dead S strain and live R strain, and the mouse died. High levels of live S strain were found in the lungs of the

dead mice. It seemed that genetic material was transferred between strains. Interestingly, Oswald Avery and coworkers prepared a mixture of dead S strain and live R strain and removed as much protein as they could from the mixture. Mice died despite the elimination of protein. The conclusion was that DNA was transferred between cells. Additional evidence to support Avery's conclusion came from Alfred Hershey and Martha Chase, who experimented with bacteriophages or viruses that infect bacteria. The virus binds to the cell's outer surface and injects its genetic material into the cell. The cell's genome is "fooled" into generating viruses until the cell ruptures. Hershey and Chase were curious to know whether viral DNA or protein was injected into the cell. To track this process, they needed to "label" both DNA and protein with radioactive isotopes. They used phosphorus ($^{32}P$) because DNA contains phosphorus and sulfur ($^{35}S$) is found in some proteins. The bacteriophage viruses were grown on a medium containing phosphorus, which was incorporated in the DNA, and a medium of Sulphur, which was incorporated in their protein coats. Then, they were allowed to infect bacteria. After infection, a large amount of the radioactive phosphorus was found in the cell and very little amount of radioactive sulfur. Hershey and Chase concluded that DNA, and not protein, was the genetic material injected into bacterial. In essence, genes, not proteins, were the unit of heredity.

50. McKinnell and DI Berardino, *The Biology of Cloning: History and Rationale*, 875–876.
51. Ibid., 876.
52. Ibid.
53. Ibid., 876.
54. Robert Briggs and Thomas J. King, "Transplantation of Living Nuclei from Blastula Cells into Enucleated Frogs' Eggs," *Proceedings of the National Academy of Sciences, USA* 38, no. 5 (May 1952): 455.
55. McKinnell and DI Berardino, *The Biology of Cloning: History and Rationale*, 876.
56. Ibid.
57. Raven *et al.*, *Biology*, 381.
58. John B. Gurdon, "Adult Frogs Derived from the Nuclei of Single Somatic Cells," *Developmental Biology* 4 (1962): 256–273. These somatic cells were either undetermined or determined by not fully developed, i.e., blastula to late gastrula or gut cells from hatched/swimming tadpoles. This is different from the present technique used today called somatic cell nuclear

transfer (SCNT), which uses somatic cells (already specialized for a specific organ/tissue) from an adult organism.

59. Robert Briggs and Thomas J. King, "Nuclear Transplantation Studies on the Early Gastrula (*Rana pipiens*)," *Developmental Biology* 2, no. 3 (January 1960): 252–270.
60. McKinnell and DI Berardino, *The Biology of Cloning: History and Rationale*, 876–877.
61. Gurdon, "Adult Frogs Derived from the Nuclei of Single Somatic Cells," 257.
62. Ibid., 69.
63. Ibid., 79.
64. Ibid., 70.
65. Ibid., 79.
66. Ibid.
67. Robert G. McKinnell, "Intraspecific nuclear transplantation in frogs," *Journal of Heredity* 53 (September–October 1962): 887–897.
68. Steven L. Stice and Carol L. Keeper, "Multiple Generational Bovine Embryo Cloning," *Biology of Reproduction* 48, no. 4 (April 1993): 715.
69. Ibid.
70. Ibid.
71. Ibid., 716.
72. Ibid., 717–718.
73. Gurdon, "The Transplantation of Nuclei between Two Species of Xenopus," 79–80. When discussing *Xenopus tropicalis* cytoplasm (fused with *Xenopus laevis laevis*), Gurdon states, "eggs contain specific substances, the presence or absence of which can prevent the normal functioning of a 'foreign' nucleus [belonging either to a different species]. These substances not only affect the functioning, but also the replication ..." In this case, I want to emphasize the *specific substances* (i.e., organelles, enzymes, etc.), if absent, might interrupt reprogramming of donor nuclei.
74. Stice and Keeper, "Multiple Generational Bovine Embryo Cloning," 718.
75. Teruhiko Wakayama, Anthony C. Perry, Maurizio Zuccoti, Kenneth R. Johnson, and Ryuzo Yanagimachi, "Full-term Development of Mice from Enucleated Oocytes Injected with Cumulus Cell Nuclei," *Nature* 394, no. 6691 (July 23, 1998): 371.
76. Ibid.
77. Ibid.
78. Ibid.

79. Ibid.
80. Ibid.
81. Teruhiko Wakayama and Ryuzo Yanagimachi, "Cloning of Male Mice from Tail-Tip Cells," *Nature Genetics* 22 (June 1999): 127.
82. Ibid.
83. Ibid.
84. Ibid.
85. Ibid.
86. Ibid.
87. Teruhiko Wakayama, Anthony C. Perry, Maurizio Zuccoti, Kenneth R. Johnson, and Ryuzo Yanagimachi, "Full-term Development of Mice from Enucleated Oocytes Injected with Cumulus Cell Nuclei," *Nature* 394, no. 6691 (July 1998): 369–374.
88. Wakayama and Yanagimachi, "Cloning of Male Mice from Tail-Tip Cells," 127.
89. Ian Wilmut, Angelika E. Schnieke, Jim McWhir, Alexander J. Kind, and Keith S. Campbell, "Viable Offspring Derived from Fetal and Adult Mammalian Cells," *Nature* 385, no. 6619 (February 1997): 811; Zhen Liu, Yijun Cai, Yan Wang, Yanhong Nie, Chenchen Zhang, Yuting Xu, Xiaotong Zhang, Yong Lu, Zhanyang Wang, Muming Poo, and Qiang Sun, "Cloning of Macaque Monkeys by Somatic Cell Nuclear Transfer," *Cell* 172, no. 4 (February 2018): 881–887.
90. Wilmut *et al.*, "Viable Offspring Derived from Fetal and Adult Mammalian Cells," 811.
91. Ibid.
92. Ibid., 812.
93. Zhen Liu, Yijun Cai, Yan Wang, Yanhong Nie, Chenchen Zhang, Yuting Xu, Xiaotong Zhang, Yong Lu, Zhanyang Wang, Muming Poo, and Qiang Sun, "Cloning of Macaque Monkeys by Somatic Cell Nuclear Transfer," *Cell* 172, no. 4 (February 8, 2018): 881–887; Rentian Wu, Zhiquan Wang, Honglian Zhang, Haiyun Gan, and Zhiguo Zhang, "H3K9me3 demethylase *Kdm4d* Facilitates the Formation of Pre-initiation Complex and Regulates DNA Replication," *Nuclei Acid Research* 45 no. 1 (January 9, 2017): 169–180; Sam Thiagalingam, Kuang-Hung Cheng, Han-Jung Lee, Nora Mineva, Aravinda Thiagalingam, and Jose F. Ponte, "Histone deacetylases: Unique Players in Shaping the Epigenetic Histone Code," *Annals of the New York Academy of Sciences* 983 (March 2003): 84–100. Histone 3 lysine 9 tri-methylation (H3K9me3) demethylase *Kdm4d* and histone deacetylase are

important olionucleotides and peptides in gene expression. To carry out gene expression, a cell must control the coiling and uncoiling of DNA around histone proteins. This is accomplished with the assistance of histone acetyl transferases, which acetylate the lysine residues in core histones, leading to a less compact and a more transcriptionally active chromatin. Histone deacetylases remove acetyl groups from lysine residues, leading to a condensed and transcriptionally silent chromatin. Histone deacetylase inhibitors, such as trichostatin A, block this action, causing hyper-acetylation of histones, facilitating reprogramming and gene expression. H3K9me3 demethylase *Kdm4d* facilitates the formation of preinitiation complex and regulates DNA replication.

94. Wilmut *et al.*, "Viable Offspring Derived from Fetal and Adult Mammalian Cells," 811.
95. Ibid.
96. Raven *et al.*, *Biology*, 381.
97. Zhen *et al.*, "Cloning of Macaque Monkeys by Somatic Cell Nuclear Transfer," 884.
98. Ibid., 883.
99. Wakayama and Yanagimachi, "Cloning of Male Mice from Tail-Tip Cells," 127.
100. Zhen Liu, Yijun Cai, Yan Wang, Yanhong Nie, Chenchen Zhang, Yuting Xu, Xiaotong Zhang, Yong Lu, Zhanyang Wang, Muming Poo, and Qiang Sun, "Cloning of Macaque Monkeys by Somatic Cell Nuclear Transfer," *Cell* 172, no. 4 (February 2018): 881–887; Deshun Shi, Fenghua Lu, Yingming Wei, Kuiqing Cui, Sufang Yang, Jingwei Wei, and Qingyou Liu, "Buffalos (*Bubalus bubalis*) Cloned by Nuclear Transfer of Somatic Cells," *Biology of Reproduction* 77, no. 2 (August 2007): 285–291; Ziyi Li, Xingshen Sun, Juan Chen, Xiaoming Liu, Samantha M. Wisely, Qi Zhou, Jean-Paul Renard, Gregory H. Leno, and John F. Engelhardt, "Cloned Ferrets Produced by Somatic Cell Nuclear Transfer," *Developmental Biology* 293, no. 2 (May 15, 2006): 439–448; Martha C. Gómez, C. Earle Pope, Angelica Giraldo, Leslie A. Lyons, Rebecca F. Harris, Amy L. King, Alex Cole, Robert A. Godke, and Besty L. Dresser, "Birth of African Wildcat Cloned Kittens Born from Domestic Cats," *Cloning and Stem Cells* 6, no. 3 (2004): 247–257.
101. Min J. Kim, Hyun J. Oh, Geon A. Kim, Erif M. N. Setyawan, Yoo B. Choi...Byeong C. Lee, "Birth of Clones of the World's First Cloned Dog," *Scientific Reports* 7, no. 15235 (November 2017): 1.

102. Ibid., 1–2.
103. Ibid., 1–2.
104. Ibid., 2.
105. William M. Rideout, Kevin Eggan, and Rudolf Jaenisch, "Nuclear Cloning and Epigenetic Reprogramming of the Genome," *Science* 293, no. 5532 (August 10, 2001): 1094.
106. Paul Shields, Alexander J. Kind, Keith H. Campbell, Ian Wilmut, David Waddington, Alan Colman, and Angelika E. Schnieke, "Analysis of Telomere Length in Dolly, a Sheep Derived by Nuclear Transfer," *Cloning* 1, no. 2 (June 1999): p.119.
107. Ibid., 12.1
108. Ibid.
109. Ibid., 123.
110. Ibid., 122.
111. Rideout *et al.*, "Nuclear Cloning and Epigenetic Reprogramming of the Genome," 1097.
112. Paul Knoepfler, *GMO Sapiens: The Life-Changing Science of Designer Babies* (Hackensack, NJ: World Scientific Publishing Co. Ptc. Ltd., 2016), 139, 145.

# Chapter 4

# Designer or Selected Babies: Self-Controlled Reproduction

For those interested in adding on eye color selection to their sex selection or genetic health procedures, we are offering a very limited substantial discount on the initial genetic testing of moms and dads with GREEN OR BLUE OR HAZEL EYES to determine if the genes you carry can actually be passed successfully and in what percentages. Because BLUE, GREEN, AND HAZEL initial testing is far more complex and expensive and is being refined by us, we are able to offer a major financial discount on initial screening for a limited time. This is a discount . . . currently available only to parents with blue, green, and hazel eyes.

— Jeffrey Steinberg, director, Fertility Institutes[1]

Soon it will be a sin of parents to have a child that carries the heavy burden of genetic disease. We are entering a world where we have to consider the quality of our children.

— Robert Edwards, British physiologist, cocreator of human IVF, and 2010 Nobel winner, 2013[2]

Chapter 1 discusses the subtle transformation after World War II of the old eugenics into the new eugenics, complete with its modified strategy of preventing the birth of diseased children in all populations by removing mutant genes from the human gene pool. In the 1960s and 1970s, the new eugenics flourished, as demonstrated by advances in genetics research and reproductive medicine aided by pharmaceutical companies and emerging biotechnology companies. As such, this chapter on the subject of designer babies could fit nicely into Chapter 1. However, this subject deserves a complete treatment.

Like human cloning and abortion, designer babies (or GM babies) provoke people's sensibilities and reignite the eternal debate on what is right or wrong in human society. At the official founding of the Kennedy Institute of Ethics on October 1, 1971, Reverend Robert J. Henle, president of Georgetown University, expressed his concerns about the then-emerging ARTs:

> Developments in medicine and biology in recent years have raised numerous ethical questions. Creating of "test tube babies" and cloning — which once were chilling concepts confined to life in Aldous Huxley's *Brave New World* — are now being studied. . . . Fertilization of human eggs in vitro and subsequent implantation into a uterus is already foreseen.
>
> Genetic engineering on human cells, whether and how much to prolong a dying patient's life, organ transplants and possible genetic abortions when tests show fetuses to have some genetic effect are other areas of concern.[3]

Jeffrey Steinberg, obstetrician-gynecologist, reproductive endocrinologist, and founder of Fertility Institutes in Encino, California, boasted that more than 200 women from foreign countries — where sex selection is illegal — have traveled to his clinics for help. "Women from countries like China and India tend to want boys," he said, exploiting the situation where pressure was placed on some of the women to adhere to their governments' one-child policy and preference for boys. "In the rest of the world, there's a slight preference for girls ... we're just happy to help them get it done."[4] The Fertility Institutes' website includes statements such as "leading IVF programs using Preimplantation Genetic Diagnosis (PGD)" and "Leading World Center for virtually 100% gender selection for family balancing" under a subheading of "Gender Selection."[5] Steinberg told Katie Moisse, science journalist, that daily injections of fertility drugs are intense for the women, but the drugs stimulate the ovaries to produce multiple eggs and prevent ovulation naturally: "They take a shot to trigger ovulation, and then they have 35 hours to get here. ... You've got to time it just right, and we'll sometimes do 24 women in two days. It's a marathon."[6] The eggs, according to Steinberg, are fertilized by the partner's sperm in the laboratory and left to develop into embryos. Subsequently, a single cell is screened for sex chromosomes and implanted into the woman's uterus.

PGD was designed to detect genetic abnormalities. According to Alan Handyside, professor of reproductive genetics at the University of Kent, and colleagues, it is a form of genetic diagnosis performed prior to implantation.[7] The patient's eggs are fertilized in vitro, and the embryos are kept in culture until a diagnosis (through blastomere or blastocyst biopsy, for example) is made.[8] These scientists stated that diagnosis can been made using several techniques, depending on the disease tested for. For example, polymerase chain reaction (PCR) methods are used for monogenic disorders, and fluorescence in situ hybridization (FISH) is used for chromosomal abnormalities and those cases in which no PCR protocol is available for an X-linked disease.[9]

Most young scientists — with their brand-new diplomas — instinctively want to go out and help humankind. Steinberg was no different; he began his work in reproductive endocrinology in the 1980s in order to help couples get pregnant. Mara Hvistendahl, Beijing-based correspondent for *Science*, wrote:

> Back in the 1980s, Steinberg tells me, he began offering in vitro fertilization to couples who were having trouble conceiving on their own. When PGD became available in the early 1990s, he started screening embryos for chromosomal disorders like Down syndrome. Then he started screening for single-gene disorders like cystic fibrosis. Then he offered sex selection to couples with a history of sex-linked disease. Finally he began offering the procedure to couples with no history of disease.[10]

The excitement over sex selection in America culminated with a *Newsweek* cover, dated January 26, 2004, of two blond babies accompanied with the following subtitle: "Girl or Boy? Now You Can Choose," which raised the ire of some experts who believed that sex selection was unethical (regardless of the fact that there were no laws against it).[11] Moisse noted a negative response to sex selection by Mark Hughes, one of the pioneers of PGD, who rejected sexual selection.[12] Hughes, a professor of molecular genetics and CEO of Genesis Genetics Institute in the Midwest, stated that he went into medicine "to diagnose and treat and hopefully cure disease. And last time I checked, your gender wasn't a disease."[13] In language that gives one example of the new eugenics, Moisse quoted Art Caplan, a medical bioethicist at New York University's Langone Medical Center, who explained that sex selection "crosses the line from treating infertility to making designer babies. ... Are we going to outlaw it? No. But should clinics try to discourage it? Yes. If we don't discourage it, what's going to happen when someone says, 'I want a tall blonde?'"[14] Other critics pointed out the high cost of sex selection ("running upwards up $20,000 out-of-pocket . . . [for] the combined cost of PGD and IVF"), which made designing a baby available only to the rich.[15]

What is troubling is that, according to Hvistendahl, the enthusiasm for sex selection is growing in the United States (unlike in Europe, where it is illegal).[16] She quoted Steinberg, who told her that "Americans are not intimidated by the technology. Nor are we averse to commercialized, made-to-order health care. Gender selection is a commodity for purchase. ... If you don't like it, don't buy it."[17] It is frightening that a respectable percentage of scientists in the United States, and around the world, have this laissez-faire attitude to using scientific technology. Since consumers drive the market and capitalism is an important aspect of the new eugenics, the same attitude might be repeated with human cloning if the technology becomes available. According to writer Jacob Brogan, there are a handful of commercial companies and institutions (many of them located in South Korea) committed to cloning pets for a price.[18] Brogan wrote that one of these companies, the US-based Viagen, charged $50,000 before taxes to clone a dog and $25,000 to clone a cat. In 2018, the world-famous singer and actress Barbara Streisand cloned her deceased dog. Brogan quoted a portion of Streisand's editorial from the *New York Times*:

> I was so devastated by the loss of my dear Samantha [the dog], after 14 years together, that I just wanted to keep her with me in some way. It was easier to let Sammie go if I knew I could keep some part of her alive, something that came from her DNA. A friend had cloned his beloved dog, and I was very impressed with that dog.[19]

When parents click "Gender Selection" on the Fertility Institutes' website, they are taken to a page that includes the phrase "Choose Your Baby's Eye Color."[20] Below this phrase is an image of a baby with accentuated blue eyes. According to the website, parents interested in adding eye-color selection to their sex selection or genetic health procedures will get a discount. Steinberg told Hvistendahl that "couples obsessed with blue or green eyes continue to call the office."[21]

In summary, advances in biotechnology are continually being pushed closer to the Rubicon by social-cultural forces. And, with preimplantation sex selection in mind, the next push that finally edges biotechnology across the Rubicon could be using the same technology to boost a child's intelligence, height, or muscle size.

## 4.1. Preimplantation Diagnosis (PGD), In vitro Fertilization (IVF), and Assisted Reproductive Technology (ART): Frontline Soldiers in the New Eugenics

Henry T. Greely, professor of law (and, by courtesy, professor of genetics) at Stanford University, used a hypothetical futuristic situation in which the present technique of PGD is made easier and cheaper through breakthroughs in the accuracy of whole-genome sequencing and the safety and effectiveness of human ES cells and iPSCs. He calls this improvement *Easy PGD*.[22] In 2016, Greely examined the possible implications of this emerging technology in terms of several issues, such as safety (including other uses of this proposed technology), equality, family, coercion, benefits, law, politics, and regulation.[23] Greely's writings are worth discussing because of his insights on new technologies and the economic, social, legal, and political forces that propel them. He states that "economic, social, legal, and political forces will combine to make this future [Easy PGD] not just achievable but, as I believe, inevitable, in the United States and in at least some other countries."[24] Greely's assessment is reasonable because the one theme of the US government when it comes to science and technology is to ensure US competitiveness. Greely's three major themes are safety and equality, regulation, and benefits.[25] This chapter focuses on safety and equality, while Chapters 10 and 11 address regulation and benefits, respectively.

In a future with Easy PGD, how safe would the technology be? Greely used the sketchy safety records of current PGD (and IVF) clinics to arrive at an answer.[26] At present, clinical procedures are not regulated by the federal government. He noted that despite the existence of published data on IVF and PGD babies (five percent of 158,000 IVF cycles employed PGD, and approximately 3,000 IVF children were born in the United States, making up 0.01 percent of all US births), there is no registry that allows IVF or PGD babies to be monitored into young adulthood to see how they do.[27] Nevertheless, the scant data available on children born as a result of IVF indicate that they are more likely to suffer from short- and long-term health problems compared to naturally conceived children. The root cause, Greely wrote, is multiple births that lead to a large number of miscarriages, stillbirths, and neonatal deaths.[27] Ironically, IVF doctors encourage parents to accept multiple births because multiple embryos increase the parents' chances of producing a baby more than one IVF cycle.[28]

Greely outlined a November 2008 report from the Centers for Disease Control and Prevention (CDC) on correlations between birth defects and IVF:

> Babies conceived with IVF have slightly higher rates of several birth defects that cause bodily malformations, such as cleft lip and palate, certain gastrointestinal system malformations, and some heart conditions. They looked at about 9,500 babies with birth defects and about 4,800 babies without; about 2.4 percent of the mothers of the babies with birth defects had used IVF, but only about 1.1 percent of the mothers of normal babies had done so.[29]

He briefly discussed two genetic diseases found in children, Beckwith-Weidemann syndrome and Angelman syndrome, which IVF children are more likely than average to have.[29] While Greely acknowledged that both IVF and natural conception come with risks, most of his discussion hinted that IVF held greater risks. He summed it up this way: "so how safe is IVF? Pretty safe — at least, safe enough to be used each year by about 160,000 American women."[30]

Easy PGD poses two main problems, according to Greely: (1) long-term health risks to children born as a result of the technique and (2) long-term risks to the PGD children and the human species due to the genetic selection (or design) possible using Easy PGD.[31] Greely made it clear that "Easy PGD is not about 'designer' babies, but about 'selected babies.'"[32] With all due respect to Greely, this is semantics. By selecting potential traits, one is essentially designing. Despite Greely's conclusion, the long-term risks to the entire species are the most frightening part of his discussion. For instance, what are the effects on the descendants of IVF babies (e.g., grandchildren and great-grandchildren)? This important question is posed elsewhere in the book in discussions of the various techniques of genetic engineering in relation to transgenerational or germ line effects and implications for the future of the human species. Chapter 3 discusses this point, for example. But the molecular geneticist Church had a different view on PGD in general, and germ line editing in particular. Paraphrasing Church, Kozubek wrote, "fewer mutations or risk are actually incurred when editing the germline than in somatic cells, since it involves fewer cells by orders of magnitude."[33]

Easy PGD may enable parents to select particular heritable genetic traits, but gene interaction is still somewhat of a black box and could play havoc in expression of traits. Because of this, Greely made an important statement about the pitfalls of genetic selection that is applicable more widely to the field of biotechnology:

> We do not know, with any certainty, what effects selecting for one variation in one particular gene might have when combined with the selection of another variation in a different gene, or in conjunction with a particular environment. It could be, when looking at two particular functions controlled by a different gene, that particular alleles of those genes might be superior for each of those functions, but that their combination would cause serious problems in some third, or fourth, or fifth function. *Picking an embryo that has genetic variations that predict both unusual height and unusually good ability at mathematics might lead to a child with a high risk for a nasty disease.*[34] (italics added)

Greeley's warning is particularly important because it brings to mind Church and his list of 10 specific genes that when edited (for gain-of-function or loss-of-function mutations) would produce "better" traits that people would desire.[35] For example, editing the gene *MSTN*, which encodes myostatin growth differentiation factor 8 and is one of genes on Church's list, results in increased muscle mass. However, a GM human with more muscle mass "could well be at increased risk for other diseases, such as cancer."[36] Chapter 15 discusses Church's list of 10 specific genes in detail.

If Easy PGD were available, parents would want to remove harmful genes from their embryos. Given that the natural environment is always changing, what if today's nonadaptive trait, encoded by a "harmful" gene, becomes tomorrow's adaptive trait? The following example is instructive. Both members of an African American couple are carriers of the sickle cell gene (for hemoglobin S). They decide to have a child, knowing that there is a 25 percent chance that the child will have sickle cell anemia and a 50 percent chance that the child will be a carrier of the mutant gene. Fortunately, Easy PGD indicates that the embryo has only one copy of the sickle cell gene. The parents decide to use Easy PGD to weed out the mutant gene. Over time, global warming generates an environment suitable for the *Anopheles* mosquitoes that carry the malaria parasite. Greely wrote that "if malaria spreads widely because of climate change or mutates in ways we cannot treat — our species could be at risk because of decisions by parents, using Easy PGD, that made sense in the short run — eliminate hemoglobin S — but not necessarily in the long run."[37] (Recall that carriers of the mutation survive better in a malarial environment because the malaria parasite cannot survive in sickled red blood cells.)

Moreover, Greely noted that with more than seven billion humans, the risk of short-term bad combinations of alleles would be small — as little as two percent of this number — if 10 percent of the population had the means to use Easy PGD and 20 percent chose combinations that resulted in harmful effects.[38] Despite only two percent of the global human population being at risk in the short-term, 120 million humans passing on mutations over time could have devastating effects on the future of the species.

As far as equality is concerned, would two percent of the global population be the only ones with access to Easy PGD? This question can also be applied to cloning, gene therapy, or genetic editing, although the answers will be different in each case. Greely's answer to this question was no.[39] However, he did expect disparities due to variations in geographical location, access, and level of Easy PGD service. Easy PGD would be classified as a clinical procedure and not a national screening program. Greely believes that health plans, private and public, would provide broad access to Easy PGD as a preventative measure to lower their costs. In connection with long-term effects, Greely made the following statement on a possible future of Easy PGD:

> Easy PGD will not lead, at least for many generations, to supermen and superwomen, but to people who do not have certain genetic diseases, who have a lower risk of getting other diseases, who have cosmetic features preferred by their parents, and who may have some marginal improvements in their behavioral traits. *It seems (to me) realistic to say that Easy PGD might produce humans that are about 20 percent healthier and, say, 10 percent both better looking and more talented.* What would follow from that? I think not much. With a 20 percent difference, the two populations would form bell curves with substantial overlap. While most Easy PGD children will be healthier, handsomer, and smarter than naturally conceived counterparts, many will not be.[40] (italics added)

The kind of society portrayed in the 1997 film *Gattaca*, in which potential children are conceived using prenatal screening to ensure they possess the best hereditary traits of their parents, is not viable in the long term and is a waste of talent, according to Greely. Most readers would agree that history has shown that human beings forget quickly and continued to make the same tragic mistakes of eugenics and genocide in the 20th century: the Nazi regime's extermination of six million European Jews in 1941–1945; Stalin's great purge in 1934–1939; Japan's Nanking massacre in 1937–1938; the Khmer Rouge regime's extermination of three million Cambodians in 1975–1979; the extermination by Rwanda's Hutu-led government and militia of approximately one million Tutsi, Twa, and moderate Hutus in 1994; and Bosnian Serbs' extermination of more than 8,000 Bosnian Muslims (and expulsion of 30,000) in 1995.[41] New divisions in society might emerge — PGD children (or GM children) versus naturally conceived children — that combine synergistically with existing divisions based on race, class, ethnicity, and religion to enhance tensions. Greely presented two examples of conflict. The first was a pair of verbal insults: "You're a dirty natural," to which the response would be, "Yeah, well so's your mother."[42] In the second example, imagine that Easy PGD proponents are referring to a Jewish couple who has a child with Tay-Sachs disease: "Why should *we* pay for the disabled children *they* chose to have?"[43] (Similar statements were used by officials in charge of the Nazi euthanasia program in 1939 to justify killing German citizens of all ages who were mentally ill or had incurable diseases.)[44] Another more fanciful but pertinent example of societal divisions which could emerge is between contemporary humans and a newly resurrected prehistoric human ancestor known as *Homo neanderthalensis*. The Neanderthal genome project published their initial draft of the genome of this species in 2010 and a more accurate version in 2013.[45] According to the initial sequences from 2010, Neanderthals share 99.7 percent of their nucleotide sequences with modern humans. If cloning or CRISPR technology (with the help of donated human eggs and brave young women to carry the embryos to term) were used to resurrect Neanderthals, they would, essentially, be GM organisms. How would we treat them? An employer might say, "We can't hire a Neanderthal for this position. They're not smart enough." The perception of them being brutish, unintelligent, and unsuccessful compared to Cro-Magnons, the earliest modern humans, has survived into the present and become a part of modern humans' psyche (the word *Neanderthal* is synonymous with *dumb* and *stupid*). The GM Neanderthal, already burdened by the name and prejudice, would be thrust into the highly charged modern environment of racial and ethnic conflict. In short, there should be strict guidelines for the use of genetic engineering to resurrect extinct species. It is troubling that a portion of the general public does not appreciate the seriousness of biotechnology but view it as an adventure. When the Neanderthal genome project was announced, for instance, Church received letters from many female applicants who volunteered to be surrogates to carry Neanderthal babies.[46]

Would parents spend more money to get more Easy PGD (for intelligence, height, eye color, hair color, etc.)? According to Greely, the answer is yes because it will involve more genes (translating into more traits from which to choose) and many more embryos to test.[47] For example, creating and testing 100 embryos might cost $10,000, while creating and testing 1,000 embryos might cost $100,000. Parents want the best for their children, and if they have the financial means, most will not pass on the deluxe version of Easy PGD. Unfortunately, biotechnological advancements in the environment of capitalism, regardless of scientists' instincts to help humankind, result in population inequality and long-term health risks for the entire human species.

Judith Daar, professor of law and clinical professor of medicine at the University of California, Irvine, generally confirmed Greely's arguments. Writing in 2017, Daar addressed the economics, equality, family, coercion, benefits, law, politics, and regulation of ART against the historical background of procreative deprivation and the American eugenics movement.[48] Daar, like Greely, noted geographical factors as a barrier to ART. She reviewed a study in which the researcher investigated the spatial disparity of ART centers based on their locations in relation to the population of reproductive age. She learned that underserved states and overserved states "had greater than 95 percent of the reproductive-age population living within sixty minutes of an ART center (including Connecticut, New Jersey, and Washington, D.C.)."[49] Daar also conducted her own search for clinics in Los Angeles County and found 13 clinics clustered in the west Los Angeles and southern San Fernando Valley areas, both inhabited by majority white populations.[50] She found no clinics in south Los Angeles or southeast Los Angeles, where Latinos and African Americans form the majority. Tethered to geographical factors are economics, because the working poor, who are either underinsured or uninsured (for this type of health care), are unable to treat their infertility. Daar made a salient statement on eugenics: "While early twentieth-century eugenicists targeted the poor as less worthy of reproducing than the rich, twenty-first century ART is more subtle but no less impactful in its sorting of prospective parents."[51] Twenty-first-century ART is part of the new eugenics, and it is indirectly targeting the poor through its cost.

But it was Daar's discussion of race and ethnicity as barriers to ART access that revealed the old eugenics hiding in plain sight in the era of new reproductive technologies. Daar, unlike Greely, delved deeper into racial and ethnic inequality by studying the myth of hyperfertility in poor women in general and women of color in particular.[52] She also discussed cultural perceptions of female infertility and mistrust of the health care system because of historical abuses.

Eugenicists in the early 20th century (in the United States and Europe) had a long-standing concern with the birth rates of different socioeconomic groups, particularly the greater fecundity of the poor compared to the middle and upper classes. In American history, women of color — particularly enslaved women of color — were "long portrayed as hyperfertile and hypersexual."[53] Thomas Jefferson recorded this stereotype in *Notes on Virginia* (1781).[54] Today, the stereotype has not changed, just merged with economics. The new label *welfare queen* appeared, and the assumption that followed is that African American women are "breeders" who will need government assistance.[55]

Daar dispelled the myth of hyperfertility in women of color.[56] Using data on incidence of infertility, Daar found the highest rates of infertility among African American women (20 percent), followed by Hispanic women (18 percent). White women had the lowest rate (7 percent). Similarly, infertility in men aged 25 to 44 years was highest among African American men and lowest among white men. These data invite the question, Why are these facts such a well-kept secret? The answers are location, cost, and the stubbornness of cultural myths.[57] First, African American women are less likely to visit a clinic, because clinic locations are not convenient. Second, African American women are less likely to receive an early diagnosis of infertility or seek medical treatment, because ART is expensive. Third, African American women have always been perceived (inside and outside the African American community) as mother figures (consider the endearing mammy during the 19th century). Consequently, the inability to conceive or carry a pregnancy to term would be embarrassing to an African American woman outside her community because she is not living up to the myth and, even more devastating, shameful within her community because she failed at motherhood.[58] Daar quoted a personal account reported by Dorothy Roberts, a lawyer and social justice advocate in reproductive health. Roberts reported one African American woman's reaction to her infertility: "Being African American, I felt that we're fruitful people and it was shameful to have this problem. That made it even harder."[59] And, quoting a minority patient, the

*New York Times* reported, "With women of color, specifically Hispanic and African-American women, the stigma attached to us is that it's not hard to have kids, and that we have a lot of kids. … And when you're the one that can't, you feel like, 'I've failed.'"[60]

There are no security borders for cultural myths or racial stereotyping. Like a colorless gas, they diffuse internally and externally, permeating everything in the environment, including physicians; physicians' attitudes and actions are consequently affected. Having internalized the African American female hyperfertility myth, a physician may consciously or unconsciously steer African American patients away from ART. Daar wrote:

> She [Roberts] cites the tendency of physicians to diagnose white professional women with infertility problems such as endometriosis that can be treated with IVF. Black patients who fail to conceive are more likely to be diagnosed as having pelvic inflammatory disease (PID), often treated with sterilization. In one study, 20 percent of black patients diagnosed with PID actually suffered from endometriosis.[61]

In short, the specter of the old eugenics, sustained by racial and economic discrimination, transformed into the new eugenics with the aid of the new reproductive technologies.

According to Daar, a further reason why infertility in women of color is such a well-kept secret and why treatment seeking is very low is historical mistrust of the health care system. Chapter 1 discusses the mandatory sickle cell screening laws established in the early 1970s. The authors of these laws failed to consider that a group already stigmatized by dark skin would be further stigmatized by sickle cell anemia. Moreover, sickle cell anemia became known among a majority of Americans as a "black" disease. Recall from Chapter 1 that this label fueled racial discrimination and even misguided eugenic-like laws, such as the New York state law ordering all persons "not of the Caucasian, Indian, or Oriental races" to be tested for the sickle cell gene before they were allowed to obtain a marriage license.

While racialized medicine contributed to the mistrust of the health care system by African Americans in the modern era, the roots of this mistrust can be found in the early 20th century, when many abuses of racial and ethnic minorities, poor people, and the mentally ill occurred and were sanctioned by state and federal government. The particular medical abuse described by Daar occurred during a clinical study called the Tuskegee syphilis experiment. In 1932, the US Public Health Service (PHS) initiated a clinical study to observe the incubation period and subsequent course of untreated syphilis in a group of African American men.[62] In the early 20th century, race and heredity were obsessions in science. Therefore, race was a major motivation in studying the progression of the disease. The PHS recruited the Tuskegee Institute, a historically black college in Alabama, for the study, and the staff in turn recruited poor African American men. The enticement for participation was free food, transportation, and minor health care. The PHS did not attempt to help "the four hundred men diagnosed with syphilis (two hundred control patients) even after the discovery of penicillin as a safe and effective treatment for syphilis in the late 1940s. More shocking to today's medical ethics sensibilities, the study persisted until 1972."[63] Men and women of color born after the 1980s might have no knowledge of the Tuskegee syphilis experiment, but they know through enculturation not to trust "the man" and the establishment's core institutions. As a result, a self-imposed barrier to ART has emerged, and scholars have perceived this crisis as one manifestation of the new eugenics.

## Notes

1. Fertility Institutes, "Choose Your Baby's Eye Color," accessed August 7, 2018, https://www.fertility-docs.com.
2. Osagie K. Obasagie, "The Eugenics Legacy of the Nobelist Who Fathered IVF," *Scientific American* (October 4, 2013), http://www.scientificamerican.com/article/eugenic-legacy-nobel-ivf/.
3. Robert J. Henle, "Bioethics Center Formed," *Chemical Engineering News* 49, no. 42 (October 11, 1971): 7.
4. Katie Moisse, "Boy or Girl? Clinic Offers Choice," *ABC News*, September 24, 2012, https://abcnews.go.com.

5. Fertility Institutes, "Baby's Eye Color"; Moisse, "Boy or Girl?"
6. Moisse, "Boy or Girl?"
7. Alan H. Handyside, Elena Kontogianni, Robert M. Winston, "Pregnancies from Biopsied Human Preimplantation Embryos Sexed by Y-Specific DNA Amplification," *Nature* 344, no. 6268 (April 14, 1990): 768.
8. Ibid.
9. Raven *et al.*, *Biology*, 339–40, 353–354. Polymerase chain reaction (PCR) is a technique used in molecular biology to amplify or copy a segment of DNA. In the first step, the two strands of DNA double helix are physically separated at high temperatures. Subsequently, the two strands become templates for DNA polymerase to selectively amplify the respective DNA segment. Polymerase is the enzyme that attaches the new bases onto template strands, and any DNA polymerase used must be able to withstand the 205 degrees F (95 degrees C) used to separate the two DNA strands. Taq polymerase, an enzyme isolated from the thermophilic bacterium *Thermophilis aquaticus* (found in hot springs), could withstand the heat required to split DNA into its two stands. Moreover, the process could be automated, and DNA could be copied continuously; Rudolf Amann and Bernhard M. Fuchs, "Single-cell Identification in Microbial Communities by Improved Fluorescent *in situ* hybridization Techniques," *Nature Reviews Microbiology* 6, no. 5 (June 2008): 339. Fluorescent in situ hybridization (FISH) is a molecular cytogenetic technique is used to detect and localize the presence or absence of specific DNA sequences on chromosomes. It enables researchers to visualize and map the genetic material in a patient's cells in order to understand a variety of chromosomal abnormalities and other mutations.
10. Mara Hvistendahl, *Unnatural Selection: Choosing Boys Over Girls, and the Consequences of a World Full of Men* (New York: PublicAffairs, 2011), 251.
11. Claudia Kalb, "Brave New Babies," *Newsweek*, January 26, 2004, http://tinyurl.com/6grvhof.
12. Katie Moisse, "Boy or Girl? Clinic Offers Choice," September 24, 2012, https://abcnews.go.com.
13. Ibid.
14. Ibid.
15. Ibid.
16. Hvistendahl, *Unnatural Selection: Choosing Boys Over Girls, and the Consequences of a World Full of Men*, 250–251.
17. Ibid., 251.

18. Jacob Brogan, "The Real Reasons You Shouldn't Clone Your Dog," March 22, 2018, https://www.smithsonianmag.com/science-nature/why-cloning-your-dog-so-wrong-180968.
19. Ibid.
20. Fertility Institutes, "Baby's Eye Color"; Hvistendahl, *Unnatural Selection: Choosing Boys Over Girls, and the Consequences of a World Full of Men*, 251.
21. Ibid.
22. Henry T. Greely, *The End of Sex and the Future of Human Reproduction* (Cambridge, MA: Harvard University Press, 2016), 3.
23. Ibid., 3–4.
24. Ibid., 2.
25. Ibid., 1–5.
26. Ibid., 210–213.
27. Ibid., 211.
28. Ibid.
29. Ibid., 212.
30. Ibid., 211.
31. Ibid., 213.
32. Ibid., 181.
33. Jim Kozubek, *Modern Prometheus: Editing the Human Genome with CRISPR — CAS9* (New York, NY: Cambridge University Press, 2016), 310.
34. Greely, *The End of Sex and the Future of Human Reproduction*, 214; Paul Knoepfler, *GMO Sapiens: The Life-Changing Science of Designer Babies* (Hackensack, NJ: World Scientific Publishing Co. Ptc. Ltd., 2016), 212–214.
35. Knoepfler, *GMO Sapiens: The Life-Changing Science of Designer Babies*, 187.
36. Ibid.
37. Greely, *The End of Sex and the Future of Human Reproduction*, 215.
38. Ibid.
39. Ibid., 137–153, 253–271.
40. Ibid., 239.
41. Adam Jones, *Genocide: A Comprehensive Introduction*, 3rd ed. (New York, NY: Routledge, 2017), 1–457.
42. Greely, *The End of Sex and the Future of Human Reproduction*, 245.
43. Ibid.

44. "A New People," the monthly magazine of the Bureau for Race Politics of the *Nationalsozialistische Deutsche Arbeiterpartei* (National Socialist German Workers' Party) or NSDAP for short. German poster (circa. 1938) reads: "60,000 koftet diefer Erbkranke die Volksgemeinfchoft ouf Lebenszeit. Volksgenoffe das ift auch Dein Geld." Translated: "60,000 Reichsmark is what this person suffering from a hereditary defect costs the People's community during his lifetime. Fellow citizens, that money is your money too."
45. Stenzel, Martin Kircher, Nick Patterson, Heng Li, Weiwei Zhai, Markus Hsi-Yang Fritz … Svante Pääbo, "A Draft Sequence of the Neandertal Genome," *Science* 328, no. 5979 (May 7, 2010): 710–712.
46. Kozubek, *Modern Prometheus: Editing the Human Genome With CRISPR — CAS9*, 311.
47. Greely, *The End of Sex and the Future of Human Reproduction*, 243.
48. Judith Daar, *The New Eugenics*: *Selective Breeding in an Era of Reproductive* (New Haven, CT: Yale University Press, 2017), 1–183.
49. Ibid., 96.
50. Ibid., 96–97.
51. Ibid., 55.
52. Ibid., 89–92.
53. Bethany Johnson and Margaret M. Quinlan, "Race, Racism, And Infertility," *Vital* (November 10, 2017), https://the-vital.com/2017/11/10/racism-infertility/.
54. Thomas Jefferson, *Writings*: *Autobiography, A Summary View of the Rights of British America, Notes on the State of Virginia, Public Papers, Addresses, Messages, and Replies, Miscellany, Letters* (London: Fitzroy Dearborn Literary Classics of the United States, 1781/1995), 75, 144–147.
55. Johnson and Quinlan, "Race, Racism, And Infertility," https://the-vital.com/2017/11/10/racism-infertility/.
56. Daar, *The New Eugenics*: *Selective Breeding in an Era of Reproductive*, 89–92, 80.
57. Ibid., 89–92.
58. Ibid., 89.
59. Ibid.
60. Ibid., 90.
61. Ibid., 91.
62. Ibid., 93–94.
63. Ibid.

# Chapter 5

# The "New World": Discovering the CRISPR System

> One sometimes finds what one is not looking for. When I woke up just after dawn on Sept. 28, 1928, I certainly didn't plan to revolutionize all medicine by discovering the world's first antibiotic, or bacteria killer. But I guess that was exactly what I did.
>
> — Alexander Fleming, 1945 Nobel winner for his 1928 discovery of benzylpenicillin from the mold *Penicillium notatum*[1]

> The people who discovered CRISPR, they weren't after genome editing, they just wanted to discover more of nature.
>
> — Rotem Sorek, Israeli molecular geneticist, 2017[2]

The CRISPR acronym is the microbiologists' explanation for an existing biological mechanism discovered in bacteria. These scientists perceive it as — to use the term originated by 15th-century explorers — the "New World," in the sense that they had never encountered the mechanism before even though it was always there. Christopher Columbus's initial goal was to find a faster route to Asia from Europe, but by chance he discovered the Americas instead (more accurately, he landed on San Salvador, an island of the Bahamas).[3] Like explorations in the 15th century, chance discoveries in science are ubiquitous. Unlike explorations in the 15th century, however, *recognizing the importance* of the chance discovery is the key factor in science. For example, during his Nobel Prize speech on December 11, 1945, Alexander Fleming stated, "The discovery of penicillin was a triumph of accident, a fortunate occurrence. ... My only merit is that I did not neglect the observation and that I pursued the subject as a bacteriologist."[4] In more recent times, Paul Berg (1980 Nobel laureate) and Janet Mertz (cancer researcher at the University of Wisconsin–Madison) — two biochemists reflecting on the emergence of recombinant DNA — wrote, "Although the emergence of recombinant DNA technology was transformational in its impact, the tools and procedures that were the keys to its development largely emerged as enhancements and extensions of existing knowledge, i.e., they were evolutionary, not revolutionary, in nature."[5] Essentially, the CRISPR system was always there, hiding in plain sight. It took microbiologists who understood existing knowledge of prokaryotic genetics to discover the CRISPR system. For example, they knew that in viral transduction (viral infection of a cell), DNA is transferred between cells by horizontal gene transfer (HGT).[6] The viruses used in gene-transfer studies are called bacteriophages (or just phages) because they specifically infect bacteria. Some of the well-known phages are the same ones that infect *E. coli.* (i.e., members of a "T" series that are all virulent or identified as lytic or causing destruction of cell: T1–T7).[7] According to Raven and colleagues, they are structurally and functionally diverse, complex, and consist of large amounts of DNA and protein. The type of transduction (generalized or specialized) and genes transferred (any gene versus a few genes) depend on the reproductive cycle of the phage.

In generalized transduction, the virus lyses (destroys) the infected host cell in which it is replicating: this is called the lytic cycle.[8] Specifically, the phage attaches to the outer surface of the bacterial cell and injects its genome into the host cytoplasm. Inside the bacterial cell, the phage hijacks the cell's replication and protein synthesis machinery to make viral components. Then mature virus particles are released through the action of enzymes that lyse the host cell. During the phage 'head packing' (i.e., replication of viral DNA in phage's head) the phage sometimes begins with bacterial DNA instead of phage DNA so that it (the phage) injects the bacterial DNA instead of viral DNA into the next cell it infects. This bacterial DNA can then be incorporated into the recipient chromosome by homologous recombination (genetic information exchanged between two chromosomes). In specialized transduction, phages carry both phage genes and chromosomal genes because they do not immediately kill the cells they infect.[9] Finding genes of phage A in phage B and vice versa was the beginning of the voyage that led to CRISPR.

On the voyage leading to CRISPR, scientists discovered some well-established cryptic prophages in the genomes of different bacterial species.[10] And some of them have altered their host cells in drastic ways, from creating virulence in a harmless bacterium to encoding functions important to cell immunity.

## 5.1. Serendipity: CRISPR in Bacteria

Since the 1960s, microbiologists have been aware of the innate immune systems of bacteria in the form of restriction endonucleases that protect them from phage infection.[11] Through natural selection, some species of bacteria and archaea (e.g., *E. coli*, *S. thermophiles*, *Yersinia pestis*, *Haloferax mediterranei*, and *Haloferax volcanii*) have acquired these enzymes. As discussed in Chapter 2, restriction enzymes recognize specific nucleotide sequences in a DNA strand, bind to the DNA on the recognition sequence, and cleave (cut) the DNA at the corresponding place in the sequence.

In 1987, molecular biologist Yoshizumi Ishino and colleagues in the Research Institute for Microbial Disease at Osaka University took advantage of the fairly new genetic engineering technology to analyze the nucleotide sequence of the *iap* gene in *E.coli*, which encodes alkaline phosphatase isozyme — a enzyme that breaks down proteins.[12] Using the technique recently perfected by geneticists Stanley Cohen of Stanford University and Herbert Boyer of University of San Francisco, Ishino and colleagues constructed a plasmid containing a minimum coding region of the *iap* gene. Then they created reconstructed plasmids by using restriction endonuclease *Eco*RI, obtained from *E. coli*, to cleave the plasmid at specific sites in order to add phage DNA. In short, they created recombinant DNA and plasmids. The reconstructed plasmids were introduced into a bacterial cell without the *iap* gene (*iap*−) and integrated into the cell's chromosome creating a high frequency recombination cell (Hfr cell).[13] Some of the *iap*− bacterial cells were transformed to $iap^+$. But the key part of this analysis is the unusual sequence found in the 3′-end flanking region (the region of DNA not transcribed into RNA) of *iap*. Essentially, Ishino and colleagues discovered a repetitive genetic code interrupted by another code that might be characterized as spacing between codes. The scientists described the "unusual structure" they found:

> Five highly homologous sequences of 29 nucleotides were arranged as direct repeats with 32 nucleotides as spacing. The first sequence was included in the putative transcriptional termination site and had less homology than the others. Well-conserved nucleotide sequences containing a dyad symmetry named REP [repetitive extragenic palindromic] sequences, have been found in *E. coli* and *Salmonella typhimurium* and may act to stabilize mRNA.[14]

Overall, Ishino and colleagues discovered (incidental to their research) these enigmatic repeat sequences whose recognition and, more importantly, significance would be officially transmitted to the scientific community in landmark papers published in 1993, 1995, and 2005 by Francisco J. M. Mojica in the Department of Physiology, Genetics and Microbiology at the University of Alicante, Spain.[15]

In 1989, Mojica was a doctoral student in the Department of Molecular Genetics and Microbiology at the University of Alicante, involved in the study of the archaeal microbe *H. meditterranei*.[16] He joined the laboratory of microbial ecologist Francisco Rodriguez-Valera, who was interested in the effect of salinity on gene expression in halobacteria. These researchers were aware that *Pstl* sites (a type II restriction endonuclease that cleaves DNA at the recognition sequence 5′-CTGCA/G-3′ and 3′-G/ACGTC-5′) in the *H. mediterranei* genome were more or less susceptible to cleavage depending on the salt concentration in which the cells were grown.[17] Based on this knowledge, their goal was to study the genome sequences that surrounded the two *Pstl* sites in order to find out which genes were expressed at different salt concentrations.

Mojica and Rodriguez-Valera selected two cloned sequences from the *H. mediterranei Pstl* genome, m122 and m61, for expression and sequence analysis. In addition to learning that the *Pstl* sites located inside the two cloned sequences were susceptible to cleavage at higher salt percentages, they found a 25 base pair alternating adenine-thymine tract within the noncoding region between open reading frames (ORFs) 122.1 and 122.2.[18] Mojica and Rodriguez-Valera presented their findings below:

> An imperfect purine-pyrimidine alternancy of at least 28 bp [base pairs] includes the *PstI* site, suggesting possible Z-DNA structures, and appears to be flanked by short direct repeats. *On the opposite side of the PstI recognition sequence, an unusually long stretch of 30-34 bp tandem repeats, spaced by 35 to 39 bp unique sequences, has been found. Each repeated sequence also includes short inverted repeats.* These tandem repeats extend for at least 900 bp and could obviously lead to the configuration of peculiar secondary structure.[19] (italics added)

Similarly, they found on opposite strands two 24-base-pair inverted repeats within the noncoding region between the starting codons of both ORF 61.1 and ORF 61.3. Like Ishino and colleagues, Mojica and Rodriguez-Valera had discovered CRISPR but were ignorant of its significance: "The presence in clone m122 of the 30 bp [base pair] tandem repeats seems to indicate a very peculiar DNA landmark. Similar structures appear in the *E. coli* chromosome located probably in non-coding regions, *but their biological role is unknown at present*"[20] (italics added). Unlike Ishino, however, Mojica's curiosity was piqued, resulting in 10 years of relentless study to learn the significance of the palindromic repeat sequences.

Building on the knowledge from previous research, such as regions of unusually long stretches of palindromic tandem repeats in *H. mediterranei*, *E. coli*, and other archaean and bacterial genomes found to be common in partitioning mechanisms of conjugate and P1 phage plasmids, Mojica and colleagues sequenced the genome of *H. volcanii* — a species closely to *H. mediterranei* — to find out if these tandem repeats also occurred in *H. volcanii* or any other haloarchaeal genomes.[21] In taking this research route, Mojica hoped to discover the biological role of the tandem repeats.

By the late 1990s, Mojica and colleagues combined their results with information on *E. coli* and learned that *E. coli*'s tandem repeats in the *sopC* region of the F plasmid (one of the genes that controls replication in *E. coli*) and chromosome were similar to the tandem repeats in *H. volcanii*, suggesting that more archaea and bacteria genomes have these repeat similarities and, more importantly, that the repeats serve a common function.[22] Finding the common function for these repeats in archaea and bacteria became Mojica's obsession.

In 2000, Mojica and colleagues embarked on a study to find out if these repeated elements occurred in prokaryotic groups phylogenetically distant from the groups where the repeats had already been detected.[23] At this point in Mojica's long research journey, he was ready to officially name this "new family of prokaryotic repeats" and describe them in detail.[24] Mojica and his colleagues described the repeats this way:

> They are repeated short elements generally occurring in clusters, but their main peculiarity is the layout: they are always regularly spaced by unique intervening sequences of constant length. *For the sake of clarity*, and ensuing from the mentioned characteristics, we will refer to the members of this family of repeats as *Short Regularly Spaced Repeats* (SRSRs).[25] (italics added)

It is notable that the term *short regularly spaced repeats* would be further modified two years later for the sake of clarity.[26]

To further elucidate the relevance of the SRSR elements, Mojica and colleagues performed a search of completed and partially completed microbial genome sequences and found the SRSRs widespread among the phylogenetic groups of all archaea and hyperthermophilic bacteria (lives in extremely hot environment), some members of the cyanobacteria and proteobacteria lineages, and two subgroups of Gram-positive bacteria.[27] Mojica and colleagues described the main features of the SRSRs as "short partially palindromic sequences of 24–40 bp [base pairs], containing inner and terminal inverted repeats of up to 11 bp and arranged in clusters of repeated units spaced by unique intervening 20–58 bp sequences … the extent of the clusters is particularly noteworthy in the Archaea."[28]

The scientists noted that the SRSR sequence was similar and conserved in members of the same phylogenetic group, indicating common origin. They found, for example, that *H. volcanii* differs from *H. mediterranei* in three out of 30 base pairs, *Pyrococcus horikoshii* differs from *Pyrococcus abysil* in two out of 29 base pairs, and *E. coli* differs from *Salmonella typhi* in only one out of 29 base pairs.[29]

The question Mojica needed to answer in light of the new data was whether this family of SRSRs had a common function in all these prokaryotes. It did not take long for other researchers to recognize the significance of these SRSRs and attempt to answer the same question.

## 5.2. Explaining CRISPR

In 2001, molecular microbiologist Ruud Jansen at Utrecht University and colleagues at the Netherlands' National Institute of Public Health were studying repetitive DNA sequences present in both domains of prokaryotic organisms.[30] The researchers described the repeats as short sequence repeats (SSRs) and grouped them into two classes: contiguous repeats and interspersed repeats. Based on their data, contiguous repeats "varied from strain to strain and this genetic heterogeneity in bacterial species may lead to phenotypic differences due to differential gene transcription or translation."[31] Interspersed repeats were less than 200 base pairs in length, noncoding, and widely dispersed throughout the genome. For instance, Jansen and colleagues found interspersed SSRs in the REP and enterobacterial repetitive intergenic consensus sequences of Enterobacteriaceae, *S. pneumonia*, *Neisseria gonorrhoeae*, and *H. influenzae*. Similarly, Mojica found interspersed SSRs in bacterial and archaeal species such as *E. coli*, *H. mediterranei*, *S. pyrogenes*, *Anabaena* sp., and *M. tuberculosis* among others.[32] Furthermore, Jansen and colleagues observed that different bacterial phyla (or even genera) had different SSR sequences.

The research teams of Mojica and Jansen were independently studying the same family of repeats but applying slightly different terminology to describe the repeats: SRSRs versus SSRs.[33] To avoid confusion, Jansen and colleagues decided to use a holistic description of the characteristic features of the family of repeats found in the genomes of various prokaryotic species. They coined the term *clustered regularly interspaced short palindromic repeats* to refer to them, and their collaborative 2002 publication first introduced the acronym *CRISPR* to the public.[34] Jansen and colleagues wrote:

> A characteristic of the CRISPRs, not seen in any other class of repetitive DNA, is that the repeats of the CRISPRs are interspaced by similarly sized non-repetitive DNA. The direct repeats vary in size from 21 bp [base pairs] in *Salmonella typhimurium* to 37 bp in *S. pyogenes*, and they are clustered in one or several loci on the chromosome.[35]

While Jansen knew much about the characteristic features of the CRISPR loci, he was unsure about their function despite Mojica having used *H. mediterranei* and *H. volcanii* to show that CRISPR might be involved in replicon partitioning. Nevertheless, in their search for the completely sequenced and partially sequenced genomes of prokaryotic species, Jansen and colleagues discovered CRISPR in more than 40 prokaryotic species (some of them already identified by Mojica).[36] Additionally, they replicated or confirmed Mojica's data concerning the narrow size range of the prokaryotic repeat sequences and spacer sequences within a given CRISPR locus (21 to 37 base pairs). But Jansen and colleagues also took a separate track in their research, focusing on several aspects of the CRISPR loci: the number of loci, which varied from a single locus in *M. tuberculosis and Neisseria meningitides*, for example, to 20 loci in *Methanococcus jannaschii*; the number of repeats within each locus, which varied from two to as many as 124 repeats (in *Methanobacterium thermoautotrophicum*); and, more interesting, the four genes associated with the CRISPR loci.[37] They identified these four genes in the flanking regions (either side) of some loci. They designated these genes CRISPR-associated genes 1–4 (*cas1*, *cas2*, *cas3*, and *cas4*).[38] They recognized that the *cas* genes in the genome of most species were present in a cluster ordered *cas3-cas4-cas1-cas2*, within a few hundred base pairs of the CRISPR locus, suggesting transcriptional organization of the *cas* genes in these species. Jansen found *cas1* gene in all species with CRISPR loci and the other three *cas* genes in some CRISPR-containing genomes; he believed that there was a functional relationship between the *cas* genes and CRISPR.[39]

Furthermore, in species with multiple CRISPR loci with the same repeat sequence, Jansen and colleagues detected *cas* genes near only one of the CRISPR loci. For example, the researchers noted that *A. fulgidus*, *P. horikoshii*, *P. multocida*, and *S. pyogenes* carried two classes of CRISPR loci with two different repeat sequences: "In these species we found two different sets of *cas* genes, indicating that each class of CRISPR has its own set of *cas* genes."[40]

The Cas proteins were analyzed, but the researchers could not deduce a clear function for some of these proteins. They did, however, find differences within and between closely related species; that is, there were no similarities between two Cas proteins from archaea species *A. fulgidus* and *P. horikoshii* and bacterial species *P. multocida* and *S. pyogenes*.[41]

Most notably, they stated that Cas1 proteins had a high isoelectronic point characteristic of proteins that bind to DNA, suggesting an "affinity of these proteins for nucleic acid."[42] Jansen, with great foresight, speculated that Cas1 proteins could have the CRISPR DNA sequences as targets.

Equally notable, the scientists predicted that Cas3 proteins were superfamily 2 helicases, enzymes which modify DNA, repair DNA, regulate transcription, segregate chromosomes, and recombine chromosomes.[43] For Cas4 proteins, they recognized similarity to the family of RecB exonucleases (cleave DNA from both ends). This predicted functionality of the Cas proteins (and *cas* genes) would become critical in later research, where the focus was to transform the CRISPR system into a technology to save people.

Jansen and colleagues suspected that the presence of multiple CRISPR loci at different locations in many prokaryotes (e.g., unrelated species such as *E. coli* and *M. avium* carry nearly identical CRISPR sequences) suggested mobility of these genetic elements and their dissemination among genetically unrelated microorganisms by lateral DNA transfer.[44] Their suspicions of lateral DNA transfer of CRISPR and the *cas* genes were based on "the absence of a phylogenetic relationship between the Cas proteins that corresponds to the phylogenetic relationship of the species."[45] Collectively, Mojica and Jansen found that the multiple CRISPR loci in archaeal and bacterial genomes each had their own unique spacer sequences. Jansen and colleagues believed that this evidence supported the idea that "the leader sequence is translocated in the genome in which it grows by duplicating repeats and generating its own unique set of spacer sequences."[46] If the spacer sequence evolved independently for each CRISPR locus in a genome, the researchers could not pinpoint the mechanism. But they believed that the Cas proteins might be involved.

## 5.3. The Biological Function of CRISPR

By December 2004, Mojica and colleagues had finally proposed a valid biological function for CRISPR in general and the spacer sequences in particular. In their landmark paper published in January 2005, Mojica and colleagues stated the following:

> We have carried out a systematic search for spacer identities, finding significant similarities to a variety of DNA molecules. The highest identities are with genetic elements, including chromosomes, bacteriophages, and conjugative plasmids, of strains closely related to the one containing the spacer. Interestingly, these targeted viruses are unable to infect the spacer-carrier cell but succeed with closely related strains lacking the specific CRISPR spacer. Likewise, plasmids efficiently transferred among various species in the same phylogenetic group cannot be stably maintained in members with a CRISPR spacer matching a sequence in the replicon. *The relationship between CRISPR and immunity against targeted foreign DNA is discussed in relation to its functional and evolutionary significance.*[47] (italics added)

The last sentence is important because it established Mojica as the first molecular microbiologist to elaborate the biological function of CRISPR (in addition to being the first biologist to recognize the importance of CRISPR).

In their search of 4,500 CRISPR spacers from 67 strains representing 36 prokaryotic genera, Mojica and colleagues found significant similarities for 88 spacers in 4 strains of archaea, 12 strains of Gram-negative bacteria, and 9 strains of Gram-positive bacteria.[48] They also learned — what would become critical information in understanding the biological function of CRISPR — that 47 out of the 88 spacers matched sequences with genes corresponding to bacteriophages, 10 within plasmid DNA, and 31 with chromosomal DNA. Subsequently, Mojica chose the workhorse *E. coli* (although no similarities to phage genes were found in its CRISPR spacers) and three other microorganisms (*S. pyogenes*, two *Sulfolobus* species, and *M. thermoautotrophicum*) on which to conduct further studies of CRISPR spacers.[49]

Beginning with the CRISPR spacers in *Sulfolobus solfataricus*, Mojica and colleagues learned that *Sulfolobus islandicus* rod-shaped viruses 1 (SIRV1) could penetrate the cell membrane of *S. solfataricus* but were unable to infect the genome that had CRISPR spacers similar to the SIRV1 sequences.[50] In contrast, simian sarcoma virus 1 (SSV1) was able to infect the genome of *S. solfataricus* because, according to Mojica, there were no CRISPR spacers similar to SSV1 sequences. Additionally, *S. tokodaii* had four CRISPR sequences similar to SIRV1 sequences,

SSV1 sequences, and conjugative plasmids. In the other archaea microbe, *M. thermoautotrophicum*, the scientists found nine CRISPR spacers showing similarity to the *Methanothermobacter* prophages: four spacers in *M. wolfeii* prophage and five spacers in *M. marburgensis* prophage.[51]

In conclusion, Mojica and colleagues confirmed their 2004 statement in which they tentatively linked CRISPR to providing immunity from infection. They stated, "Indeed, the preferential occurrence of CRISPR spacers derived from genetic elements that fail to infect the corresponding spacer-carrier strain ... strongly suggests a relationship between CRISPR and such immunity."[52] In essence, the interruption of the phage genes responsible for critical functions (plasmid transference, DNA replication, virus assembly, replicon partitioning, phage integration, and incision among others) could prevent infection. And Mojica believed that this activity could be performed by "CRISPR-RNA molecules, acting as regulatory RNA that specifically recognizes the target through the homologous RNA-spacer sequence, similarly to the eukaryotic interference RNA."[53] Two years later, in 2007, another group of researchers would continue the voyage of discovery, focusing their attention exclusively on understanding the workings of CRISPR-RNA molecules in conjunction with the Cas system.[54] Within five years, they would successfully engineer a technology whereby this natural system could be used in genetic editing.

Meanwhile, other research teams were independently studying CRISPR structures, replicating Mojica and Jansen's findings, and generating new information about CRISPR. One of these research teams was led by molecular microbiologist Alexander Bolotin at the Institut National de la Recherche Agronomique, Paris. In a May 2005 paper, Bolotin proposed — building on the foundation laid down by Mojica — that the "widespread presence of CRISPRs in bacterial genomes may be due to their protective function against foreign DNA invasion."[55] Bolotin and his team used computational biology to analyze the vast amounts of molecular microbiology data, finding another cluster of three genes associated with CRISPR that they called *cas1B*, *cas5*, and *cas6*. (The Cas 5 family group consists of large proteins that cause cellular death by introducing double-stranded breaks into DNA, and Cas 6 consists of short protein of high isoelectronic point.)[56] They also found that some spacers had similar genes to phage sequences and extrachromosomal elements, confirmed by sequence analysis of CRISPR structures from 24 strains of *S. thermophilus* and *S. vestibularis*. Bolotin and his colleagues found a strong correlation between the number of spacers in *S. thermophiles* and its resistance to phage infection; they believed that the mechanism involved in phage resistance might work via an "anti-sense RNA inhibition of phage gene expression."[57]

Simultaneously, molecular microbiologist Christine Pourcel at the Institut de Génétique et Microbiologie, Université Paris-Sud and colleagues analyzed three CRISPR elements in several variant strains of *Yersinia pestis*, commonly known as bubonic plague and found worldwide, and nine strains of *Yersinia pseudotuberculosis*.[58] They found one of these CRISPR elements, designated YP1, located near the replication terminus, where levels of recombination are high, and it was and associated with *cas1* and *cas3* genes.[59] Pourcel and colleagues noted that a 28-base-pair repeated sequence was found interspaced with spacers approximately 33 base pairs in length in all three CRISPR elements. They believed that the majority of these spacers were similar to fragments of a prophage. The researchers wrote, "The observation that *Y. pestis* CRISPR loci acquire new spacers from a prophage DNA is quite striking. Most of these phage sequences are also present in the *Yersinia enterocolitica* genome separated 41–186 million years ago."[60] They reasoned that while the prophage was present in an ancestor, the descendant *Y. pestis* CRISPR loci have acquired a new sequence pattern.

Like Bolotin, Pourcel and colleagues exploited computational biology to propose a mechanism by which CRISPRs could acquire phage DNA. They performed a search to find regions of local similarity between gene sequences for the *S. pyogenes* spacer sequences against five sequenced genomes; only Group A Streptoccocal (GAS) strain of *S. pyogenes* had CRISPR according to the researchers.[61] Interestingly, apart from the GAS strain, they found that seven out of the nine spacers had phage-associated sequences present in their genomes. The scientists believed that there might be a correlation between the presence of CRISPR and the absence of a corresponding prophage, because CRISPR is able to use pieces of foreign DNA as part of its defense mechanism.[62] In the end, Pourcel and colleagues concluded, like Mojica and Bolotin, that "CRISPRs may represent a memory of past 'genetic aggressions.' "[63]

By 2007, a great deal of information about CRISPR appeared in the literature, and researchers had begun to exploit the CRISPR mechanism for applications in medical genetics. But molecular microbiologists Rodolphe Barrangou and Philippe Horvath, senior researchers at DuPont-Danisco Food Processors and Manufactures in Wisconsin, along with Hélène Deveau, of the Department of Biochemistry and Microbiology, Université Laval, Québec, were more concerned with gathering more conclusive data on the function of the spacers in CRISPR loci.[64] Barrangou and Horvath's study of CRISPR appears to have been rooted in two important facts: (1) from an industrial perspective, domesticated bacteria used in the formation of dairy cultural systems for the production of yogurt and cheese were often susceptible to phage attack; and (2) from an evolutionary perspective, the universal progressive proliferation of phages results in bacteria evolving defense strategies (i.e., blocking absorption, restricting incoming DNA, or preventing DNA infection) to counter phage infection.[64] The research model for their analysis was *Streptococcus thermophilius* strains because the regular use of these strains in dairy processing had "soured the milk" due to the presence of phages.[65] The analysis of the spacers in the CRISPR loci revealed sequence homology with bacteriophages and plasmid sequences. Before Barrangou and colleagues began their analysis, they knew about the working hypothesis from other research that CRISPR and the *cas* genes provided immunity against phage infection by a mechanism based on RNA interference (iRNA), in which animal and plant cells form RNA double helixes to prevent viral infection.[66] Their strategy was threefold: (1) perform effect-on-function and gain-of-function studies on spacers in CRISPR loci to test the working hypotheses using the CRISPR1 locus in *S. thermophiles*; (2) investigate the origin and function of additional spacers in phage-resistant mutants; and (3) further understand the role of CRISPR1 in phage-host interactions in *S. thermophiles*.[67]

In first part of their project, the scientists analyzed the CRISPR sequences of different *S. thermophile* strains (industrial and phage-resistant variants) and found several interesting features: spacers were clustered overwhelmingly at the CRISPR1 locus, and the spacers of phage-resistant strains, while nearly identical to the parental strains, had additional spacers. They deduced "a potential relationship between the presence of additional spacers and the differences observed in the phage sensitivity of a given strain."[68]

To test their hypothesis that the CRISPR loci of the strains that were resistant to phage attack were modified, Barrangou and colleagues created what they called a "phage-host model system," in which a phage sensitive wild-type *S. thermophilus* strain used in the dairy industry, DGCC7710, would be challenged by two virulent bacteriophages: phage 858 and phage 2972 (isolated from products in the dairy industry).[69] Initially, the scientists challenged the wild type with phage 858 and phage 2972 independently. Then they used both phages simultaneously. In the end, they generated nine phage-resistant mutants with variable CRISPR1 loci.[70] For instance, the mutants had one to four additional spacers inserted next to the original 32 spacers present in the wild type toward one end of the CRISPR1 loci. And, more importantly, sequence analysis of the additional spacers in the phage-resistant mutants revealed similarity to phage 858 and phage 2972 genome sequences. The scientists concluded that "on becoming resistant to bacteriophages, the CRISPR1 locus was modified by the integration of novel spacers, apparently derived from phage DNA. The presence of a CRISPR spacer identical to a phage sequence provides resistance against phages containing this particular sequence."[71]

On further examination of the spacers, the scientists found other interesting characteristics. For instance, single-nucleotide polymorphism between spacer and phage sequence resulted in no resistance to phage, and the more spacers added, the greater was the resistance attained.[72] In short, driven by phage exposure, some bacterial species adapt rapidly in order to survive and reproduce. At this point in their analysis, the scientists decided to conduct effect-on-function experiments by adding and deleting spacers in the CRISPR1 loci and then testing the sensitivity of the resulting strains to phages.

In the end, the researchers agreed that iterative addition of spacers in *S. thermophiles* strains occurred regularly and resulted in higher levels of phage resistance. Barrangou and colleagues presented some key conclusions based on their results: (1) in CRISPR-mediated phage resistance, a bacterial strain acquires a new spacer specific to the phage genome without damage to the strain, and with each phage challenge, new generations of multiresistant *S. thermophiles* strains are formed; and (2) the position of the spacer (always inserted at the leader end) might serve as a memory of a previous phage attack.[73]

## 5.4. CRISPR Cascade Complex

As the focus on the spacers in CRISPR loci continued, other scientists (studying Jansen's data) tried to further elucidate the role of the *cas* genes. Molecular microbiologist Stan J. Brouns of Wageningen University in the Netherlands, and collaborators from the University of Sheffield in the United Kingdom and the NIH's National Library of Medicine analyzed the *E. coli* K12 CRISPR/*cas* system and identified the Cascade complex, a CRISPER-associated protein complex composed of five Cas proteins (CasA, CasB, CasC, CasD, and CasE) which they believed were responsible for antiviral defense.[74] In the next step of their research, Brouns and colleagues performed effect-on-function studies, such as deleting a *cas* gene to find out the effects on formation of transcripts of the CRISPR region in *E. coli* K12. Analysis revealed that a small CRISPR-RNA (crRNA) product was present in the control strains of the wild type and a non-*cas* gene knockout.[75] Interestingly, Brouns and colleagues found crRNA in much greater quantities in the *casA*, *casB*, and *casC* knockout strains (and it was absent from strains lacking the *casE* gene).[76] In essence, crRNA originates from a crRNA precursor (pre-crRNA), and *casE* is essential for generation of crRNA.

To test whether crRNA combined with Cascade promotes phage resistance, Brouns and colleagues designed two artificial CRISPRs against four essential genes in phage λ.[77] First, the coding CRISPR ($C_{1\text{-}4}$) generated crRNAs complementary to both the mRNA and the coding strand of the four genes. Then the template CRISPR ($T_{1\text{-}4}$) generated crRNA complementary to the template strand of the same protospacer regions. Brouns and colleagues subsequently performed plaque assays with virus-infected *E. coli* and learned that the introduction of $C_{1\text{-}4}$ or $T_{1\text{-}4}$ anti-λ-phage CRISPRs in the *E. coli* strain expressing only Cascade did not result in reduced sensitivity to virulent phage λ.[78] Remarkably, they found that *E. coli* strains that expressed Cascade *and* Cas3 were less sensitive to phage infection.[79] Brouns wrote, "The template CRISPR rendered the strain insensitive to the phage at the highest phage titer tested

($>10^7$-fold), whereas the coding CRISPR reduced the sensitivity $10^2$-fold) and produced plaques with a diameter of the standard λ plaque."[80] In the end, the scientists learned that both Cascade and Cas3 were critical to phage resistance (as with the omission of Cas3, phage resistance was lost when Cascade was omitted).

Brouns and his collaborators added a major piece of the puzzle in the continuing race to further elucidate the function of the CRISPR/Cas system. They stated:

> Our results demonstrate that a complex of five Cas proteins is responsible for the maturation of pre-crRNA to small crRNAs that are critical for mediating antiviral response. ... The Cascade-bound crRNA serves as guide to direct the complex to viral nucleic acids to mediate an antiviral response. We hypothesize that crRNAs target virus DNA, because anti-λ CRISPRs of both polarities lead to a reduction of sensitivity to the phage. ... We conclude that the transcription of CRISPR regions — and the cleavage of pre-crRNA to mature crRNA by Cas proteins — is the molecular basis of the antiviral defense stage of the CRISPR/cas system, which enables prokaryotes to effectively prevent phage predation.[81]

Other scientists, such as molecular geneticist Jason Carte at the University of Georgia and colleagues from the Department of Chemistry and Biochemistry Florida State University, came to similar conclusions a few months later. In a November 2008 paper, Carte and colleagues identified *Pyrococcus furiosus* Cas6 as a new endoribonuclease (an enzyme that cleaves double-stranded or single-stranded RNA) that cleaved CRISPR RNAs within the repeat sequences to release individual invader targeting RNAs.[82]

The finding that Cascade-bound crRNA serves as a guide to direct and mediate antiviral responses, coupled with knowledge of *cas* genes and spacers in CRISPR loci, set the stage for the next act in the CRISPR discovery saga, the conversion of this mechanism into technology for genetic engineering (through gene therapy or genetic editing) to help people. As stated several times in this book, the desire to help humankind is a potent stimulus behind the success of emergent technologies. This practical effort is ubiquitous in past history. For example, Eli Whitney's cotton gin was used to make linens and clothing; Alfred Nobel's dynamite was used in mining and drilling; J. Robert Oppenheimer's atomic bomb was used to force Japan to surrender during World War II; and Tim Berners-Lee's World Wide Web is used to connect people instantly, but all have also led to unintended consequences.[83] Continuing the example, (1) the invention of the cotton gin caused a massive growth in the production of cotton in the antebellum South accompanied by a need for plantations and slave labor; (2) dynamite increased devastation to the human body and property in warfare; (3) the atomic bomb led to the present global proliferation of advanced nuclear weapons; and (4) the World Wide Web has spawned social networks that are exploited by bad actors to spread misinformation and, ironically, divide people. CRISPR is another emergent technology (not unlike other biotechnologies discussed in this book) that has been, and still is, revolutionizing biotechnology in positive ways. But unintended consequences are also emerging, such as embryo editing, designer babies, and elective (nontherapeutic) enhancement, all echoes of the old eugenics resonating in this age of the new eugenics. Additionally, there is the problem of dual use, because terrorists might acquire cheap and efficient gene-editing technology such as CRISPR "to create potentially harmful biological agents or products," according to the worldwide threat assessment report released by the US intelligence community in 2016.[84]

## Notes

1. Calyampudi Radhakrishna Rao, *Statistics and Truth: Putting Chance to Work* (Singapore: World Scientific, 1989), 23.
2. Tracy Vence, "Notable Quotes from the American Association for the Advancement of Science Annual Meeting," *Speaking of Science Policy*, February 21, 2017, http://www.the-scientist.com.
3. Lawrence Bergreen, *Columbus: The Four Voyages, 1492–1504* (New York, NY: Penguin Books, 2012), 3–423.
4. Alexander Fleming, "Penicillin: Nobel Lecture," in *Nobel Lectures: Physiology or Medicine 1942–1962* (1964), ed., Nobel Foundation Amsterdam, Netherlands (Amsterdam, NL: Elsevier Publishing Company, 1964), 83.

5. Paul Berg and Janet E. Mertz, "Personal Reflections on the Origins and Emergence of Recombinant DNA Technology," *Genetics* 184, no. 1 (January 2010): 9.
6. Peter Raven, George B. Johnson, Kenneth A. Mason, Jonathan B. Losos, Susan R. Singer, *Biology*, 9th ed. (New York: McGraw-Hill, 2008), 516.
7. Ibid., 558, 533–534.
8. Ibid., 533–534.
9. Ibid., 558, 533–534.
10. Francisco Juan Martinez Mojica, César Diez-Villaseñor, Jesús Garcia-Martínez, and Elena Soria, "Intervening Sequences of Regularly Spaced Prokaryotic Repeats Derive from Foreign Genetic Elements," *Journal of Molecular Evolution* 60, no. 2 (February 2005): 174–182; Ruud Jansen, Jan. D.A. van Embden, Wim. Gaastra, and Leo. Schouls, "Identification of Genes That Are Associated with DNA Repeats in Prokaryotes," *Molecular Microbiology* 43, no. 6 (January 2002),: 1565–1575; Alexander Bolotin, Benoit Quinquis, Alexei Sorokin, and S. Dusko Ehrlich, "Clustered Regularly Interspaced Short Palindrome Repeats (CRISPRs) Have Spacers of Extrachromosomal Origin," *Microbiology* 151, no. 8 (May 2005): 2551–2561; Christine Pourcel, Grégory, and Gilles Vergnaud, "CRISPR Elements in *Yersinia pestis* Acquire New Repeats by Preferential Uptake of Bacteriophage DNA, and Provide Additional Tools for Evolutionary Studies," *Microbiology* 151, no. 3 (January 2005),: 653–663; Rodolphe Barrangou, Christophe Fremaux, Hélène Deveau, Melissa Richards, Patrick Boyaval, Sylvain Moineau, Dennis A. Romero, and Philippe Horvath, "CRISPR Provides Acquired Resistance Against Viruses in Prokaryotes," *Science* 315, no. 5819 (March 23, 2007): 1709–1712; Hélène Deveau, Rodolphe Barrangou, Josaine E. Garneau, Jessica Labonté, Christophe Fremaux, Patrick Boyaval, Dennis A. Romero, Philippe Horvath, and Sylvain Moineau, "Phage Response to CRISPR-Encoded Resistance in *Streptococcus thermophiles*," *Journal of Bacteriology* 190, no. 4 (February 2008): 1390–1400.
11. Raven *et al.*, *Biology*, 327–328; Francisco Juan Martinez Mojica, C. Ferrer, Guadalupe Juez, and Francisco Rodriguez-Valera, "Long Stretches of Short Tandem Repeats Are Present in the Largest Replicons of *The Archaea Haloferax mediterranei and Haloferax volcanii* and Could Be Involved in Replicon Partitioning," *Molecular Microbiology* 9, no. 1 (March 1995): 89.
12. Yoshizumi Ishino, Hideo Shinagawa, Kozo Makino, Mitsuko Amemura, and Atsuo Nakata, "Nucleotide Sequence of the *iap* Gene, Responsible for Alkaline Phosphatase Isozyme Conversion in *Escherichia coli*, and

Identification of the Gene Product," *Journal of Bacteriology* 169, no. 12 (December 1987): 5429–5433.

13. Raven *et al.*, *Biology*, 555; Ishino *et al.*, "Nucleotide Sequence of the *iap* gene," 5431.
14. Ishino *et al.*, "Nucleotide Sequence of the *iap* gene," 5432.
15. Francisco Juan Martinez Mojica and Francisco Rodriguez-Valera, "Transcription at Different Salinities of *Haloferax mediterranei* Sequences Adjacent to Partially Modified *PstI* Sites," *Molecular Microbiology* 9, no. 3 (April 1993): 613–621; Mojica *et al.*, "Long Stretches of Short Tandem Repeats," 85–93; Mojica *et al.*, "Intervening Sequences of Regularly Spaced," 174–182.
16. Jim Kozubek, *Modern Prometheus: Editing the Human Genome with CRISPR-CAS9* (New York, NY: Cambridge University Press, 2016), 10–11.
17. Mojica *et al.*, "Transcription at Different Salinities," 613.
18. Ibid., 617.
19. Ibid.
20. Ibid., 619.
21. Mojica *et al.*, "Long Stretches of Short Tandem Repeats," 85–93.
22. Ibid., 91.
23. Ibid.; Francisco Juan Martinez Mojica, Cesar Diez-Villaseñor, Elena Soria, and Guadalupe Juez, "Biological Significance of a Family of Regularly Spaced Repeats in the Genomes of Archaea, Bacteria and Mitochondria," *Molecular Microbiology* 36, no. 1 (January 2000): 244–246.
24. Ibid., 244.
25. Ibid.
26. Ibid.
27. Ibid., 89.
28. Ibid.
29. Ruud Jansen, Jan. D.A. van Embden, Wim. Gaastra, and Leo. Schouls, "Identification of Genes That Are Associated with DNA Repeats in Prokaryotes," *Molecular Microbiology* 43, no. 6 (January 2002): 1567.
30. Jansen *et al.*, "Identification of Genes," 1565.
31. Ibid.
32. Ibid.
33. Ibid.
34. Ibid., 1567.
35. Ibid.
36. Ibid.

37. Ibid.
38. Ibid.
39. Ibid.
40. Ibid., 1569.
41. Ibid.
42. Ibid., 1573.
43. Ibid.
44. Ibid.
45. Ibid., 1572.
46. Ibid.
47. Mojica *et al.*, "Intervening Sequences," 174–175.
48. Ibid., 175.
49. Ibid.
50. Ibid.
51. Ibid., 177–178.
52. Ibid., 180.
53. Ibid., 181.
54. Jennifer A. Doudna and Samuel H. Sternberg, *A Crack in Creation: Gene Editing and the Unthinkable Power to Control Evolution* (New York: Houghton Mifflin Harcourt, 2017), 60–85.
55. Bolotin *et al.*, "Clustered Regularly Interspaced," 2551.
56. Ibid., 2553.
57. Ibid., 2560.
58. Pourcel *et al*, "CRISPR elements in *Yersinia*," 653–654.
59. Ibid., 654.
60. Ibid., 661.
61. Ibid.
62. Ibid.
63. Ibid.
64. Barrangou *et al.*, "CRISPR Provides Acquired Resistance," 1709–1712.
65. Ibid., 1709–1710.
66. Ibid., 1710.
67. Ibid.
68. Ibid., 1710–1711.
69. Barrangou *et al.*, "CRISPR Provides Acquired Resistance," 1710; Deveau *et al.*, "Phage Response to CRISPR-Encoded," 1390–1400.
70. Barrangou *et al.*, "CRISPR Provides Acquired Resistance," 1710.
71. Ibid., 1710–1711.

72. Ibid., 1710.
73. Ibid., 1710–1711; Deveau *et al.*, “Phage Response to CRISPR-Encoded,” 1395, 1399.
74. Barrangou *et al.*, “CRISPR Provides Acquired Resistance,” 1711; Deveau *et al.*, “Phage Response to CRISPR-Encoded,” 1399.
75. Stan J. Brouns, Matthijs M. Jore, Magnus Lundgren, Edze R. Westra, Rik J. Slijkhuis, Ambrosius P. Snijders, Mark J. Dickman, Kira S. Makarova, Eugene V. Koonin, and John van der Oost, “Small CRISPR RNAs Guide Antiviral Defense in Prokaryotes,” *Science* 321, no. 5891 (August 15, 2008): 960–963.
76. Ibid., 961.
77. Ibid.
78. Ibid., 963.
79. Ibid.
80. Ibid.
81. Ibid.
82. Jason Carte, Ruiying Wang, Hong Li, Rebecca M. Terns, and Michael P. Terns, “Cas6 Is an Endoribonuclease That Generates Guide RNAs for Invade Defense in Prokaryotes,” *Genes & Development* 22, no. 24 (December 2008): 3489.
83. Bernard Grun, *The Time Tables of History: A Horizontal Linkage of People and Events* (New York, NY: Simon & Schuster, 2005), 375, 523, 537, 551, 699, 723.
84. News at a Glance, “CRISPR Dubbed a WMD,” *Science* 351, no. 6275 (February 19, 2016): 793, http://sciencemag.org/content/351/6275/792.

# Chapter 6

# Genome Editing: Rewriting the Fundamental Code of Life

For the birth of something new, there was to be a happening. Newton saw an apple fall; James Watt watched a kettle boil; Roentgen fogged some photographic plates. And these people knew enough to translate ordinary happenings into something new.

— Alexander Fleming, 1945 Nobel winner for the 1928 discovery of benzylpenicillin from the mold *Penicillium notatum*[1]

I lie in bed almost every night and ask myself that question. … When I'm ninety, will I look back and be glad about what we have accomplished with this technology? Or will I wish I'd never discovered how it works?

— Jennifer Doudna, American biochemist, 2015[2]

Scientific discovery is not the product of one individual. It is the product of many individuals past and present. As the British philosopher of science Eric Scerri wrote, "Science is a cumulative, incremental collective effort. Fierce competition among individuals is inevitable, and it may serve to develop better science in the short run, but overall, even heroic individual achievements are simply not as important as the ever-evolving whole."[3] This belief suggests that any accolades or prizes offered should be shared among several scientists rather than heaped upon one scientist. For instance, Erwin Chargaff, the Austro-Hungarian-born American biochemist, should have been (in my opinion) included in the 1962 Nobel Prize shared by James Watson, Francis Crick, and Maurice Wilkins "for their discoveries concerning the molecular structure of nucleic acids."[4] Through careful experimentation with nucleic acids, Chargaff discovered two rules: (1) the proportion of A always equals that of T, and the proportion of G always equals that of C; and (2) there was always an equal proportion of purines (A and G) and pyrimidines (C and T).[5] These rules were instrumental in the discovery of the double helix. In another example, Martha Chase, one of a handful of early female geneticists, conducted a career-defining experiment with lead researcher Hershey that proved that DNA, and not protein, was the genetic material. However, Chase was not included in the 1969 Nobel Prize for the discovery shared by Hershey, microbiologist Salvador Luria, and biophysicist Max Debrück.[6] One wonders if chemist and x-ray crystallographer Rosalind Franklin would have shared in the 1962 Nobel Prize if she had not died of ovarian cancer in 1958.[7] After all, it was her x-ray diffraction image of crystallized DNA that revealed a double helix structure and enabled Watson and Crick to construct a model of this DNA molecule.[8] There are many examples in history (and in the present) of scientists unintentionally (it is hoped) passed over for an important prize. Scientists themselves might be the remedy to this problem, in that they must continue to publicly acknowledge one another's work to counter any unintentional lapse in judgment by administrative committees. Scientists can learn a lesson in this type of professionalism from the actions of Charles Darwin. In June 1858, Darwin received a 20-page letter from British naturalist Alfred Russel Wallace, who outlined the theory of natural selection.[9] This revelation forced Darwin to publish his book on the subject — the draft of which he had initially completed in 1844 — in 1859. But before Darwin published his book, he sent Wallace's letter with his (Darwin's) contribution to the Linnaean Society of London as a composite paper on how natural selection accounts for the evolution and variety of species.[10] Ironically, neither Darwin nor Wallace attended the meeting at which the paper was presented. In October 1858, Wallace wrote a letter to the botanist Joseph Hooker, expressing his surprise and gratitude:

> I beg leave to acknowledge the receipt of your letter of July last, sent me by Mr. Darwin, & informing me of the steps you had taken with reference to a paper I had communicated to that gentleman. … I cannot but consider myself a favoured party in this matter, because it has hitherto

> been too much the practice in cases of this sort to impute *all* the merit to the first discoverer of a new fact or a new theory, & little or none to any other party who may, quite independently, have arrived at the same result a few years or a few hours later.[11]

In more recent times, 2020 Nobel laureate Jennifer Doudna, in the prologue of the book *A Crack in Creation* that she wrote with colleague and biochemist Samuel Sternberg, recognized all scientists involved in CRISPR without parsing out how much one did or who did what and when.[12] She wrote, "Finally, we humbly acknowledge the countless scientists who have played crucial and invaluable roles in the study of CRISPR and gene editing, and we apologize to the many colleagues whose work we didn't have space to mention."[13] Additionally, Doudna acknowledged the body of research conducted with ZFN and transcription activator-like effector nuclease (TALEN) proteins after her research team had successfully constructed the CRISPR/Cas9 technology.

This leads to the second part of the CRISPR discovery saga, in which this natural mechanism was exploited as a gene-editing technology to help people. This transition would not have been possible without Jansen and Bolotin's discovery of CRISPR-associated genes and double-stranded breaks, caused by *cas*5, that result in cellular death.[14] Or Brouns analysis of *E. coli* K12, resulting in the discovery of the Cascade complex of five Cas proteins.[15] Brouns believed that this complex was a defense against viral infection. His research team's effect-on-function studies revealed several critical facts: (1) small crRNA combined with Cascade promoted phage resistance; (2) *E. coli* strains that expressed Cascade and Cas3 were less sensitive to phage infection; and (3) Cascade-bound crRNA served as a guide to direct and mediate antiviral responses.

## 6.1. Events Enabling the Launch of the Research Leading to the CRISPR/Cas9 Technology

Putting genes to therapeutic use did not begin with CRISPR. As discussed in Chapter 11, geneticists in the 1970s were trying to insert elements of simian virus 40 (SV40 — a polyomavirus found in both humans and monkeys) into eukaryotic cells to investigate signals that controlled the expression of genes.[16] They eventually engineered SV40 to transfer and express foreign genes into cells. In addition to these recombinant retroviruses, scientists also designed cells to target molecules associated with specific tumors. These early forms of gene therapy had some success, but they were also accompanied by severe tragedies as a result of an overproduction of immune cells due to the immunogenic properties of the viruses. Knowing that gene therapy was ineffective for a wide range of genetic conditions which are not caused by missing genes (for example, Huntington's disease, where addition of a normal copy of the gene would not make a difference because the mutant gene overrides the effect of the healthy copy of the gene), Doudna fantasized early in her career about finding a way to *repair* mutant genes rather than replacing them.[17] She wrote that this was probably perceived as science fiction in the 1980s:

> One idea we kicked around was the possibility that RNA molecules, those intermediaries between DNA and proteins in cells, could be edited to fix mutations they carried over from DNA. Occasionally we also discussed another possibility: editing the source code of such defective RNAs — that is, the actual DNA of the genome. This would be game changing, we agreed. The question was, would it ever be anything but a pie-in-the-sky idea?[18]

Other scientists modified DNA in the laboratory and injected it into the fertilized eggs of mice, and the cells themselves (rather than recombinant viruses) merged the foreign DNA into their genomes.[19] Scientists could then observe gene expression or effect on function. Doudna's interest continued from the 1980s into the early 2000s, when scientists demonstrated methods to transform mammalian cells, which culminated in a love affair with CRISPR once she was introduced to this mechanism.

Doudna believed that the fundamental question relating to the nonviral-based DNA transfer used in the 1980s was "how exactly was DNA finding its way into the genome?"[20] Molecular geneticist Mario Capecchi, 2007 Nobel winner who was at the time a professor in the Department of Human Genetics at the University of Utah, provided the answer: homologous recombination.[21] That is, recombination as in meiosis I, when homologous chromosomes exchange parts and, therefore, genetic information. While the phenomenon of homologous recombination had been well known since the early 20th century, Capecchi was nevertheless surprised that it occurred in laboratory-cultured mammalian cells. In a 1982 article, he wrote, "It will be interesting to determine whether we can exploit [the enzymes involved] to 'target' a gene by homologous recombination to a specific chromosomal location."[22] This statement seems to foreshadow CRISPR. Interestingly, Capecchi's study of replacing human cells derived from bladder tumors with new cells grown in the laboratory (with recombinant DNA) worked very well. It was the first instance of gene editing. As Doudna recalled, Capecchi did not stop there; he also proposed that homologous recombination could be used to inactivate genes to study effect on function and simultaneously learn the gene's function.[23] This pushing of the envelope, going further and further to learn more and more, is the normal process in science. The question, however, is, how far can scientists go before they cross the line? British biologists Martin Evans, of the Department of Genetics at Cambridge University, and Matthew Kaufman, professor of anatomy at the University of Edinburgh, went further than anyone in the 1980s. They created live, genetically modified (or designer) mice.[24] Specifically, they cultured and targeted genes in mouse embryonic cells and injected the modified stem cells back into mouse embryos. Evans shared a Nobel Prize with Capecchi in 2007 for creating gene modifications in mice (including the "knockout" mouse) and related gene-targeting technology.[25]

In the late 1980s, the promise of homologous recombination as a gene-editing technology was not fulfilled, because of the overwhelming problem of nonhomologous recombination (which occurred when the modified DNA randomly integrated into the genome and was not delivered to the target sequence).[26] Regardless, Jack Szostak, professor of genetics at Harvard Medical School, was obsessed with understanding how gene targeting and homologous recombination actually worked. Doudna explained:

> He wanted to understand how two strands of DNA from one chromosome could merge with two matching strands of DNA from a second chromosome, exchange information during some kind of fused intermediate stage, and then separate again to re-form the individual chromosomes after cells divide.[27]

Szostak proposed the *DNA double-strand break model* based on the results of his genetics experiments with yeast.[28] A double-stranded break occurs when both strands in the DNA double helix are severed, which causes genome rearrangements and possibly cellular death. However, there is a repair process, since DNA is frequently exposed to radiation and chemicals without destruction or mutation, and it requires homologous recombination. Szostak's double-standed break model prompted speculation that a defective gene could be replaced by a copy engineered in the laboratory by inducing a double-stranded break at the location of the defective gene.[29] The cell would then attempt to repair the break by finding a matching copy, which would happen to be the engineered sequence with the corrected gene.

Doudna discussed the gene-editing experiment conducted in 1994 by developmental biologist Maria Jasin, investigator at the Memorial Sloan Kettering Cancer Center in New York, in which Jasin simultaneous introduced a nuclease and a piece of synthetic DNA to mammalian cells to both induce a double-stranded break and facilitate subsequent repair (the synthetic DNA matched the DNA sequence that had been cut).[30] To induce a double-strand break at a specific location in the genome and avoid randomly causing cellular death, Jasin needed a special endonuclease (enzymes that cuts DNA inside the strand). She chose the endonuclease designated I-*Sce*I found in the yeast strain *Saccharomyces cerevisiae*.[31] This homing endonuclease is special because it recognizes a specific 18-base-pair sequence (TAGGGATAACAGGGTAAT) to cut a given segment.[32] Since the gene for the I-*Sce*I endonuclease is not found in the mouse (or human) genome, Jasin had to splice a copy of this sequence into the mouse genome so that the endonuclease had a region to cut. While Jasin succeeded in inducing only 10 percent of the cells to precisely repair, Doudna believed that Jasin had great success (in the sense that no one else in the mid-1990s had achieved *precise repair of mutant genes by homologous recombination*).[33] The downside to all of this, however, was that the I-*Sce*I endonuclease was too specific to be used therapeutically, because various disease-associated genes had both more than and less than 18 base pairs combined with different base pair sequences.[34] In essence, the one factor that was needed to take this gene-editing technology to the next level was a reprogrammable endonuclease that could target and cut different DNA sequences. This problem could not be solved by modifying I-*Sce*I endonuclease, because it was too complex, and other nucleases were too short to work properly. The solution to this problem was provided by biomedical engineer Srinvasan Chandrasegaran, a newly hired assistant professor at Johns Hopkins Bloomberg School of Public Health in Baltimore. He proposed finding pieces of proteins that exist naturally and combining them so that they could be reprogrammed to target and cut different DNA sequences.[35]

Chandrasegaran chose the restrictive enzyme FokI (naturally found in the bacterium *Flavobacterium okeanokoites* that infects freshwater fish) because it could induce the double-strand breaks in DNA and was not burdened by a preference for a specific sequence.[36] Initially, he used fruit fly homeodomain protein to do the targeting with some success. Then he turned to zinc-finger proteins (ZFN), which are a family of naturally occurring proteins built of multiple repeated segments arranged in tandem, with each segment binding to a specific DNA sequence to turn genes on and off.[37] To obtain zinc-finger proteins, Chandrasegaran went to see another Johns Hopkins scientist, Jeremy Berg, a biological chemist who was designing new zinc fingers. Berg gave Chandrasegaran plasmids containing zinc-finger proteins which bound to a specific sequence.[38] Chandrasegaran subsequently combined the DNA-recognition module from a zinc-finger protein with cleavage domain of FokI. The hybrid or designer endonuclease cut exactly the DNA he expected, despite the fact that he had fused two protein components from completely different sources (i.e., FokI and zinc-finger protein). This was a significant breakthrough in bioengineering that led to a publication in 1996 detailing the steps in the engineering of the first ZFN.[39] With this initial success in the laboratory, the next step was putting ZFN to practical use.

Chandrasegaran joined forces with biochemist Dana Carroll, professor in the Department of Biochemistry at the University of Utah School of Medicine. As *Nature* writer Monya Baker told the story, Carroll had had trouble inducing double-strand breaks. According to Baker, Carroll said, “We were puttering along not getting anything dramatic, and we saw the *PNAS* paper.”[40] Immediately, Carroll made a phone call to Chandrasegaran, proposed a collaboration, and explained that he (Carroll) had a way to detect DNA repair in frog eggs. These two men had never met before

1996, but together they showed that ZFNs could cut DNA and trigger homologous recombination within eggs.

Carroll wanted to create new ZFNs that could knock out a gene in an organism. Baker wrote that he consulted another colleague at the University of Utah named Kent Golic, an expert in fruit fly genetics, who suggested an experiment targeting a gene encoding the body pigment called *yellow*.[41] When this gene (*yellow*) is disrupted in fly larvae, it causes the adult fly's brown cuticle to be flecked with amber specks. The experiment worked; Baker recalled that tiny amber specks were visible through the microscope.[42] In her book, Doudna wrote enthusiastically that "this was a profoundly significant development for gene editing. Not only were ZFNs practical enough to use in animals but, more important, they could be redesigned to target new genes."[43] In the example of *CCR5* and AIDS, *CCR5* is the means of entry of the HIV virus into the T cell to cause infection. Chandrasegaran believed that gene therapy could potentially knock out both copies of the *CCR5* gene, preventing or treating HIV/AIDS in high-risk healthy people.[44]

Carroll and Chandrasegaran's breakthrough brought excitement to the young field of biotechnology, as had cloning, recombinant DNA, and gene therapy before. Researchers and biotechnology companies alike wanted a piece of the action. One scientist-physician, Matthew Porteus, a medical doctor specializing in pediatric hematology and oncology in the combined Boston Children's Hospital / Dana Farber Cancer Institute program, was, like most scientists in the biological sciences, motivated by curing the sick.[45] Baker reported that Porteus encountered patients with extreme sickle cell anemia during his career but could not help them.[46] He became hopeful when he read about the creation of the knockout mouse, because the same gene-editing technique could be applied to fix the hemoglobin mutation responsible for sickle cell anemia. As a result, Porteus applied to the laboratory of molecular biologist and 1975 Nobel winner David Baltimore at the Massachusetts Institute of Technology and California Institute of Technology to complete his fellowship and postdoctoral studies "in developing homologous recombination as a strategy to correct disease causing mutations in stem cells as definitive and curative therapy for children with genetic diseases of the blood, particularly sickle cell disease."[47] It is interesting to note that the initial lack of enthusiasm for this project (to genetically modify human cells) among the researchers in Baltimore's laboratory was not because of the dangers of germ line (inherited) modification but because it was deemed impossible. Porteus noted the general impression in the laboratory after other scientists read his proposal: "Engineering human cells was just not something that was considered possible."[48] But Baltimore was willing to give Porteus a chance, and in 2003, Porteus was the first to show that a gene in human cells could be cleaved at specific sequences by an engineered ZFN.[49] Subsequently, Porteus and collaborators at Sangamo Biosciences (now Sangamo Therapeutics) — an American biotechnology company based in Richmond, California — reported the use of engineered ZFNs to edit a gene in human cells.[50]

Doudna explained that ZFNs were adopted by scientists interested in engineered crops and animal models, which suggested that DNA double-strand breaks induced highly efficient homologous recombination in many types of cells (including nonmammalian cells): "Concurrently, papers reporting that ZFNs had been used to modify genes in zebrafish, worms, rats, and mice began trickling out. *This work was intriguing and caught my attention in publications and at conferences due to its exciting potential*"[51] (italics added). Doudna was not alone; a growing number of biotechnology companies in the 1980s and 1990s were also intrigued, although their interest was specifically to use ZFNs to turn genes on and off and not for the more controversial application of modifying DNA.[52] In 1995, Edward Lanphier, CEO of Sangamo Biosciences, consolidated the intellectual property and know-how obtained by retaining British chemist and 1982 Nobel winner Aaron Klug and, as chief science officer, the biochemist Carl Pabo, professor at the Massachusetts Institute of Technology, thereby facilitating the creation of zinc fingers.[53] The new landscape was now prepared: Sangamo Biosciences had total control over the knowledge for constructing workable ZFNs.

But the promise of ZFNs was not totally fulfilled, according to Doudna. She explained that, in theory, designing ZFNs was easy — particularly for researchers who had experience in protein engineering and used it extensively.[54] But in practice, she added, it was difficult for inexperienced researchers to design. For example, many newly designed ZFNs did not recognize the DNA sequences they were targeting; some ZFNs targeted similar sequences, damaging cells; and others recognized the correct sequence but did not cut. Baker used an interesting analogy to describe zinc-finger binding: "Just as fingers on a hand have a certain order — pinky, ring, middle, index, thumb — zinc-finger proteins also fit together in a certain way. … A finger could bind tightly to one triplet of DNA when it is in the 'index' position, but have less-secure grip when it is in other positions."[55] In short, ZFNs would not burst upon the scene like recombinant DNA, PCR, or the soon-to-emerge CRISPR. ZFNs, because of the problems detailed above, were not easily reprogrammable to be useful for editing different DNA sequences.

Science is an endless journey of discovery. Therefore, it was not surprising when a simpler way was found to bind DNA. This technology, according to Doudna, derived from studies of novel types of proteins found in bacteria of the genus *Xanthomonas* that infect and cause black rot in more than 200 crop species, including staples such as peppers, tomatoes, cotton, soybeans, and rice.[56] The bacteria inject into a plant a number of effector proteins, including transcription activator-like effectors (TALEs), which stimulate the bacterial infection.[57] As Kozubek explained it, a TALE can bind to DNA in a plant, facilitating bacterial infection by switching on plant genes that lower the plant's innate immunity.[58]

According to Doudna, TALEs were more widely adopted than ZFNs because their construction was straightforward. For instance, whereas each finger of a zinc-finger protein recognizes a codon in DNA, each segment in a TALE recognizes just a single nucleotide.[59] This difference "allowed scientists to easily deduce a code for what segment would recognize a given letter of DNA, and then they simply arranged those segments, one after the other, to recognize a longer sequence of DNA within a gene."[60] Geneticists Ulla Bonas, Thomas Lahaye, and Jens Boch of the Institute at Martin Luther University in Halle-Wittenberg, Germany, deduced this code by conducting detailed experiments on the regulatory parts of plant genes activated by bell pepper pathogens to locate where the proteins bound.[61] Based on their findings, Bonas and colleagues were able to accomplish several things: (1) find binding sites for seven other TALEs; (2) construct artificial promoters to match particular TALEs; and (3) construct artificial TALEs that bound predicted sequences of DNA.[62] Like the discovery of the function of CRISPR, many researchers were simultaneously investigating TALEs. For instance, plant pathologist Adam Bogdanove at Iowa State University studied two rice pathogens, each of which made 20 TALEs.[63] Baker described Bogdanove's work, noting that the proteins from each pathogen regulated different sets of rice genes despite the fact that their structures were similar. However, Bogdanove discovered one section in the middle of the genome sequence that was highly repetitive and composed of three dozen amino acids, each unique only for two amino acid positions or *residues* — a term coined by plant pathologist Frank White at Iowa State University.[64] White had developed a classification system for these differing residues called *repetitive variable di-residues* (RVDs).[65] Bogdanove noticed that the number of repeats in the TALE that infected the bell pepper was equal to the number of nucleotides to which it bound. As Bogdanove recalled, "I thought, 'Could it be so simple?' "[66]

At this point in 2007, Bogdanove had a working hypothesis that each RVD in a TALE repeat bound to a single nucleotide. He consulted bioinformatician Matthew Moscou, a colleague at the University of Utah. Their plan was to find regions where the protein unit and nucleotide aligned by running RVDs along promoter regions. Their plan worked: "Within a week, they had found positions on the plant promoters where the RVDs locked in … the association between nucleotide and RVD was constant over some 20 TALE-promoter pairings. The code was clear."[67] What came next were simultaneous publications in *Science* in 2009 by the Bogdanove and Bonas teams on how TALEs recognized nucleotides and the possible applications in the real world — critical information for additional funding.

Bogdanove collaborated with plant genomics expert Daniel Voytas, director of the Center for Precision Plant Genomics at the University of Minnesota, who had pioneered using ZFNs in plants.[68] Initially, Voytas expected a slow and difficult road in engineering TALEs for practical applications, based on his experience engineering ZFNs. However, he was pleasantly surprised. They fused a new TALE to the DNA-cleaving domain of *FokI*, the same restriction enzyme used by Chandrasegaran to create ZFN. After repeating the experiment several times, with positive results each time, the new gene-editing tools known as TALENs came on to the scene.[69] Molecular biologist Keith Joung (pioneer in techniques for designing gene-editing nucleases at Massachusetts General Hospital) said that the demand for this technology — from scientists ready to try their hand at gene editing and biotechnology companies eager to commercialize it — was remarkable.[70] But unlike all of those scientists who were dazzled by TALENs — which were still the craze two years after being announced to the world — Joung remained cautious. He believed that it was too soon to declare TALENs the gene-editing technology of choice, because they were still brand-new.[71] It turned out that Joung was right about waiting before crowning TALENs the "king" of gene-editing technology, because they were supplanted very quickly (just as the hydrogen bomb supplanted the atomic bomb) by CRISPR, the present and true king of gene editing.

***

Doudna was primed to be one of the key scientists to transform CRISPR into a technology because of her two deep desires: to study unexplored natural phenomena and to apply the knowledge obtained to the real world.[72] In the mid-1990s, she gave a television interview about her RNA research to determine the molecular structure of ribozymes. She speculated that RNA might be used in editing DNA: " 'One possibility,' I had said, 'is that we might be able to cure or treat people that have genetic defects. ... We hope that [this discovery] will provide some clues as to how we might be able to modify the ribozyme so that it can act like a molecular repair kit and repair defective genes.' "[73] While ribozyme-based therapies never proved effective against genetic diseases, according to Doudna, her background in RNA research would become invaluable in grasping the workings of CRISPR.[74]

In 2006, Doudna was introduced to CRISPR by Jillian Banfield, a fellow professor at the University of California, Berkeley, who specialized in geomicrobiology in the Department of Environmental Science, Policy, and Management.[75] In the beginning, according to Doudna, Banfield was investigating what Doudna heard as *crisper* and wanted to find any molecular geneticist who knew something about iRNA. Banfield wanted to know if there was a connection between CRISPR and iRNA.[76] From basic investigation before their meeting and after Banfield explained and sketched CRISPR on paper, Doudna gained a basic understanding of the mechanism:

> Finally, the words that the acronym stood for — clustered regularly interspaced short palindromic repeats — began to make sense to me. The diamonds were the short repeats, the squares were interspacing sequences that regularly interrupted the repeats, and these diamond-square arrays were clustered in just one region of the chromosome, not randomly distributed throughout. (When I inspected the repeating DNA sequences more closely, back in my office, the *P* in the acronym also became clear: the sequences were nearly the same when read in either direction, just like a palindrome such as "senile felines."[77]

Immediately, Doudna said, her desires kicked in, and she was hooked.[78] Initially, she had read Mojica's 1990s papers on the genome sequences of *H. mediterranei* and *H. volcani* and his seminal 2005 paper on CRISPR (see Chapter 5). She also read Jansen's 2002 paper, in which he identified *cas* genes which encoded specialized enzymes adjacent to CRISPR loci; Bolotin's paper on CRISPR spacers; Pourcel's paper explaining the origin of spacers in CRISPR elements; and Barrangou's papers on CRISPR and *cas* genes providing resistance against phages. Since one of her research interests was iRNA, she mined papers written by biologists Eugene Koonin and Kira Makarova (staff scientists at the National Center for Biotechnology Information, National Library of Medicine, Bethesda, Maryland) on iRNA. She commented on this in her book:

> Combining the results of a number of earlier studies with an expert analysis of the prevalence of CRISPRs in different species, they had pieced together an intriguing new hypothesis that suggested RNA was a key participant in the immune system of single-celled microorganisms like bacteria — and that this system might be functionally similar to one of my research interests, RNA interference.[79]

As the existing research data on CRISPR settled in Doudna's mind, she wondered if CRISPR could be one of several bacterial antiviral defense mechanisms.[80] Between Koonin's and Jansen's papers, Doudna had charted a course for her laboratory research, and, like the scientific explorer she was, Doudna was ready to make new discoveries.

## 6.2. The Research Producing CRISPR/Cas9

The research that transformed CRISPR into a technology began in early 2007. Doudna's working hypothesis was that "CRISPRs and *cas* genes were parts of the same antiviral immune system and that RNA was employed by this system to detect viruses."[81] The model organism chosen by Doudna and her postdoctoral researcher, Blake Widenheft, was *Pseudomonas aeruginosa* — a Gram-negative bacterium that causes disease in animals, including humans and plants, and contained CRISPR.[82] The plan was to infect *P. aeruginosa* with *Pseudomonas* phages to test the ability of Cas proteins (beginning with Cas1 the most widely found in CRISPR) encoded in *P. aeruginosa* to recognize or destroy viral DNA.

However, Doudna and her colleagues were one or two years behind in reading the literature in this area of research.[83] Recall that in 2005, Bolotin and colleagues (after sequencing CRISPR structures from twenty-four strains of *S. thermophiles* and *S. vestibularis*) learned that phage sensitivity of an *S. thermophilus* strain correlated with the number of spacers in the CRISPR locus the strain carries (see Chapter 5). Bolotin hypothesized that the spacer elements provide cell immunity against phage infection by coding an antisense RNA. Recall also that Barrangou and colleagues published their findings building on the work of Bolotin (see Chapter 5) around the time that Doudna launched her CRISPR research in 2007. They analyzed 30 phage-insensitive mutants of *S. thermophiles* and found that the addition of a new spacer in CRISPR1 occurred frequently after a phage challenge, and, more remarkably, they demonstrated two critical features of CRISPR as a result of their analyses: (1) the iterative addition of spacers increased resistance of the host to phage infection, and (2) the newly added spacer must be identical to a region in the phage genome to confer bacterial resistance (see Chapter 5).[84]

Doudna absorbed the data from these studies. However, some details of CRISPR were still unclear, such as the system for targeting phages: "It is clear that this immune system was targeting the phages' genetic material for destruction — but how? What part of the cell was doing the targeting?"[85] Soon, the answer to her question was provided by the molecular microbiologist Brouns (see Chapter 5). Recall that Brouns and colleagues showed how Cas proteins used phage-derived sequences in CRISPRs to mediate an immune response. Mature crRNAs, formed from pre-crRNAs, served as small guide RNAs that enabled helicase Cas3 to disrupt phage infection. Doudna believed that the basic functions of RNA, such as being able to pair with a matching single RNA strand (i.e., RNA–RNA double helix) or pairing with a matching single DNA strand (RNA–DNA double helix), allowed it to play a key role in the CRISPR immune system.[86] With this versatility, she added, "CRISPR RNA molecules could single out both DNA and RNA molecules from invading phages for attack during an infection by pairing with any that they matched and initiating some sort of immune response in the cell."[87] This new revelation about the versatility of RNA was heaven to Doudna, simply because she began her career studying RNA. At this point, she wondered if CRISPR might be similar to iRNA.[88]

This answer would come from molecular microbiologist Luciano Marraffini and molecular biophysicist Erik Sontheimer, who were, at the time, in the Departments of Molecular Biology and Cell Biology at Northwestern University. In their important paper, published in 2008, they showed that CRISPR interference prevents conjugation and plasmid transformation, limiting horizontal gene transfer and the spread of resistant genes in pathogenic bacteria.[89] Their work was influenced by the rise of antibiotic resistant strains of *Staphylococcus aureus* and *Staphylococcus epidermis* as a result of acquiring resistant genes through horizontal gene transfer (within and between species) via plasmid conjugation.

Based on all the data generated from their numerous experiments, Marraffini and Sontheimer showed that neither RNA nor iRNA (RNA interference) is similar or a target during CRISPR interference. In fact, they showed that CRISPR interference acts at the DNA level: "A DNA targeting mechanism for CRISPR interference implies a means to prevent its action at the encoding CRISPR locus itself, as well as other potential chromosomal loci such as prophage sequences."[90] This work might have been the foundation on which Doudna, molecular biologist and 2020 Nobel laureate Emmanuelle Charpentier (director of infection biology at the Max Planck Institute), and colleagues transformed CRISPR into a technology. In fact, Marraffini and Sontheimer were prescient about the practical utility of the CRISPR system. They emphasized the following:

> From a practical standpoint, the ability to direct the specific, addressable destruction of DNA that contains any given 24 — 48 nucleotide target sequence *could have considerable functional utility, especially if the system can function outside of its native bacterial or archaeal context. ... If CRISPR interference could be manipulated in a clinical setting, it would provide a means to impede the ever-worsening spread of antibiotic resistance genes and virulence factors in staphylococci and other bacterial pathogens.*[91] (italics added)

Doudna wrote that after first meeting with Banfield in 2006 and reading more about CRISPR, she became hooked.[92] Two years later, after learning more about CRISPR by reading numerous articles, including the brand-new article published in 2008 by Marraffini and Sontheimer detailing their elegant experiments with CRISPR interference, she and her colleagues began the task of "cracking the code."[93]

## 6.3. Type I CRISPR-Cas Immune System

Scientists in the throes of their research do not contemplate, or are unwilling to contemplate, the negative or dual-use impact of their work outside the controlled environment of the laboratory; Doudna was not immune to this malady. The CRISPR/Cas9 technology she helped to create was so simple to use, effective, and, therefore, susceptible to misuse (e.g., for germ line editing or elective enhancements) that in 2015 she felt compelled to join other colleagues in calling for a moratorium on using CRISPR/Cas9 in genetic editing.[94] While CRISPR/Cas9 has become a potent weapon in the arsenal of the new eugenics, this technology, in some cases, beckons the spirit of the old eugenics. However, in 2008, when Doudna and her team began laboratory analyses to further elucidate the CRISPR mechanism, dual-use or questionable applications were not part of their thinking, which centered on helping people.

It would still take a few years to produce the gene-editing technology now ubiquitously known in the world of bioengineering as CRISPR/Cas9. And, according to Doudna, it began with three critical questions: (1) By what mechanism do bacteria cleave short segments of phage DNA during an infection and precisely integrate these segments into the CRISPR array so that the defense system can target the viral DNA? (2) How are the long strands of crRNA produced inside the cell and converted to short segments of virus-matching sequences? (3) How does a segment of RNA pair with its phage DNA counterpart and cause its destruction?[95] This "weapons system," as she called it, would hone her team's focus toward the biochemical aspects of the mechanism in terms of the *cas* genes that flank the CRISPR regions of bacterial genomes.[96] Doudna wrote:

> The proteins encoded by these *cas* genes, we reasoned, must work intimately with CRISPR DNA — or perhaps CRISPR RNA molecules, or even phage DNA. One thing seemed certain: we'd need to find out how these genes worked and discern the biochemical functions of the proteins they produced before we could understand the CRISPR immune system as a whole.[97]

Widenheft, a member of Doudna's team, chose the two CRISPR-positive bacterial species *E. coli* and *P. aeruginosa*, built plasmids from the individual *cas* genes based on these choices, and then finally generated Cas proteins.[98] After this laborious process, Widenheft conducted experiments to gauge the function of these proteins and discovered two proteins that the team called Cas1 and Cas6. They learned that Cas1 was responsible for cutting and inserting segments of phage DNA into CRISPR during the immune system's memory-forming stage, and Cas6 cleaved the crRNA sequence into shorter segments, which were subsequently used to target phage DNA.[99] With this piece of the puzzle in place, so to speak, they were able to tentatively answer the question regarding how bacteria cleave short segments of phage DNA during an infection and precisely integrate these segments into the CRISPR array so that the defense system can target the viral DNA. Doudna believed that they could develop a practical diagnostic tool to detect specific disease viruses and cut them out.[100] Soon, public and private grant money flowed in, which is traditionally the fuel that propels emergent technologies forward so fast that warning signs are missed. These soon-to-be-commercialized Cas proteins could be used to detect the presence of viral or bacterial DNA in body fluids.

Doudna and Widenheft discovered what she called "the actual weapon in the attack": Cas3, another protein in the CRISPR system, which shifted their focus to the destruction phase of CRISPR.[101] They also discovered the specificity with which the Cascade complex locked on to viral DNA targets that were identical to crRNA (rather than its own DNA, which would result in detrimental effects). This accuracy increased its potential for therapeutic use. However, there was still a long way to go, even though the important pieces of the puzzle were in place. Doudna learned quickly that her results and conclusions were only relevant for the Type I CRISPR-Cas immune system and not for the other two types (and possible subtypes within those two types).[102] For instance, she described the Type I systems — found in *E. coli* and *P. aeruginosa* — as shredding DNA using the Cas3 enzyme, a contrast from the Type II system found in *S. thermophiles*, which she described as more restrained and precise: "the cutting in *S. thermophiles* operated more like a pair of scissors, clipping the DNA apart at exactly the site where the letters of the viral genome matched the letters of the CRISPR RNA."[103] Unfortunately, Doudna and her team did not know how the Type II enzymes worked together with crRNA to target a specific region to cut. Discovering this mechanism was their next challenge.

***

In 2008, Brouns and colleagues identified the Cascade complex and demonstrated that this complex was responsible for the maturation of pre-crRNA to small crRNAs, which are [small crRNAs] critical in preventing phage infection.[104] In fact, they noted that Cascade-bound crRNA and Cas3 serve as a guide to disrupt viral proliferation. Using this data, Charpentier, molecular biologists Elitza Deltcheva and Krzysztof Chylinski of the Umeå Centre for Microbial Research at Umeå University in Sweden and the Max Perutz Laboratories at the University of Vienna, and other colleagues designed experiments to uncover the process that directed the maturation of crRNAs.[105]

Charpentier, "obsessed with how bacteria cause disease and working with the right models" and hoping that the CRISPR system "might give us new ways of tackling *Streptococcal* infections saving countless lives," chose *Streptococcus pyogenes* for analysis because it was a human pathogen that caused pharyngitis (commonly known as strep throat) and it happened to have new RNA molecules.[106] Remarkably, Charpentier's team's analysis of *S. pyogenes* genome sequences uncovered two different subtypes of CRISPR/Cas loci, each containing distinct sets of repeats and *cas* genes: CRISPR01 and CRISPR02.[107] On further analysis, the scientists found similarities between the CRISPR spacers and the prophage sequences, indicating that lysogenic phages are targeted by the CRISPR/Cas systems of *S. pyogenes*.

To confirm that CRISPR01 and CRISPR02 loci were active, Charpentier and her team examined *S. pyogenes* (strain SF370) in vivo by differential RNA sequencing (dRNA-seq) and detected six crRNAs (from CRISPR01) originating from a sequence of pre-crRNA.[108] Additionally, the scientists detected abundant RNA species they called *trans*-encoded crRNA (tracrRNA). Using Northern blot (RNA blot) probing, the scientists detected four different forms of tracrRNA — all present throughout the growth of the bacterial strain — having nucleotide lengths of 171, 89, 75, and 65. Additionally, they learned that both the 171- and 89-nucleotide tracrRNAs contained a segment of 25 nucleotides which were complementary to all repeats in CRISPR01, indicating to Charpentier that they probably paired with pre-crRNA.[109] Charpentier and her team expressed a prediction: "Moreover, the tracrRNA and pre-crRNA processing sites detected by dRNA-seq fell in the putative RNA duplex region, indicative of co-processing of the two RNAs upon pairing."[110] Therefore, tracrRNA was essential for processing pre-crRNA. In essence, tracrRNA directs maturation of pre-crRNA to generate crRNA (active species).[111] But the work did not end here; maturation of crRNA required other complex enzymes.

New experiments showed that ribonuclease III (RNaseIII) and Cas protein Csn1 were required for crRNA maturation.[112] In Barrangou's research (see Chapter 5), Csn1 (referred to by Barrangou as Cas5 and COG3513) was essential for CRISPR-mediated immunity in *S. thermophilus*. Its inactivation in a specific wild-type strain resulted in loss of phage resistance, and it "acts as a nuclease, because it contains an HNH-type nuclease motif."[113] (An HNH motif bears the amino acid sequence H-N-H and is found in nucleases.) Continuing with Charpentier, her team found that both the 171-nucleotide and 89-nucleotide tracrRNAs promoted RNase III cleavage within the repeat to produce intermediate crRNA species. If there were mutations in the complementary regions of tracrRNA or pre-crRNA interrupting RNase III cleavage, the scientists found that RNase III cleavage was restored when tracrRNA and crRNA mutants were combined. Taken together, Charpentier and colleagues noted that "RNase III serves as a host factor in tracrRNA-mediated crRNA maturation, and constitutes the first example of a non-Cas protein that is recruited to CRISPR activity."[114] Additionally, they believed that the Cas proteins facilitated the coprocessing of the RNA duplex in vivo, because deletion of the *csn1-cas1-cas2-csn2* operon disrupted the processing of both tracrRNA and pre-crRNA.

Furthermore, Charpentier and colleagues proposed — in one of their numerous models — that Csn1 facilitates the base pairing of tracrRNA and pre-crRNA for subsequent recognition and cleavage of pre-crRNA repeats by RNase III.[115] They added the following predictions: (1) Csn1 might also mediate the second cleavage to occur at a specific distance within the spacers because it has RuvC-like and McrA/HNH nuclease motifs (special proteins and restriction endonuclease), and (2) Csn1 might protect tracrRNA and pre-crRNA from other RNases, as suggested by the reduced tracrRNA in the absence of *csn1*.[116] Overall, the results of Charpentier and colleagues showed that the alternative pathway of CRISPR01 crRNA maturation is achieved by "the concerted action of three novel factors, a trans-encoded small RNA, a host-encoded RNase and a Cas protein previously not implicated in pre-crRNA cleavage" in the absence of Cse3 (CasE), Cas6, and Csy4 — found in many other CRISPR/Cas subtypes — that cleave (as endoribonucleases) within the repeat sequences of pre-crRNA to produce mature crRNAs.[117]

Doudna became aware of Charpentier's work on CRISPR01 Type II in *S. pyogenes* at the annual conference of the American Society for Microbiology, where Charpentier presented the findings from her 2011 collaborative paper that had been published in *Nature* a few weeks earlier.[118] Charpentier and her collaborators wanted to know more about the biochemistry of the *csn1* gene, and Doudna and her team had the motivation and, more importantly, the expertise to help. Additionally, Doudna had a bias for Type II CRISPR systems.[119] Now that Charpentier, Barrangou, Horvath, and other scientists had identified the Type II systems lacked the Cascade and Cas3 proteins, the more intriguing challenge for her was to investigate Csn1's role in targeted DNA destruction.[120] This new research path meant a change in the model organism from *S. thermophilus* to *S. pyogenes* because of its Type II CRISPR immune system and the practical medical benefit of eliminating its pathogenicity.[121]

Martin Jinek, Czech molecular biologist and a postdoctoral researcher in Doudna's laboratory at the time, accepted her invitation to lead the collaboration with Charpentier.[122] The Csn1 enzyme, or Cas9 as it was permanently referred to after the summer of 2011, "was likely to be a key player in the DNA destruction phase of the immune response in Type II CRISPR systems," according to Doudna.[123] The initial step in the project was to purify the Cas9 and test its ability to cleave a plasmid DNA or short DNA target segment consisting of twenty nucleotides.[124] This short segment is located adjacent to a protospacer motif (PAM) of two to six base pairs and a sequence complementary to a mature crRNA consisting of 42 nucleotides (including the 20 complementary nucleotides). They then conducted biochemical tests to determine how Cas9 and crRNA interacted during an immune response. The Doudna-Charpentier research team learned that mature crRNA alone could not direct Cas9 to cleave plasmid DNA without the addition of tracrRNA (essential to crRNA maturation and target DNA recognition).[125] The tracrRNA paired with crRNA and triggered Cas9 cleavage of plasmid DNA; cleavage was prevented when the crRNA-DNA match was imperfect. At this point, the team knew that tracrRNA not only triggered pre-crRNA processing by RNase III but also activated crRNA-guided, site-specific DNA cleavage by Cas9.[126] In essence, the Doudna-Charpentier research team identified "a DNA interference mechanism involving a dual-RNA structure that directed a Cas9 endonuclease to introduce site-specific double-stranded breaks in target DNA."[127] Specifically, the tracrRNA:crRNA-guided Cas9 protein made use of HNH and RuvC endonucleases to simultaneously cleave the complementary and noncomplementary strands of DNA. Additionally, several other important issues were addressed: (1) whether the entire length of the tracrRNA was necessary for site-specific Cas9-catalyzed DNA cleavage;

(2) the protospacer sequence requirements for type II CRISPR/Cas immunity in bacterial cells; and (3) the role of the PAM in target DNA cleavage by the Cas9-tracrRNA:crRNA complex.[128] First, they learned that shortening the length of the crRNA by 10 nucleotides from the 3′-terminal end still triggered Cas9-catalyzed cleavage, but only in the presence of tracrRNA. However, a 10-nucleotides deletion from the 5′-terminal end of crRNA stopped cleavage by Cas9. Second, the scientists believed that a "seed" region located at the 3′ end of the protospacer sequence was crucial for interaction with crRNA and later cleavage by Cas9. Third, they learned that mutations introduced at the 5′ end of the protospacer and in the region close to the PAM and the Cas9 cleavage sites resulted in reduced plasmid cleavage efficiency. Finally, they observed that mutations of the PAM motif did not affect cleavage of target single-stranded DNA, suggesting that the PAM motif was required only in target double-stranded DNA and formation of an R-loop structure.[129]

Now that the mechanism of CRISPR/Cas systems in adaptive immunity against phages and plasmids had been elucidated, particularly Cas9's role in target DNA destruction, the research team tackled a question involving potential practical applications: "If bacteria could program Cas9 to cut up other DNA sequences, could we, the researchers, program Cas9 to cut up other DNA sequences — viral or not — as we suspected?"[130] In other words, could a single RNA-guided Cas9 be engineered to target and cleave any double-stranded DNA sequence of interest (matching crRNA)? Doudna and Charpentier believed that the tracrRNA:crRNA duplex made it possible without the limitations of ZFN and TALEN discussed earlier in the chapter. They subsequently engineered five 20-nucleotide fragments of single chimeric RNA, fusing the 3′ end of crRNA to the 5′ end of tracrRNA (mimicking the dual-RNA structure required to guide site-specific DNA cleavage by Cas9), and tested their efficacy against a plasmid carrying the gene encoding green-fluorescent protein.[131] Each 20-nucleotide fragment was complementary to the protospacer sequence in the target DNA. In all five cases, they found that cleavage by Cas9 programmed with these chimeric RNAs was site-specific: all the green-fluorescent DNA was sliced apart at the intended sites.[132] In a statement rife with hyperbole, Doudna wrote, "Out of this fifth *bacterial weapons system*, we had built the means to *rewrite the code of life*"[133] (italics added). A few years later, that statement — particularly the part about "rewriting the code of life" — would give Doudna nightmares as the CRISPR/Cas9 technology, thanks to its low cost and ease of use, quickly spread unchecked throughout the global community.

## 6.4. Type II CRISPR-Cas Immune System

Scientists frequently reach similar conclusions independently at about the same time. This happened in 2012 when Lithuanian biochemists Giedrius Gasiunas and Virginijus Šikšnys (chief scientist) at Vilnius University's Institute of Biotechnology and colleagues Barrangou and Horvath, employing their favorite microbe model *S. thermophilus* (DGCC7710), identified the CRISPR3/Cas system (see Chapter 5) to isolate the Cas9-crRNA complex.[134] They showed that this complex functioned as an RNA-guided endonuclease that used RNA to find the respective DNA site to be cleaved and Cas9 to do the cutting — as demonstrated by the Doudna-Charpentier team a few months before. Doudna stated in her book that "they failed to uncover the crucial role of the second RNA (called tracr-RNA), which we had demonstrated was an essential component of the DNA-targeting and DNA cutting reaction."[135]

In 2013, researchers around the world were engineering type II CRISPR/Cas systems to function in human and mouse cells with custom single-guide RNA or different single-guide RNAs.[136] For instance, Church and colleagues at Harvard Medical School and Boston University's Department of Biomedical Engineering constructed two custom guide RNAs (gRNAs), T1 and T2, in human embryonic kidney (HEK 293Ts) cells with a genomically integrated coding sequence disrupted by a 68-base-pair genomic fragment that rendered the expressed protein fragment nonfluoresent.[137] The T1 and T2 gRNAs targeted the disrupting fragment and edited it with correction rates of three percent and eight percent respectfully, with the first fluorescent cells appearing 20 hours after fragment insertion (they first appeared 40 hours after insertion using TALENs). Similarly, genetic editing occurred when Church and colleagues used the T1 and T2 gRNAs to simultaneously target HEK 293Ts, human chronic myelogenous leukemia K562 cells, and p-glycoprotein 1 (PGP1) human iPSCs.[138] The correction rates obtained with the two gRNAs were 10 percent and 25 percent in 293Ts, 13 percent and 38 percent in K562s, and 2 percent and 4 percent in PGP1.

Another parallel study had independently confirmed Church's work on the efficiency of CRISPR-mediated gene targeting in mammalian cells. American biochemist Feng Zhang of the Broad Institute of the Massachusetts Institute of Technology and Harvard University and colleagues codon-optimized the *S. pyogenes Cas9* (*SpCas9*) and *RNase III* (*SpRNase III*) genes in human 293FT cells and attached nuclear localization signals to target SpCas9 to the nucleus.[139] They subsequently reconstituted the noncoding RNA components (of the *S. pyogenes* Type II CRISPR system) by generating an 89-nucleotide tracrRNA under the RNA polymerase III U6 promoter to form pre-crRNA. Zhang and colleagues transfected 293FT cells with different combinations of the four CRISPR components (SpCas9, SpRNase III, tracrRNA, and pre-crRNA) and achieved targeted cleavage of the protospacer.[140] They also found that some chimeric RNA designs were more effective than others in cleaving their genomic targets.

Doudna discussed the excitement at the birth of the CRISPR/Cas9 technology and experiencing firsthand the positive results of CRISPR/Cas9 being used to edit and replace the mutated β-globin gene responsible for sickle cell anemia.[141] This excitement transformed into the launching by her and her colleagues of companies to develop CRISPR-based therapies. And there was no shortage of financing to support the explosion of these fledging biotechnology companies.[142] Unlike the Cohen-Boyer patents, which were filed jointly by their academic institutions (and not by the scientists themselves), no institution had induced the numerous CRISPR scientists — primary or secondary developers of the technology — to explore their entrepreneurial instincts. However, unlike the Cohen-Boyer experience, a major patent war would break out.[143] As stated elsewhere in this book, the motivation of scientists to correct disorders, in alliance with the market forces which make possible continuous flow of materiel and production, sustain the new eugenics.

## 6.5. The CRISPR/Cas9 Technology

The unknown causes of off-target nuclease activity induced by CRISPR/Cas9 (or ZFN and TALEN) systems; the march toward inevitable germ line edits; the user-friendly nature of CRISPR; and the differing ethical standards from country to country resulted in a call in 2014 for a temporary moratorium to discuss the path forward for genomic engineering.[144] Despite this, the march toward germ line editing allegedly reached its destination four years later.

In November 2018, Chinese biophysics researcher Jiankui He, a former associate professor at the Southern University of Science and Technology in Shenzhen, China, claimed to have used CRISPR to alter both copies of the gene for the protein CCR5 in the embryos of twin girls, making them resistant to HIV infection.[145] If this is true, it would be the first known human-genome-editing experiment on embryos that were implanted and brought to term. This is the nightmare scenario of the new eugenics: heritable changes in the human genome. And there are also shadows of the old eugenics in terms of the goodwill of a scientist wanting to prevent future illness by altering the present generation.

In contrast, Canquan Zhou, chief scientist of the Center for Reproductive Medicine at Sun Yat-sen University in Guangzhou, China, and his team were cognizant of the ethical concerns in germ line editing, which prompted them to use tripronuclear zygotes (i.e., fertilized by two sperm cells and therefore nonviable), generally those discarded in clinics.[146] Zhou and his team made the following statement in the introduction to a 2015 paper: "Because ethical concerns preclude studies of gene editing in normal embryos, we decided to use tripronuclear (3PN) zygotes, which have one oocyte nucleus and two sperm nuclei."[147] Despite this adherence to a largely universal ethical standard, the editors of *Nature* and *Science* had previously rejected Zhou's collaborative manuscript because of the use of embryos, ignoring the fact that these embryos were nonviable.[148] The editors probably believed that Zhou's experiment was an opening that would lead, eventually, to the use of viable embryos; they did not want to be complicit in such an endeavor.

The scientists used CRISPR/Cas9 to edit the mutated human ß-globin (HBB) gene (responsible for ß-thalassemia) and used Δ-globin (HBD), with its similar sequence, as a repair template.[149] They designed three gRNAs (G1, G2, and G3) that targeted different regions of the HBB gene and transfected the gRNA-Cas9 into human 293T cells. They found that G1 and G2 gRNAs were efficient in cleavage activities. In the analysis of the specificity of gene targeting, for instance, the scientists used PCR amplification of G1 and gRNA target regions. They learned that 28 embryos were cleaved by Cas9, indicating an efficiency of 52 percent, and 14.3 percent (four out of the 28 Cas9-cleaved embryos) were edited using single-strand DNA as a repair template.[150] However, along with cleavage and repair, there was off-target cleavage activity. They explained, "While G2 gRNA showed very low off-target cleavage activity in the intergenic region, gRNA G1 did not exhibit detectable off-target cleavage at the top seven predicted off-target sites."[151] Further analyses demonstrated that CRISPR/Cas9 had notable off-target effects in human 3PN embryos.

Approximately a year after Zhou completed the gene editing of nonviable embryos, an international team of researchers led by American reproductive biologist Shoukhrat Mitalipov (who heads the Center for Embryonic Cell and Gene Therapy at the Oregon Health & Science University in Portland) used CRISPR/Cas9 to correct a mutation in the gene called myosin binding protein C, cardiac type (*MYBPC3*) in dozens of viable human embryos *not destined for implantation.*[152] Mutations in the *MYBPC3* gene can cause a condition called hypertrophic cardiomyopathy in which the heart muscle thickens, causing sudden death, particularly in young athletes.[153] In addition to correcting the mutant gene, Mitalipov and his team monitored off-target edits causing unwanted mutations and the risk of generating mosaics — embryos with normal and abnormal cells. They attempted two experiments to reduce the risk of both of these problems.[154] First, they injected CRISPR/Cas9 directly into the cells (rather than the normal approach of inserting DNA encoding CRISPR/Cas9 into cells). Since Cas9 degraded faster than the DNA that encoded it, there was less time for Cas9 to cleave if the guide was off target. In the second experiment, the team injected both CRISPR/Cas9 and sperm (to fertilize the egg) simultaneously in an attempt to reduce mosaics.

Out of 58 embryos fertilized with sperm carrying the *MYBPC3* mutant gene, 16 of them — if implanted — would have developed into male offspring with the disease.[155] However, these embryos were edited with high efficiency with the generation of just a single mosaic, compared to 13 mosaics when CRISPR/Cas9 was injected 18 hours after fertilization. The fact that mosaics were generated is a serious problem, and Joung cautioned that there may be off-target mutations that are not perceptible at this early stage in development.[156] The potential for these and other unintentional alterations to be inherited fuels the international apprehension of human germ line editing.

In science, answering one question leads to many others. Doudna and Charpentier had put together many key pieces of CRISPR puzzle: how bacteria can steal short segments of DNA from a phage genome in the midst of an infection and insert these segments precisely in the CRISPR array during the immune system's memory-forming stage; how crRNA molecules are produced inside the cell and converted to short, single-virus-matching sequences; and how a segment of RNA can form a complementary bond with its phage DNA counterpart and cause that DNA to be destroyed. Together, the answers to these questions formed the Cas proteins and tracrRNA:crRNA-guided Cas9. But this was not the end of it; the structural basis for gRNA recognition and DNA targeting by Cas9 was still unknown.[157] Zhou's work highlighted the problem of off-target cleavage and mutations at nontarget sites in the edited embryos. These critical problems have resulted in calls for caution in clinical applications until more studies on the structural and biochemical aspects of CRISPR/Cas9 are completed: "Further investigation of the molecular mechanisms of CRISPR/Cas9-mediated gene editing in human model is solely needed. In particular, off-target effect of CRISPR/Cas9 should be investigated thoroughly before any clinical application."[158] Therefore, it is folly for synthetic biologists or biochemists to say that they have created life or can rewrite the code of life. This is pure hyperbole! Life is complex — as noted by Venter and colleagues during their synthesis of a 531 kilobase synthetic genome they named *Mycoplasma mycoides* JCV1-syn3.0, which generated more questions about the underlying mechanisms. Additionally, biological organisms do not live in a vacuum; they interact with and compete with other organisms. Scientists must check their hubris lest they make irreversible modifications to life with unknown consequences for the entire ecosystem.

## Notes

1. André Maurois, *The Life of Sir Alexander Fleming Discoverer of Penicillin* (New York, NY: E.P. Dutton, January 1, 1959), 167.
2. Michael Specter, "The Gene Hackers: A Powerful New Technology Enables Us to Manipulate Our DNA More Easily Than Ever Before," *New Yorker*, November 8, 2015, 20.
3. Eric R. Scerri, "Forget Genius. Science Is the Product of Less-Than-Brilliant Minds," *Los Angeles Times,* February 20, 2017, 4, http://www.latimes.com/opinion/op-ed/la-oe-scerri-science-is-not-about-brillant-breakthroughs-20170220-story.html.
4. Nobel Media AB, "The Nobel Prize in Physiology or Medicine 1962," accessed March 21, 2019, http://old.nobelprize.org/nobel_prizes/medicine/laureates/1962/.
5. Peter Raven, George B. Johnson, Kenneth A. Mason, Jonathan B. Losos, Susan R. Singer, *Biology*, 9th ed. (New York: McGraw-Hill, 2008), 260.
6. Milly Dawson, "Martha Chase dies," *Genome Biology* 4, no. spotlight-20030820-01 (August 20, 2003), https://doi.org/10.1186/gb-spotlight-20030820-01.
7. Brenda Maddox, "The Double Helix and the Wronged Heroine," *Nature* 421, no. 6921 (January 23, 2003): 407–408.
8. Ibid.
9. Ronald K. Wetherington, "The Age of Darwin, II: After the Voyage" in *Readings in the History of Evolutionary Theory: Selections from Primary Sources*, edited by Ronald K. Wetherington (New York, NY: Oxford University Press, 2012), 102.
10. World History Project, "July 1 1858: Charles Darwin First Goes Public On Views of Evolution," accessed March 23, 2019, https://worldhistoryproject.org/1858/7/1/charles-darwin-first-goes-public-on-views-of-evolution.
11. Wetherington, "The Age of Darwin," 102.
12. Dr. Eric Lander is president and founding director of the Broad Institute of MIT and Harvard University. He might be described as brilliant, being skilled in mathematics and molecular genetics and applying those skills in the Human Genome Project as one of the principal leaders. In 2016, he published a short article in the high-impact scientific journal *Cell* titled

"The Heroes Of CRISPR," where he is accused of bias in his parsing out of recognition — assigning more recognition to scientists employed at the Broad Institute. Others have accused Lander of sexism for giving male scientists "top billing" and presenting female scientists' contribution as equivalent to laboratory assistants. The suggestion was that women scientists were being turned into handmaidens. Jim Kozubek, in his book *Modern Prometheus*, presents Johns Hopkins historian of science Nathaniel Comfort's deconstruction of Lander's essay. Comfort begins with the premise that a Nobel Prize and millions of dollars resulting from patents might be at stake: "Who claims them will be decided in part by what version of history becomes accepted as 'the truth ...' " Good writers know how rhetoric can be used to persuade. Does Lander use writing techniques to advance a self-interested version of history? On the first read, Lander's piece seems eminently fair, even generous. It "aims to fill in [the] backstory" of Crispr, Lander writes; "the history of ideas and the stories of pioneers — and draw lessons about the remarkable ecosystem underlying scientific discovery ... By turning his lens on such unsung heroes, laboring away at universities well beyond the anointed labs of Harvard, MIT, UCSF, Johns Hopkins, and the like, Lander creates the impression of inclusiveness, of the sharing of credit among all the 'heroes' of Crispr. But when he reaches Doudna and Charpentier's chapter in the story, the generosity becomes curiously muted. Though Lander maintains his warm, avuncular tone, Doudna and Charpentier enter the story as brave soldiers, working shoulder to shoulder with others ... Charpentier reported in a note on Pubmed: 'I regret that the description of my and collaborators' contributions is incomplete and inaccurate. The author did not ask me to check statements regarding me or my lab. I did not see any part of this paper prior to its submission by the author. And the journal did not involve me in review process" (28–29).

13. Jennifer A. Doudna and Samuel H. Sternberg, *A Crack in Creation: Gene Editing and the Unthinkable Power to Control Evolution* (New York: Houghton Mifflin Harcourt, 2017), xx.
14. Ruud Jansen, Jan. D.A. van Embden, Wim. Gaastra, and Leo. Schouls, "Identification of Genes That Are Associated with DNA Repeats in Prokaryotes," *Molecular Microbiology* 43, no. 6 (January 2002): 1565–1575; Alexander Bolotin, Benoit Quinquis, Alexei Sorokin, and S. Dusko Ehrlich, "Clustered Regularly Interspaced Short Palindrome Repeats (CRISPRs) Have Spacers of Extrachromosomal Origin," *Microbiology* 151, no. 8 (May 2005): 2551–2561.

15. Stan J. Brouns, Matthijs M. Jore, Magnus Lundgren, Edze R. Westra, Rik J. Slijkhuis, Ambrosius P. Snijders, Mark J. Dickman, Kira S. Makarova, Eugene V. Koonin, and John van der Oost, "Small CRISPR RNAs Guide Antiviral Defense in Prokaryotes," *Science* 321, no. 5891 (August 15, 2008): 960–963.
16. Jim Kozubek, *Modern Prometheus: Editing The Human Genome With CRISPR-CAS9* (New York, NY: Cambridge University Press, 2016), 106–110; Robert L. Sinsheimer, interview by Shelley Erwin, May 30, 1990, and March 26, 1991, California Institute of Technology Oral History Project (Pasadena, CA: Caltech Archives, 1992), 50–52.
17. Doudna and Sternberg, *A Crack in Creation*, 21.
18. Ibid., 21–22.
19. Ibid., 22.
20. Ibid.
21. Ibid., 23.
22. Ibid.
23. Ibid., 24–25.
24. Ibid., 25.
25. The Nobel prize in Physiology or Medicine 2007. Mario R. Capecchi, Martin J. Evans, Oliver Smithies, "for their discoveries of principles for introducing specific gene modifications in mice by the use of embryonic stem cells." Nobel Media AB 2019. Sunday, October 27, 2019, https://www.nobelprize.org/prizes/medicine/2007/summary/.
26. Doudna and Sternberg, *A Crack in Creation*, 26.
27. Ibid., 27.
28. Ibid., 27–28.
29. Ibid.
30. Ibid., 28.
31. Ibid., 28–29.
32. Doudna and Sternberg, *A Crack in Creation*, 29; Yohanns Bellaiche, Vladic Mogila, and Norbert Perrimon, "I-*Sce*I Endonuclease, a New Tool for Studying DNA Double-Strand Break Repair Mechanisms in Drosophilia," *Genetics* 152, no. 4 (July 1, 1999): 1037–1044.
33. Doudna and Sternberg, *A Crack in Creation*, 29.
34. Ibid., 29–30.
35. Ibid., 30–31.
36. Ibid., 31; David A. Wah, Jurate Bitinaite, Ira Schildkraut, and Aneel K. Aggarwal, "Structure of FokI Has Implications for DNA Cleavage,"

*Proceedings of the National Academy of Sciences* U.S.A. 95, no. 18 (September 1, 1998): 10564; Yang-Gyun Kim, Jooyeun Cha, and Srinivasan Chandrasegaran, "Hybrid Restriction Enzymes: Zinc Finger Fusions to FokI Cleavage Domain," *Proceedings of the National Academy of Sciences* U.S.A. 93, no. 3 (February 6, 1996): 1156.

37. Doudna and Sternberg, *A Crack in Creation*, 29; Bellaiche *et al.*, "I-*Sce*I Endonuclease," 1037–1044; Monya Baker, "Gene-editing Nucleases," *Nature Methods* 9 (January 2012): 23–26; Kim *et al.*, "Structure of FokI," 1156.
38. Kozubek, *Modern Prometheus*, 286.
39. Kim *et al.*, "Structure of FokI," 1156–1160.
40. Baker, "Gene-editing Nucleases," 24.
41. Kozubek, *Modern Prometheus*, 289.
42. Ibid., 290.
43. Doudna and Sternberg, *A Crack in Creation*, 32.
44. Kozubek, *Modern Prometheus*, 293–295.
45. Ibid., 58, 209; Matthew H. Porteus, Professor of Pediatrics (Stem Cell Transplantation), Bio, accessed March 28, 2019, https://profiles.stanford.edu/matthew-porteus.
46. Baker, "Gene-editing Nucleases," 24; Porteus, https://profiles.stanford.edu/matthew-porteus.
47. Porteus, https://profiles.stanford.edu/matthew-porteus.
48. Baker, "Gene-editing Nucleases," 24.
49. Ibid.
50. Kozubek, *Modern Prometheus*, 290; Baker, "Gene-editing Nucleases," 24.
51. Doudna and Sternberg, *A Crack in Creation*, 33.
52. Baker, "Gene-editing Nucleases," 23–24.
53. Ibid., 24.
54. Doudna and Sternberg, *A Crack in Creation*, 33.
55. Baker, "Gene-editing Nucleases," 24.
56. Doudna and Sternberg, *A Crack in Creation*, 33; Baker, "Gene-editing Nucleases," 24.
57. Doudna and Sternberg, *A Crack in Creation*, 33–34; Kozubek, *Modern Prometheus*, 292; Baker, "Gene-editing Nucleases," 24–25.
58. Kozubek, *Modern Prometheus*, 292.
59. Doudna and Sternberg, *A Crack in Creation*, 34.
60. Ibid.
61. Baker, "Gene-editing Nucleases," 24–25.

62. Ibid.
63. Ibid.
64. Ibid.
65. Kozubek, *Modern Prometheus*, 292; Baker, "Gene-editing Nucleases," 24–25.
66. Baker, "Gene-editing Nucleases," 24–25.
67. Baker, "Gene-editing Nucleases," 25.
68. Ibid.
69. Kozubek, *Modern Prometheus*, 292; Baker, "Gene-editing Nucleases," 24–25.
70. Baker, "Gene-editing Nucleases," 26.
71. Doudna and Sternberg, *A Crack in Creation*, 38.
72. Ibid.
73. Ibid.
74. Ibid., 39.
75. Ibid., 39–40.
76. Ibid., 41.
77. Ibid., 42.
78. Ibid., 44.
79. Ibid., 45.
80. Ibid., 52.
81. Ibid., 53.
82. Ibid., 43, 52–54.
83. Hélène Deveau, Rodolphe Barrangou, Josaine E. Garneau, Jessica Labonté, Christophe Fremaux, Patrick Boyaval, Dennis A. Romero, Philippe Horvath, and Sylvain Moineau, "Phage Response to CRISPR-Encoded Resistance in *Streptococcus thermophiles*," *Journal of Bacteriology* 190, no. 4 (February 2008): 1399.
84. Doudna and Sternberg, *A Crack in Creation*, 57.
85. Ibid., 58.
86. Ibid.
87. Ibid.
88. Luciano A. Marraffini and Erik Sontheimer, "CRISPR Interference Limits Horizontal Gene Transfer in Staphylococci by Targeting DNA," *Science* 322, no. 5909 (December 19, 2008): 1843–1845.
89. American Type Culture Collection (ATCC): The ATCC Bacterial collection is a nonprofit organization that collects, stores, and distributes for research standard reference microorganisms and cell lines. It has a diversified

collection of prokaryotes, containing more than 18,000 strains in over 750 genera; Marraffini and Sontheimer, "CRISPR Interference Limits," 1843.

90. Marraffini and Sontheimer, "CRISPR Interference Limits," 1843.
91. Ibid., 1843–1844.
92. Ibid., 1844.
93. Ibid.
94. Ibid.
95. Ibid.
96. Doudna and Sternberg, *A Crack in Creation*, 58–59.
97. Ibid., 60.
98. The National Academies of Sciences, Engineering and Medicine, International Summit on Human Gene Editing: A Global Discussion, December 1–3, 2015 (Washington, DC: The National Academies Press), 1–8; Nicholas Wade, "Scientists Seek Ban on Method of Editing the Human Genome," *New York Times*, March 19, 2015, https://www.nytimes.com/2015/03/20/science/biologists-call-for-halt-to-gene-editing-technique-in-human.html.
99. Doudna and Sternberg, *A Crack in Creation*, 61.
100. Ibid., 58.
101. Ibid., 63.
102. Ibid.
103. Ibid., 64–65.
104. Ibid., 65.
105. Ibid., 67.
106. Ibid., 68.
107. Ibid., 69.
108. Stan J. Brouns, Matthijs M. Jore, Magnus Lundgren, Edze R. Westra, Rik J. Slijkhuis, Ambrosius P. Snijders, Mark J. Dickman, Kira S. Makarova, Eugene V. Koonin, and John van der Oost, "Small CRISPR RNAs Guide Antiviral Defense in Prokaryotes," *Science* 321, no. 5891 (August 15, 2008): 960.
109. Elitza Deltcheva, Krzysztof Chylinski, Cynthia M. Sharma, Karine Gonzales, Yanjie Chao, Zaid A. Pirzada, Maria R. Eckert, Jörg Vogel, Emmanuelle Charpentier, "CRISPR RNA Maturation by *Trans*-encoded Small RNA and Host Factor RNase III, *Nature* 471, no. 7340 (March 31, 2011): 602–607.
110. Kozubek, *Modern Prometheus*, 17; Doudna and Sternberg, *A Crack in Creation*, 73.

111. Deltcheva *et al.*, "CRISPR RNA Maturation," 602–603.
112. Ibid., 602.
113. Ibid., 602–603.
114. Deltcheva *et al.*, "CRISPR RNA Maturation," 603.
115. Ibid., 602.
116. Ibid.
117. Rodolphe Barrangou, Christophe Fremaux, Hélène Deveau, Melissa Richards, Patrick Boyaval, Sylvain Moineau, Dennis A. Romero, and Philippe Horvath, "CRISPR Provides Acquired Resistance Against Viruses in Prokaryotes," *Science* 315, no. 5819 (March 23, 2007): 1711; Peter Raven, George B. Johnson, Kenneth A. Mason, Jonathan B. Losos, Susan R. Singer, *Biology*, 9th ed. (New York: McGraw-Hill, 2008), 345.
118. Deltcheva *et al.*, "CRISPR RNA Maturation," 604.
119. Ibid.
120. Ibid.
121. Ibid.
122. Doudna and Sternberg, *A Crack in Creation*, 70–71.
123. Ibid., 72.
124. Ibid., 72–75.
125. Ibid., 72–73.
126. Ibid., 74.
127. Ibid., 75.
128. Ibid., 76.
129. Martin Jinek, Krzysztof Chylinski, Ines Fonfara, Michael Hauer, Jennifer A. Doudna, Emmanuelle Charpentier, "A Programmable Dual-RNA-Guided DNA Endonuclease in Adaptive Bacterial Immunity, *Science* 337, no. 6096 (August 17, 2012): 816.
130. Ibid.
131. Ibid., 820.
132. Ibid., 818–819.
133. An R-loop is a three-stranded nucleic acid structure, composed of a DNA:RNA hybrid and the associated nontemplate single-stranded DNA. R-loops may be created by hybridization of mature mRNA with double-stranded DNA forming DNA-RNA hybrid.
134. Doudna and Sternberg, *A Crack in Creation*, 81.
135. Jinek *et al.*, "A Programmable Dual-RNA," 820; Doudna and Sternberg, *A Crack in Creation*, 81–84.
136. Doudna and Sternberg, *A Crack in Creation*, 83.

137. Ibid., 84.
138. Giedrius Gasiunas, Rodolphe Barrangou, Philippe Horvath, and Virginijus Siksnys, "Cas9-crRNA Ribonucleoprotein Complex Mediates Specific DNA Cleavage for Adaptive Immunity in Bacteria," *Proceedings of the National Academy of Sciences* U.S.A. 109, no. 39 (September 4, 2012): 2579–2586.
139. Doudna and Sternberg, *A Crack in Creation*, 91.
140. Ibid., 95–100.
141. Prashant Mali, Luhan Yang, Kevin M. Esvelt, John Aach, Marc Guell, James E. DiCarlo, Julie E. Norville, and George M. Church, "RNA-Guided Human Genome Engineering Via Cas9," *Science* 339, no. 6121 (February 15, 2013): 823.
142. Ibid., 823–825.
143. Le Cong, F. Ann Ran, David Cox, Shuailiang Lin, Robert Barretto, Naomi Habib, Patrick D. Hsu, Xuebing Wu, Wenyan Jiang, Luciano A. Marraffini, and Feng Zhang, "Multiplex Genome Engineering Using CRISPR/Cas Systems," *Science* 339, no. 6121 (February 15, 2013): 819–823.
144. Ibid., 820.
145. Ibid., 820–821.
146. Doudna and Sternberg, *A Crack in Creation*, 87–89.
147. Ibid., 89.
148. Jim Kozubek, *Modern Prometheus: Editing the Human Genome With CRISPR-CAS9* (New York, NY: Cambridge University Press, 2016), 24–30, 49–51, 306; Jon Brooks, "Making Sense of the CRISPR Patent Dispute Between the University of California and Broad," accessed May 3, 2019, https://www.kqed.org. Kozubek and Brooks does an excellent job in detailing the long-running CRISPR patent dispute between the University of California at Berkeley and Harvard/MIT Broad Institute. Detailed in Chapter 6, Luciano Marraffini and Erik Sontheimer at Northwestern University showed in 2008 how the CRISPR system targeted DNA. They were probably the first scientists to realize the potential of CRISPR as a technology for genetic engineering. Before CRISPR became a common acronym in the field of biotechnology, Marraffini and Sontheimer filed a patent application on CRISPR as a technology, but the application was disallowed on lack of description to enable to deploy it. In 2012, Doudna's patents filed under the Regents of the University of California, Berkeley, the University of Vienna and Emmanuelle Charpentier was registered as a provisional patent filed seven months before Feng Zhang filed under the Broad

Institute, Inc., MIT, and president and fellows of Harvard College in 2013. In 2014, the Broad Institute paid for an expedited review, receiving a key patent (for editing the DNA in *mouse and human cells*) while University of California, Berkeley waited for approval on its own filing. University of California, Berkeley went to court to challenge the Broad Institute's patent approval, arguing that the Broad Institute patents "interfered" with the patent for which they (UC Berkeley) had applied. In 2017, they lost the court battle; the Patent Office ruled that the Broad Institute's patent was sufficiently different compared to UC Berkeley. This decision was upheld by the US Court of Appeals in the fall of 2018. Finally, in late 2018 / early 2019, University of California, Berkeley was granted a foundational CRISPR/Cas9 patent (for editing the DNA in *all types of cells*). With both institutions owning CRISPR patents, the seeds of future conflict — as to who will get compensated when the technology is used therapeutically — were sown.

149. Wade, "Scientists Seek Ban," https://www.nytimes.com/2015/03/20/science/biologists-call-for-halt-to-gene-editing-technique-in-human.html.
150. Jon Cohen, "What Now for Human Genome Editing: Claimed Creation of CRISPR-edited Babies Triggers Calls for International Oversight?" *Science* 362, no. 6419 (December 7, 2018): 1090–1092; Dennis Normile, "Government Report Blasts Creator of CRISPR Twins," *Science* 363, no. 6425 (January 25, 2019): 328.
151. Puping Liang, Yanwen Xu, Xiya Zhang, Chenhui Ding, Rui Huang, Zhen Zhang, Jie Lv *et al.*, "CRISPR/Cas9-mediated Gene Editing in Human Tripronuclear Zygotes," *Protein Cell* 6, no. 5 (May 2015): 364; Doudna and Sternberg, *A Crack in Creation*, 214–216.
152. Liang *et al.*, "CRISPR/Cas9-mediated Gene Editing," 364.
153. Doudna and Sternberg, *A Crack in Creation*, 216.
154. Liang *et al.*, "CRISPR/Cas9-mediated Gene Editing," 364.
155. Ibid., 366.
156. Ibid., 364.
157. Heidi Ledford, "CRISPR Fixes Embryo Error: Gene-editing Experiment in Human Embryos Pushes Scientific and Ethical Boundaries," *Nature* 548, no. 7665 (August 2, 2017): 13–14.
158. Saul Winegrad, "Cardiac Myosin Binding Protein C," *Circulation Research* 84, no. 10 (May 28, 1999): 1117–1126.
159. Ledford, "CRISPR Fixes Embryo Error," 13–14.
160. Ibid.
161. Ibid., 14.

162. Fuguo Jiang, Kaihong Zhou, Linlin Ma, Saskia Gressel, and Jennifer A. Doudna, "A Cas9-guide RNA Complex Preorganized for Target DNA Recognition," *Science* 348, no. 6242 (June 26, 2015): 1477–1481; Liang *et al.*, "CRISPR/Cas9-mediated Gene Editing," 368.

# Chapter 7

# Crossing the Rubicon: Therapy versus Elective Enhancement

If we begin to medicalize what we now consider normal traits, enhancement and trait-selection will become more "legitimate" because they will be understood as part of medical treatment, driving people toward using these technologies.

— Sonia Suter, professor of law, George Washington School of Law, 2007[1]

Researchers have developed a synthetic gene that when injected into the muscle cells of mice makes muscles grow and prevents them from deteriorating with age. So the question arises, should this become possible in human beings? What are the proper uses of genetic alteration of muscles? Should it just be to cure muscular dystrophy and the atrophy of muscles that come with age? Or should athletes be able to use it to bulk up with steroids.

— Michael Sandel, 2004[2]

In this age of light speed advances in biotechnology, particularly genetic engineering, some bioethicists have recalled Nazi doctors and the singular nature of the Nazi "medicine" that they practiced. But many scholars have also invoked horror stories and their fictional characters, such as Frankenstein and his monster or Dr. Moreau and his half-human/half-animal creatures, like yellow traffic lights indicating caution. They also want to emphasize to the public the dangers posed by misguided or unregulated experiments or technology. A good approach is that taken by British science writer Jon Turney, who framed the problem so that everyone can make sense of their relationship with technology: "My premise is that fictional representations matter, that the science and technology we ultimately see are partly shaped by the images of the work which exist outside the confines of the laboratory report or the scientific paper."[3] Put simply, science does not occur in a vacuum; it interacts with the greater culture, both shaping and being shaped by public perceptions.

In *Frankenstein*, British author Mary Shelley told the story of young medical student Victor Frankenstein, who is obsessed with uncovering the "elixir of life."[4] Sharing his unfortunate story with Mr. Walton, captain of the ship that has rescued him from the Arctic ice, Frankenstein remarks:

> Life and death appeared to me ideal bounds, which I should first break through, and pour a torrent of light into our dark world. ... Pursuing these reflections, I thought, that if I could bestow animation upon lifeless matter, I might in the process of time ... renew life where death had apparently devoted the body to corruption.[5]

Prior to this time, to test his hypothesis, Frankenstein constructs a human body from parts of dismembered corpses and animates it with electricity using lightning. On seeing the corpse come to life, Frankenstein is horrified. Regretting his mistake — "the wretch — the miserable monster whom I had created" — Frankenstein runs away, rejecting his creation.[6] Obviously, Shelley was knowledgeable of the new science of the Enlightenment. And her complex narrative could have been a caution to scientists that they had a moral responsibility to use their science to do no harm and not violate nature. Ultimately, she believed that scientists should not use their knowledge to play God.

Another British author, H. G. Wells, told a story with similar philosophical themes, including moral responsibility and violation of nature.[7] Wells told the story — through his protagonist Edward Prendick — of the obsessed, immoral, formerly eminent British physiologist Dr. Moreau whose gruesome experiments in gross anatomy (known as vivisection in pre-20th-century literature) are publicly exposed.[8] As a result, Moreau is forced to flee to an island in the South Pacific. Prendick survives a shipwreck and is rescued by, and becomes a guest of, the captain of a ship. He is subsequently transported to "the island" and, in turn, becomes a reluctant guest of Mr. Montgomery, Moreau's assistant.[9] Wandering alone on the island, Prendick is shocked when he encounters half-human/

half-animal creatures. Eventually, he surmises that Moreau has been performing painful experiments, based on the loud cries he frequently hears coming from Moreau's laboratory.[10] Subsequently, he assumes that Moreau has performed experiments to convert humans to animals. However, Moreau makes a point of correcting Prendick, expressing the fact that he [Moreau] has been striving to make a complete transformation of animal to human.[11]

It is probably not fair to equate these science fiction novels and their flawed characters with modern scientists. As stressed throughout this book, most scientists are honorable men and women working very hard to help humankind and obtain some fame and fortune in the process. But it is interesting that, initially, Frankenstein (like modern scientists) has honorable intentions: "I entered with the greatest diligence into the search of the philosopher's stone and the elixir of life. But the latter obtained my most undivided attention … *but what glory would attend the discovery, if I could banish disease from the human frame, and render man invulnerable to any but a violent death*!"[12] (italics added). And Dr. Moreau, driven by the successes in the final stage of his research, believes that the pain he has inflicted has been insignificant to the final goal: the benefits of the work outweigh the risks.[13] These attitudes are common among scientists, and in countries where regulation is lax, some of these scientists might be tempted to cross the line.

There are many therapeutic treatments in medicine, some of which are available to healthy individuals as elective enhancements. As such, the concern is for therapeutic and nontherapeutic enhancements that impact animal and human physiology or genetics. These enhancements, as discussed at the beginning of this book, usher in a new type of eugenics. Ultimately, the concern here is any research (e.g., animal-human hybrids, chimeras, xenotransplantation, organic bioelectronics, gene editing, gene therapy, or complex chemicals that alter body physiology) that crosses the boundary into the realm of the unknown, invoking the monsters of Frankenstein and Moreau and leading to the questioning of human identity.

The mixing of human and animal genetic material has a long history in biology.[14] For instance, the fusion of human and animal cells to create somatic cell hybrids was a technique first used in the 1970s and 1980s to illuminate the interactions between nuclear and mitochondrial genomes.

At about the same time, the United Kingdom's HFEA licensed the transfer of human sperm to hamster eggs, thus creating a hybrid, as a diagnostic test for the quality of human sperm.[15] Today, scientists argue that the hostility of the public toward embryo destruction, coupled with the decreasing availability of human eggs, has compelled them to find alternative means, such as interspecies cloning (merging human somatic cells with eggs from other species or transplanting human iPSCs in animal embryos to advance studies in regenerative medicine).[16] The gut reaction of the general public to the suggestion of mixing human and animal material has been fear and disgust. Obviously, most lay members of the public have not been aware that mixing human and animal genetic material has been going on since the 1970s. Once the public was informed about the details of the science, especially the potential benefits of creating human-animal hybrids (e.g., cures for a diseases), the experiments became acceptable to some.[17] Would the public really sacrifice safety and accept the risk of heritable diseases and mutations through the use of human-animal hybrids and chimeras for *potential* benefits? What if the benefits were more tangible, such as a ready-made replacement for a diseased heart or liver generated by implanting human stem cells in an animal embryo so that it grows specific human organs? This would eliminate the wait for donor organs.[18] Would the public feel more compelled to accept the risk? Unfortunately, there are no definitive answers to these questions. A wise course would be to exercise caution and hold a series of international scientific conferences to formulate tentative rules of regulation that could be tailored to the current laws of different countries.

Nevertheless, a team of researchers led by developmental biologist Jun Wu of the Salk Institute for Biological Studies in La Jolla, California, claimed that interspecies blastocyst complementation enabled organ-specific enrichment of xenogenic iPSC derivatives.[19] They explained that interspecies blastocyst complementation might allow human organ formation in animals whose organ size, anatomy, and physiology were close to those of humans.[20] Wu and colleagues decided to use pigs to create human-pig chimeras, because pigs were similar in body size and muscles to humans, and their litter sizes would allow faster production of organs. The scientists injected more than 1,400 pig embryos with three types of iPSCs: normal cells, cells primed to develop into tissue, and cells that were not primed to develop into tissue.[21] They also modified the human cells for ease of identification within the chimera. Ethics regulations required Wu and his team to not grow chimera embryos beyond 28 days.[22] Consequently, at 28 days, the embryos were destroyed.

Notably, the chimeras injected with the stem cells not primed to develop into tissue contained the largest proportion of human cells, approximately one in 100,000 cells, but this was not enough to serve a useful purpose in organ donation, according to Niromitsu Nakauchi, a stem cell researcher at Stanford University.[23] A possible solution that caused some people to shudder was *to make pigs into human-organ donors* by using CRISPR to knock out pig proteins that could cause an immune reaction in humans.[24] But this solution would weaken the taxonomic barrier that makes *Homo sapiens* a "good species" in the sense of not allowing hybridization. Experiments creating human-rabbit and human-cow hybrids were not immune to this charge of contaminating a "good species." It also did not help the case for this type of research when the reported successes of these experiments from 1998 to 2003 were questioned.[25]

The goal of developmental biologists in the early 21st century has been to create a new source of ES cells, because, against the backdrop of the abortion debate, destroying embryos to harvest ES cells (each a potential life to the pro-life movement) was unacceptable to some people. But one research team, led by developmental biologist Hui Zhen Sheng of Shanghai Second Medical University in China, used donor cells from the foreskins of a five-year-old boy and two adult men and facial tissue from an adult woman to create more than 400 hybrid human-rabbit embryos, of which 100 survived to the blastocyst stage.[26] This team reported in a little-known peer-reviewed bimonthly scientific journal called *Cell Research* (affiliated with the Shanghai Institute of Cell Biology and the Chinese Academy of Sciences) that they had had great success generating ES cells with the ability to transform into many tissue types. Many researchers at the time questioned the lack of details in this report.[28] Robert Lanza, currently head of Astellas Global Regenerative Medicine in Massachusetts, remarked, "If this is true, it's important. But their results are very hard to believe."[29] The consensus among scientists in this field was that Sheng's team did not demonstrate that the ES cells reproduced indefinitely.

Several observations pertinent to all human-hybrid research are warranted. For instance, any rabbit proteins present will cause an immune reaction in humans; remnants of the rabbit's mtDNA in the cytoplasm of the egg might not be compatible with human nDNA.[30] More controversially, the human-animal hybrid embryos may have to receive human rights. In fact, the associate director of Pro-Life Activities for the United States Conference of Catholic Bishops, Richard Doerflinger, believed strongly that the human-rabbit embryos were human enough to deserve protections: "I think because all the nuclear DNA is human … we'd consider this an organism of the human species."[31]

The human-cow hybrid interspecies clone came into being with the added burden that the somatic cells used were donated by a research scientist working on the project.[32] In 1998, Jose Cibelli, formerly at the biotechnology company Advanced Cell Technologies in Massachusetts, donated cells from the inside of his cheek. These cells were subsequently fused with a cow egg.[33] Reprogramming occurred, and cleavage began to the cell stage 32, at which point the human-cow embryo was destroyed. This experience is reminiscent of the behavior of Moreau or Frankenstein. If the interspecies clone was allowed to continue beyond implantation, it would have developed as Cibelli's identical twin (the mitochondrial genes would have belonged to the cow): this is a boundary that no respectable scientist or open-minded bioethicist would cross. But there are those who warn against panicking. R. Alta Charo, professor of law and bioethics at the University of Wisconsin–Madison, wrote: "Short of putting one of these embryos into a woman's body for development to term, I don't think this work harms anyone alive."[34] And Arthur Caplan, professor of bioethics at New York University's Langone Medical Center, noted that while he understood the fear of interspecies cloning and stem cell research, he believed the following:

> There is no risk of making monsters this way. The biology will not work. Nor is the intent of any of these experiments anyway, so I don't think that fear is justified. I come down on the side that says if you can make great gains by making embryo hybrids in preventing premature death and understanding disease then a limited amount of such research is morally justified.[35]

With respect to Caplan, each year boundaries that should not be violated often are in the race to understand development, disease, and the prevention of death. The public, and regulators to some extent, are unaware of these breaches of ethics and only hear news of the final research outcomes. For example, Dolly was born on July 6, 1996, but was introduced to the world in February 1997, when she was seven months old. This means that the cloning techniques were being perfected in 1995 or before, because two lambs, Megan and Morag, were derived from a nine-day-old embryo in 1995, two years before the researchers took the next step to clone Dolly.[36] This might simply be a situation, as occurs in many research laboratories, where scientists wanted to protect their techniques to generate publications and file patents. Nevertheless, it is frightening to think of the unsanctioned research that scientists around the world are engaged in, guided by the overriding belief that the ends justify the means.

## 7.1. Creating Human-Animal Hybrids in the Name of Health: Violating Species Boundaries and/or Proliferation of Zoonoses (Animal Diseases)?

In trying to find an alternative approach to existing pharmacotherapy for treating Parkinson's disease, Yale School of Medicine professor of psychiatry D. Eugene Redmond Jr., in collaboration with colleagues, created a disease-relevant model using new-world monkeys (*Clorocebus sabaeus*), a species with a complex brain and a set of relevant behaviors similar to those of humans.[37] In 2007, the scientists treated these monkeys with 1-methyl-4-phenyl-1,2,3,6-tetrahydropyridine (MPTP) — a potent neurotoxin that degenerated dopaminergic neurons in each monkey's substantia nigra — providing a model for Parkinson's disease to test new cell-based therapeutics (during which millions of human neural stem cells were injected into the monkeys' brains).[38] According to Hemraj Dodiya, Jeffrey Kordower, and Dustin Wakeman, formerly of the Department of Neurological Sciences at the Rush University Medical Center in Chicago, Parkinson's disease is a debilitating neurodegenerative disorder affecting nearly 1,700,000 Americans and increasing in incidence as the average age increases.[39] They wrote that "loss of dopamine-containing neurons in the substantia nigra (SN) and the resulting deficit in sustained dopaminergic tone in the striatum and other areas of the brain appear to be responsible for the characteristic motor symptoms (tremor, rigidity, askinesia, shuffling gait, and postural instability) of PD [Parkinson's disease]."[40] It is not difficult to understand why scientists might explore the boundaries, and then go beyond them, to alleviate the pain and suffering of people with debilitating diseases. Redmond and colleagues performed a recombinant adeno-associated virus (rAAV2)-mediated gene transfer of glial cell line-derived neurotropic factor (GDNF) to the striatum of MPTP-treated monkeys.[41] Surprisingly, the primary fetal ventral mesencephalon dopamine-precursor cells survived and began directional growth of axons; the nigrostriatal reconstruction had begun in Parkinson's disease.[42] It was also revealed at this time that human dopaminergic neuroblasts had been grafted into the substantia nigra of adult mice lesioned with 6-hydroxydopamine (OHDA). Anders Björklund and colleagues at the Department of Experimental Sciences, Lund University, Sweden, wrote that these mice "retain the capacity to follow endogenous instructional cues and reconstruct a new functionally relevant nigrostriatal pathway."[43]

Near the sacred boundary of never implanting a human-animal hybrid embryo into a human female is the boundary of never injecting human neural cells into the brains of animals. Redmond and his team crossed this boundary when they used monkeys. Insoo Hyuh, professor of bioethics and philosophy at Case Western Reserve University, had concerns about this research. He recently addressed the human-hybrid embryo ethical dilemma:

> If you up the biological contribution of the human stem cells, are you also somehow turning them morally into a human-like thing with human rights? It's so difficult to know how you would actually address that. It's not measurable. I would argue, *as long as you are avoiding the brain, or there's not a significant change to the brain, of structure and possible functioning,* then I don't think you even need to go down than philosophical path.[44] (italics added)

Despite Hyuh's statement, Redmond had no ethical qualms. He adamantly believed that "if the use of monkeys leads to the cure of Parkinson disease for the 500,000 people in the United States (and millions more around the world) some of whom suffer, suffocate, and die each year, it is an acceptable moral price to pay."[45] He specifically noted why the monkey with MPTP-induced Parkinson disease is the best model:

> The central nervous system and higher brain functions are sufficiently different that monkey experiments are often essential for progress with neuropsychiatric and brain-related problems. Parkinson disease represents a research problem for which monkey studies can be justified. … Accidental exposures of humans to MPTP simulate Parkinson disease almost completely, confirming that monkeys exposed to MPTP are a reasonable model for studying the condition in humans.[46]

It is commendable, and simultaneously frightening, that most of the best research scientists in the world have these convictions. And if one asks seasoned and new medical scientists why they entered the medical field, their responses would be near identical to Redmond's, or at least have the same theme: "I became a physician in order to cure, alleviate, and understand diseases and to 'do good' if possible."[47] The unanswerable question is how will scientists, raised and educated in different social environments that shaped their characters, maintain their honorable intentions, not cross certain boundaries, and pursue more distant benefits of their research?

Alternatively, scientists' backgrounds may play a minor role in their impulses to cross boundaries and ride the momentum of science "that makes you want to go further and further."[48] Developmental scientists are not only implanting human stem cells in animal embryos to grow specific human organs, they are also violating species boundaries by creating hybrid interspecific chimeras. It is well known in ecology that hybridization occurs naturally in animals (sibling species) and plants (more so in plants, causing polyploidy and subsequent duplication of the genomes of the two species, resulting in allopolyploidy), causing chaos in nature.[49] Most hybrids are sterile because their parents' reproductive systems are not totally compatible, the result of being different species. In essence, the hybrid is an evolutionary waste. The chaos derives from the fact that some hybrids are fertile and better able to survive (heterosis) in new environments than the parental species.

Humans have been involved in artificial selection and hybridization of plants and animals since the agricultural revolution. So, what is the problem here? The problem is that humans are for the first time directly violating species boundaries — through genetic engineering — to study genetic and behavioral differences between sibling species. In 1980, for instance, developmental biologist Janet Rossant, professor in the Departments of Molecular Genetics and Pediatrics at the University of Toronto, and the late William Frels of the Roswell Park Cancer Institute in Buffalo, New York, successfully generated live chimeras between sibling species *Mus musculus* (an albino laboratory mouse) and *Mus caroli* (Ryukyu mouse from East Asia).[50] Interestingly, this was not the first time scientists performed this type of experiment. In fact, in 1973 British embryologists Richard Gardner and Martin Johnson created rat-mouse chimeras by a new technique at the time called blastocyst injection.[51] The rat-mouse (intergenera) embryonic tissues developed into few offspring, which developed abnormally.

*M. musculus* and *M. caroli* do not interbreed in nature, and, predictably, research scientists Rossant and Frels found differences in genes, preimplantation development (*M. caroli* completes cleavage to gastrulation approximately 16 hours before *M. musculus* does), and gestation (M. *caroli* has shorter gestation).[52] They created chimeras by injecting the inner cell mass of *M. caroli* into *M. musculus* blastocysts. The resulting embryos were transferred into the uterus of *M. musculus* (specifically, the nonagouti mice homozygous for the *b* allele of the glucose phosphate isomerase [GPI] gene, in order to identify chimeric cells of the two tissue types). Out of nine fetuses dissected to study fetal tissues, six were identified as chimeric using GPI analysis, indicating that the *M. caroli* and *M. musculus* cells were mostly compatible.[53] In summary, out of 48 offspring, 38 were determined to be chimeric by hair and eye pigmentation and GPI analysis: "Coat and eye pigmentation, as well as the percentage of *M. caroli* GPI enzyme in the blood, varied in each chimera."[54] This quote highlights the complexities of these experiments and the many unknowns — the causes for the variation of coat and eye pigmentation in each chimera — that indicated the need for less celebration and more caution.

Caution is also needed in xenotransplantation — the process of grafting or transplanting organs, tissues, or living cells between members of different species.[55] This approach, more than any other procedures discussed so far, brings to mind Dr. Moreau. The scarcity of organs for life-saving transplants has made xenotransplantation an attractive option, especially when compared to injecting human iPSCs into an early animal embryo to create chimeric animals (such as a pig that can grow human organs to order).[56] But decades of scientific failures, struggles to suppress the host immune response, the problem of zoonotic diseases (animal diseases that affect humans), and associations with the darker side of science and experimentation portrayed in fiction like *Frankenstein* and the *Island of Dr. Moreau* and films like 1978's *Coma* have forced xenotransplantation into obscurity with the label of pseudoscience.[57] Xenotransplantation has been resurrected recently by results in which healthy monkeys survived with pig organs. Kelly Service — writing for *Science* — indicated that some researchers at a meeting of the International Xenotransplantation Association were hoping for human trials soon.[58] Members of this organization feel confident enough to retake their position in mainstream science. Better immunosuppressant drug regimens and CRISPR/Cas9 have given xenotransplantation experts added confidence and encouragement.[59] With CRISPR, for example, antigens on pig cells causing an immune response or harmful viral sequences (e.g., from porcine reproductive respiratory syndrome virus) can be edited out.

Service discussed the work of research scientists at Emory University who transplanted the kidney of a pig genetically engineered without antigens — that caused an immune response — into a rhesus macaque monkey.[60] This kidney sustained the monkey for more than 400 days before the monkey's immune system rejected it. Equally notably, a research team in Germany announced a survival record of 90 days for a pig's heart transplanted into a baboon.[61] Service quoted University of Munich cardiac surgeon and team member Paolo Brenner speaking on the baboon's progress: "The baboon was still in very good condition when the experiment was stopped after 3 months."[62] Nevertheless, researchers are fighting a constant battle with the pig's resilient immune system, which presents a significant problem for transplant organs susceptible to inflammation, such as lungs. This problem of resistance should be no surprise to researchers in the biological fields. Biological systems are constantly evolving and adapting in changing environments, and no matter how much genetic engineers tinker with human or animal genomes for human health and safety, any quick fix today might be detrimental tomorrow because the environment is constantly changing.

## 7.2. Benefits and Risks of Elective Biological Enhancement

The blurred lines between therapeutic enhancements and elective cosmetic enhancements, discussed elsewhere in this book, become sharper when consumers apply pressure for cosmetic enhancement that will make them phenotypically attractive and cognitively and physically above average. And the brilliant scientists employed by multinational corporations to support their research file patents in order to make more money as consumer demand increases.

In athletics, for example, some of the common drugs used for bigger muscles, better performance, and heightened energy are anabolic steroids, EPO, human growth hormone, androstenedione, amphetamines, and testosterone.[63] A few of these drugs will change the physiology of adult cells of the individual but not impact potential offspring. However, the drug EPO, while it provides more energy and endurance for the individual, it also makes the blood become thick so that the heart has to work harder to pump it through the vessels, which can lead to a stroke.[64] In the competitive arena of sports, athletes are pushed to become better, stronger, and faster. In the environment of the new eugenics, this will come with a risk to human biology and human culture.

Most parents want their children to be healthy and intelligent. And most human beings would not reject the chance to be an Einstein. Recall from Chapter 1 the hypothetical memory-enhancing drug used for Alzheimer's patients that not only cured them but made them function beyond the capacity of the average human. Or imagine that a completely new "smart" pill is created. What will be the risk to human cognition and culture? Greely suggested that neuroscientists face technical and societal challenges when working with the brain:

> We understand the brain much less than we understand the insulin system or heart rhythms ... although you can do great damage with other

> implants, you are rarely going to change, directly at least, the patient's personality. Even with DBS [deep brain simulation] for Parkinson's disease, we see patients whose personality and behavior changes dramatically with regard to gambling or to sex.[65]

The questions posed are the following: What is the correlation between the amount of stimulation the machine can deliver and changes in the brain? (This question could be applied to any neurotechnology or chemical therapy.) And can the technology or therapy be used for enhancement, such as brain manipulation of soldiers or elective stimulation by private citizens so that they continue to function without injury, pain, and fatigue?

Michael Bess, chancellor's professor of history at Vanderbilt University in Nashville, in his 2015 book *Our Grandchildren Redesigned*, provided a short historical outline of research in augmented cognition, learning, and memory:

> 1990: the neuroscientist Eric Kandel identifies a brain chemical known as CREB [cAMP response element binding protein] as a key factor in the processes of memory formation.
>
> 1994: the neuroscientists Tim Tully and Jerry Yin alter the way CREB operates in a strain of fruit flies. The CREB-boosted flies learn new behaviors at rates up to ten times faster than their unmodified brethren.
>
> 1999: Princeton neuroscientists Joe Tsien genetically engineers a strain of mice possessing extra copies of the $NR_2B$ gene [N-methyl D-aspartate receptor subtype 2B], which regulates the activity of synaptic receptors in the brain. The $NR_2B$-enhanced mice perform up to five times better than unmodified mice in tests of memory and learning.[66]

Many people would jump at the chance to retain more memories and become more intelligent. Some examples are parents with dreams of their child becoming a Nobel laureate, an employee who wants to work twice as fast with little sleep and multitask to obtain a promotion, or a college engineering student who needs to master calculus. The risk would probably be an overload of information. Bess argued that there might be a good reason why people occasionally forget things throughout their lives in the absence of any mental pathology. He said:

> When we examine the functioning of memory as a practical component in a person's daily life, we find it is just as important to be able to selectively *lose* information as it is to retain it. Without this ability, we would rapidly find ourselves drowning in a sea of trivial details, impressions, emotions, and images.[67]

With a nervous breakdown in the individual, society is duped by superficial research promising "superintelligence"; it is like a get-rich-quick scheme.

Similarly, humans also jump at the chance to slow down or stop aging. Recall the discussion on telomeres, which become progressively shorter with age. At a specific time, the cell stops dividing and enters a state of senescence.[68] Is there a drug or enzyme that could be inserted into human cells to reverse the processes of telomere shortening and senescence? Researchers Hamayoun Vaziri and Samuel Benchimol at the Ontario Cancer Institute in the Department of Biophysics at the University of Toronto answered this question in the late 1990s. They began by testing the telomere hypothesis, which proposed that "critical shortening of telomeric DNA due to the end-replication problem is the signal for the initiation of cellular senescence."[69] To test this hypothesis, they used a critical catalyst called telomerase reverse transcriptase (hTERT), which is a complementary DNA encoding the catalytic subunit of human telomerase (an RNA-protein complex that synthesizes telomeric repeats).[70] Vaziri and Benchimol expressed hTERT in healthy human fibroblasts, which lacked telomerase activity, to determine whether telomerase activity could extend the life span of the cells. In their results, they noted that "retroviral-mediated expression of hTERT resulted in functional telomerase activity in normal aging human cells. Moreover, reconstitution of telomerase activity *in vivo* led to an increase in the length of telomeric DNA and to extension of cellular lifespan."[71] While these results might imbue one with a sense of awe at this potential "fountain of youth," there is a serious problem in all of this, which is the unintended insertion of retroviruses into a target cell's genome, resulting in disastrous effects (as discussed in Chapters 2 and 11).

Other scientists in search of the fountain of youth use *C. elegans*.[72] Recall that *C. elegans* is one of the model animals used in biology to study development and regulation. This nematode is also widely used as a model animal in the study of aging and longevity-related processes.[73] In 2005, geneticist Juilius Halaschek-Weiner and colleagues at the British Columbia Cancer Agency and the Molecular Biology Program at the University of Missouri used serial analysis of gene expression (SAGE) — a method that efficiently quantifies large numbers of mRNA transcripts by sequencing — to identify a longevity-related *daf-2* gene in a long-lived *C. elegans* that encoded the insulin/insulin-like growth factor 1 receptor.[74] Halaschek-Weiner and colleagues stated that reduction of *daf-2* signaling in these mutant worms led to a doubling in mean life span. They prepared *C. elegans* SAGE libraries from one-, six-, and ten-day-old adult *daf-2* mutants and six-day-old control adults. They reported:

> Differences in gene expression between daf-2 libraries representing different ages and between daf-2 versus control libraries identified not only single genes, but whole gene families that were differentially regulated. These gene families are part of major metabolic pathways including lipid, protein, and energy metabolism, stress response, and cell structure. Similar expression patterns of closely related family members emphasize the importance of these genes in aging-related processes. Global analysis of metabolism-associated genes showed hypometabolic features in mid-life daf-2 mutants that diminish with advanced age. Comparison of our results to recent microarray studies highlights sets of overlapping genes that are highly conserved throughout evolution and thus represent strong candidate genes that control aging and longevity.[75]

If this gene were harnessed and synthesized for human trials, there might be complications because of the complexities of biological mechanisms of different animals coupled with their varied interactions with the environment. Bess discussed this very well in his book:

> The more we learn about the functioning of genes, the more salient becomes the role played by the mediating factors that regulate their activation, deactivation, and transcription. Many of our twenty thousand genes, it turns out, are constantly being switched on and off in complex combinations, as well as in still more complex chronological *sequences*

of combinations. The activity is profoundly bidirectional, in the sense that genes both regulate, and are regulated by, the chemical processes going on in the cell and broader organism that surrounds them.[76]

In essence, there is much still to learn about how genes are expressed when transferred from one organism to another. The *daf-2* mutant gene might be expressed adversely in humans due to our biological complexity.

## Notes

1. Sonia M. Suter, "A Brave New World of Designer Babies," *Berkeley Technology Law Journal* 22, no. 2 (March 2007): 936.
2. Michael Sandel, "The Pursuit of Perfection: A Conversation on the Ethics of Genetic Engineering," March 31, 2004, 4, http://www.pewforum.org/2004/03/31/the-pursuit-of-perfection-a-conversation-on-the-ethics-of-genetic-engineering.
3. Jon Turney, *Frankenstein's Footsteps: Science, Genetics, and Popular Culture* (New Haven, CT: Yale University Press, 1998), 3.
4. Mary W. Shelley, *The Complete Frankenstein* (1818; repr., Washington, DC: Literary Folio, 2017), 32.
5. Shelley, *The Complete Frankenstein*, 47.
6. Shelley, *The Complete Frankenstein*, 52.
7. Herbert G. Wells, *The Island of Dr. Moreau* (1896; repr., London: Harmondsworth, Penguin, 1946), 243–248.
8. Ibid., 127–146.
9. Ibid., 45.
10. Ibid., 54.
11. Ibid., 127–146.
12. Shelley, *The Complete Frankenstein*, 32.
13. Wells, *The Island of Dr. Moreau*, 127–146.
14. Sarah Taddeo and Jason S. Robert, " 'Hybrids and Chimeras: A Report on the Findings of the Consultation,' by the Human Fertilisation Embryology Authority in October 2007" in Embryo Project Encyclopedia (City, ST: Publisher, 2014), http://embryo.asu.edu/handle/10776/8240.
15. Ibid.
16. David Robson, "The Birth of Half-Human, Half-Animal: The Quest to Create Animals with Human Organs Has a Long History — And It Is Now

Becoming a Reality. Has Science Taken a Step Too Far?" January 5, 2017, http://www.bbc.com/earth/story/20170104-the-birth-of-the-human-animal-chimeras.

17. Taddeo and Robert, "'Hybrids and Chimeras: A Report on the Findings of the Consultation,' by the Human Fertilisation Embryology Authority in October 2007," http://embryo.asu.edu/handle/10776/8240.
18. Robson, "The Birth of Half-Human, Half-Animal," http://www.bbc.com/earth/story/20170104-the-birth-of-the-human-animal-chimeras.
19. Jun Wu, Aida Platero-Lugengo, Masahiro Sakurai, Atsushi Sugawara, Maria Antonia Gil, Takayoshi Yamauchi, Keiichiro Suzuki … Juan Carlos Izpisua Belmonte, "Interspecies Chimerism with Mammalian Pluripotent Stem Cells," *Cell* 168, no. 3 (January 26, 2017): 473.
20. Ibid.
21. Ibid.; Sara Reardon, "Hybrid Zoo: Introducing Pig — Human Embryos and a Rat-Mouse: Chimaeras Could Pave the Way for Growing Human Organs in Other Animals," January 26, 2017, http://www.nature.com/news/hybrid-zoo-introducing-pig-human-embryos-and-a-rat-mouse-1.21378.
22. Ibid.
23. Reardon, "Hybrid Zoo: Introducing Pig — Human Embryos And a Rat-Mouse"; http://www.nature.com/news/hybrid-zoo-introducing-pig-human-embryos-and-a-rat-mouse-1.21378.
24. Ibid.
25. Taddeo and Robert, "'Hybrids and Chimeras: A Report on the Findings of the Consultation,' by the Human Fertilisation Embryology Authority in October 2007," http://embryo.asu.edu/handle/10776/8240.
26. Philip Cohen, "Human-Rabbit Embryos Intensify Stem Cell Debate," *New Scientist* (August 15, 2003, Daily News): 1–3.
27. Ibid.
28. Ibid.
29. Ibid., 1.
30. Rick Weiss, "Cloning Yields Human-Rabbit Hybrid Embryo," August 14, 2003, http://www.washingtonpost.com/archive/politics/2003/08/14/cloning-yields-human-rabbit-.
31. Ibid.
32. Patrick Dixon, "Cow/Human Clone Hybrid — Cow and Human Mixed Together," accessed September 6, 2018, https://www.globalchange.com/humancow.htm.
33. Ibid.

34. Rick Weiss, "Cloning Yields Human-Rabbit Hybrid Embryo," August 14, 2003.
35. Maggie Fox, "Human-Cow Hybrid Embryos Made in Lab," April 3, 2008, http://www.abc.net.au/science/articles/2008/04/03/2206835.htm.
36. Ian Wilmut, "Cloning for Medicine," *Scientific American* (December 1998): 58–63.
37. D. Eugene Redmond Jr., J.D. Elsworth, R. H. Roth, C. Leranth, T.J. Collier, B. Blanchard, K.B. Bjugstad, R.J. Samulski, P. Aebischer, J.R. Sladek, "Embryonic Substantia Nigra Grafts in the Mesencephalon Sent Neurites to the Host Striatum in Non-Human Primate After Overexpression of GDNF," *The Journal of Comparative Neurology* 515, no. 1 (July 1, 2009): 31.
38. Dustin Wakeman, Hemraj R. Dodiya, and Jeffrey H. Kordower, "Cell Transplantation and Gene Therapy in Parkinson's Disease," *Mount Sinai Journal Medicine* 78, no. 1 (January–February 2011): 126.
39. Ibid., 127.
40. Ibid.
41. Redmond *et al.*, "Embryonic Substantia Nigra Grafts in the Mesencephalon Sent Neurites to the Host Striatum in Non-Human Primate After Overexpression of GDNF," 31–32.
42. D. Eugene Redmond Jr., J.D. Elsworth, R. H. Roth, C. Leranth, T.J. Collier, B. Blanchard, K.B. Bjugstad, R.J. Samulski, P. Aebischer, J.R. Sladek, "Embryonic Substantia Nigra Grafts in the Mesencephalon Sent Neurites to the Host Striatum in Non-Human Primate After Overexpression of GDNF," *The Journal of Comparative Neurology* 515, no. 1 (July 1, 2009): 31–40; J.D. Elsworth, D. Eugene Redmond Jr., C. Leranth, K.B. Bjugstad, J.R. Sladek Jr., T.J. Collier, S.B. Foti, R.J. Samulski, K.P. Vives, R.H. Roth, "AAV2-mediated Gene Transfer of the Striatum of MPTP Monkeys Enhances the Survival and Outgrowth of Co-implanted Fetal Dopamine Neurons," *Experimental Neurology* 211, no. 1 (May 2008): 255–256.
43. L. H. Thompson, S. Grealish, D. Kirk, A. Björklund, "Reconstruction of the Nigrostriatal Dopamine Pathway in the Adult Mouse Brain," *European Journal of Neuroscience* 30, no. 4 (August 2009): 625–638.
44. Susan Scutti, "First Human-Pig Embryos Made, Then Destroyed," CNN, January 30, 2017, https://www.cnn.com/2017/01/26/health/human-pig-embryo/index.html.
45. D. Eugene Redmond, "Using Monkeys to Understand and Cure Parkinson Disease," *Hastings Center Report* 42, no. S1 (November–December 2012): S10, https://animalresearch.thehastingscenter.org.

46. Ibid., S8.
47. Ibid.
48. David Robson, "The Birth of Half-Human, Half-Animal: The Quest to Create Animals with Human Organs Has a Long History — And It Is Now Becoming a Reality. Has Science Taken a Step Too Far?" January 5, 2017, http://www.bbc.com/earth/story/20170104-the-birth-of-the-human-animal-chimeras.
49. Raven *et al.*, *Biology*, 477–480.
50. Janet Rossant and William I. Frels, "Interspecific Chimeras in Mammals: Successful Production of Live Chimeras Between *Mus musculus* And *Mus caroli*," *Science* 208, no. 4442 (1980): 419–421; Nicole Newkirk, "'Interspecific Chimeras in Mammals: Successful Production of Live Chimeras between *Mus musculus* and *Mus caroli*,' (April 25, 1980), by Janet Rossant and William I. Frels," in *Embryo Project Encyclopedia* (City, ST: Publisher, 2007), http://embryo.asu.edu/handle/10776/1712.
51. Richard L. Gardner and Martin H. Johnson, "Investigation of Early Mammalian Development Using Interspecific Chimeras between Rat and Mouse," *Nature New Biology* 246, no. 151 (November 21, 1973), 86–89; Newkirk, "Interspecific Chimeras in Mammals," http://embryo.asu.edu/handle/10776/1712.
52. Rossant and Frels, "Interspecific Chimeras in Mammals," 420; Newkirk, "Interspecific Chimeras in Mammals," http://embryo.asu.edu/handle/10776/1712.
53. Rossant and Frels, "Interspecific Chimeras In Mammals," 420.
54. Ibid.
55. Food and Drug Administration, "Xenotransplantation," accessed September 15, 2018, https://www.fda.gov.
56. Kelly Service, "Xenotransplant Advances May Prompt Human Trials: Record Survival of Primates with Pig Organs Raises Hopes," *Science* 357, no. 6358 (September 29, 2017): 1338.
57. Ibid.
58. Ibid.
59. Ibid.
60. Ibid.
61. Ibid.
62. Ibid.
63. Michael J. Sandel, *The Case Against Perfection: Ethics in the Age of Genetic Engineering*, (Cambridge, MA: The Belknap Press of Harvard University Press), 25–38; Jim Kozubek, *Modern Prometheus: Editing the Human*

*Genome with CRISPR-CAS9* (New York, NY: Cambridge University Press, 2016), 161–163.

64. Kozubek, *Modern Prometheus*, 161–163.
65. Emily Underwood, "Brain Implant Trials Raise Ethical Concerns: NIH Workshop Delves into Challenges of Testing Invasive Neuromodulation Technology," *Science* 348, no. 6240 (June 12, 2015): 1187.
66. Michael Bess, *Our Grandchildren Redesigned: Life in the Bioengineered Society of the Near Future*, (Boston: Beacon Press, 2015), 16–17. CREB (Camp response element binds to certain DNA sequences called Camp response elements (CRE) to increase or decrease the transcription of genes. Since the reduction of CREB is implicated in the pathology of Alzheimer's disease, increasing the expression of CREB might prevent the formation of misfolded proteins (prions); N-methyl D-aspartate receptor subtype 2B ($NR_2B$) is a protein subunit related to learning, memory processing, and feeding behaviors.
67. Bess, *Our Grandchildren Redesigned*, 17.
68. Ibid.; Raven *et al.*, *Biology*, 273.
69. Hamayoun Vaziri and Samuel Benchimol, "Reconstitution of Telomerase Activity in Normal Human Cells Leads to Elongation of Telomeres and Extended Replicative Lifespan," *Current Biology* 8, no. 5 (February 26, 1998): 279.
70. Ibid.
71. Ibid.
72. Juilius Halaschek-Weiner, Jaswinder S. Khattra, Sheldon McKay, Anatoli Pouzyrev, Jeff M. Stott, George S. Yang, Robert A. Holt, Steven J. M. Jones, Marco A. Marra, Angela R. Brook-Wilson, Donald L. Riddle, "Analysis of Long-lived *C. elegans* daf-2 Mutants Using Serial Analysis of Gene Expression," *Genome Research* 15, no 5 (May 2005): 603–615.
73. Ibid., 603.
74. Ibid., 603–604.
75. Ibid., 604.
76. Bess, *Our Grandchildren Redesigned*, 41.

# Chapter 8

# Playing God: Synthetic Biology and the Attempt to Control the Machinery of Life

Remember, that I am thy creature: I ought to be thy Adam; but I am rather the fallen angel, whom thou drivest from joy for no misdeed.

— Frankenstein's monster addressing Victor Frankenstein in Mary Shelley's *Frankenstein*, 1818[1]

We're moving from reading the genetic code to writing it.

— J. Craig Venter, CEO of the J. Craig Venter Institute, 2005[2]

In May 2000, members of the British all-party Parliamentary and Scientific Committee charged that scientists were "playing God" by modifying the genetic essence of life.[3] In response, James Watson — the provocative 1962 Nobel laureate and a special guest of the committee — responded rhetorically, "But then, in all honesty, if scientists don't play God, who will?"[4] Most scientists would not display such hubris in my opinion. Applied philosopher Henk van den Belt of the Department of Social Science at Wageningen University in the Netherlands believed that this charge might have been premature, since synthetic biologists take on different postures when the situation calls for it. He wrote, "It seems to me that it might be helpful to consider the two contrasting postures, arrogance and humility, as *different* registers from the *same* rhetorical repertoire, which scientists can play according to the demands of the situation. If the situation demands that critics be silenced, scientists will indeed play the register of humility."[5] Playing the humility card might be the right strategic choice because it keeps professional critics at bay and stifles any accusations of blasphemy by the public. It is interesting to compare the responses to accusations of playing God of three prominent synthetic biologists: J. Craig Venter, venture capitalist and CEO of the J. Craig Venter Institute; Church (the Harvard Medical School molecular geneticist mentioned in previous chapters); and Andrew D. Endy, professor of bioengineering at Stanford University. Their humble responses tentatively calmed the religious factions of the public but simultaneously caused concern among the informed public, because the synthetic biologists referred in their responses to *manipulating* or *modifying* life.[6] For instance, Venter insisted that he was not playing God; he was simply "*modifying life* to come up with new life forms"[7] (italics added). Church went into more detail, emphasizing the words *engineering* and *designing* in connection with life and making it clear that scientists are not *creating* it:

> We're acting as *engineers*, possibly as intelligent *designers*. The religiously-inclined would not put humans in the same league with the "Intelligent Designer," or God. As creative as we become, and as industrious and as good as we are at designing and manufacturing living things, which we've been doing since the stone age — no matter how good we get at that, it's like calling a candle a supernova. A candle is not a supernova; it's not even in the same league. And we, as intelligent designers, are not in the same league as the "Intelligent Design" forces that started the whole shebang. We're not designing sub-atomic particles from scratch; we're not designing galaxies. We're really not even designing the basic idea of life; we're just *manipulating* it.[8] (italics added)

And Endy replied with an explanation similar to Church's, with equal confidence, regarding what synthetic biologists do:

> I don't view [my research] projects as creating life, but rather [as] construction projects. For me as an *engineer*, there is a big difference

> between the words creation and construction. Creation implies I have unlimited power, perfect understanding of the universe, and the ability to *manipulate* matter at a godlike level. That's not what I have. I have an imperfect understanding, a budget, limited resources, and I can only manipulate things quite crudely. In that context, with those constraints, I'm a more humble constructor.[9] (italics added)

Venter, Church, and Endy were correct when they said that they were not creating life. But the long-term consequences of modifying life to create new life forms are worrying, regardless of the transformative, marketable, and beneficial products, such as gene therapies that obliterate leukemia or biodegradable plastics synthesized from sugars.[10] What are the risks to the environment and living organisms, for example, if these biodegradable plastics were to get into natural and artificial water and food systems? Sadly, these risk assessments are not available.

The European Commission was also concerned about the potential risks associated with synthetic biology research. In 2014, its scientific committees on consumer safety, emerging and newly identified health risks, and health and environmental risks published their opinion on the definition of synthetic biology and the adequacy of risk assessment methods, including research priorities on risk assessment.[11] They defined synthetic biology as "the application of science, technology and engineering to facilitate and accelerate the design, manufacture and/or modification of genetic materials in living organisms."[12] This is an accurate description of the current research in biotechnology: a race to modify or repair biological organisms that is the foundation of the new eugenics. Systems biologist Rainer Breitling and synthetic biologist Eriko Takano — both professors at the University of Manchester — and their colleague Timothy Gardner, CEO of Riffyn, strongly supported the wording in this definition because it treats synthetic biology and engineering or modification as fundamentally similar.[13] As such, guidelines for genetic engineering should also apply to synthetic biology.

In their draft opinion, the European Commission's scientific committees discussed several questions, three of which are noted — in one form or another — in this book: (1) Will the continual acceleration of biotechnologies outweigh the benefits (or overburden current risks assessments)?

(2) Will genetically engineered organisms display aberrant or normal behavior? (3) Is there a standard cellular or physiological system for genetically modified organisms beyond which substantial divergence or mutation poses danger to natural organisms? In the end, the committees recommended advanced safety systems ("safety locks") built into products of synthetic biology.[14]

Returning to the theme of playing God, every generation has experienced a revolution in science in which the public has been exposed to new technologies through university education, medical treatment, or buying the new technology in the free market. A good example of this was in the late 18th century, when the Industrial Revolution was in its infancy. The public was fascinated with electricity.[15] The Italian biologist and physicist Luigi Galvani discovered that the muscles of *dead* frogs' legs twitched when struck by an electric spark.[16] This established the connection between electricity and life. It was the first of many studies in bioelectricity undertaken by Galvani and later his nephew, Giovanni Aldini, once Galvani's health declined. Aldini introduced the connection between electricity and life to the British people through public demonstrations of electrical stimulation (using physicist Alessandro Volta's new battery) applied to the bodies of recently executed criminals.[17] Turney wrote that "contemporary accounts suggested that during these experiments, 'the body became violently agitated and even raised itself as if about to walk, the arms alternately rose and fell and the forearm was made to hold a weight of several pounds, while the fists clenched and beat violently the table upon which the body lay.' "[18] While these accounts were more or less rife with exaggeration, one could see how the 18th-century man or woman would be fascinated by this spectacle.

Humphry Davy also harnessed electricity; but he did not stimulate corpses. He studied the interaction between electricity and chemical elements.[19] Using the Voltaic battery, Davy isolated several elements: potassium, sodium, chlorine, boron, strontium, magnesium, and calcium. Davy was amazed with the results of electrochemistry and the "*powers which may be almost called creative*"[20] (italics added). In Shelley's novel *Frankenstein*, details of the creature's animation are surprisingly superficial, but there are hints that electricity is involved.[21] This meant that, unlike Georgian (a period in British history between 1714 to 1830 named after king George I to George IV) women in general and women of her age and class in particular, Shelley had been exposed to the science of electricity and was both amazed and terrified by its awesome power. Reacting to the storm and grieving for his murdered brother, Victor Frankenstein says, "While I watched the storm, so beautiful yet terrific ..."[22] The real terror, in Shelley's view, was scientists using new technology to play God. For instance, she wrote the following in the introduction to the 1831 version of *Frankenstein*: "Frightful must it be; for supremely frightful would be the effect of any human endeavor to mock the stupendous mechanism of the Creator of the world."[23] In one of the letters written by Frankenstein's monster to Mr. Walton, Shelley again added the theme of playing God: "He had played god, creating life and then abandoning it. I had also played god, but in creation's stead, I took away life."[24]

In the middle of the 20th century, Robert J. Oppenheimer — one of the "fathers of the atomic bomb" — was terrified after seeing the awesome power of the weapon during the Trinity test and viewing photographs of the subsequent devastation of Hiroshima and Nagasaki, the two Japanese cities the bomb was dropped on in 1945.[25] He was reminded of a phrase from the *Bhagavad Gita*: "Now I am become Death, the destroyer of worlds."[26]

In more recent times, scientists whose research involved engineering a new strain of influenza H5N1 (commonly known as bird flu) to be readily transmissible between mammals, cloning Dolly the sheep, or using CRISPR/Cas9 to edit genomes have all been viewed as having "Promethean" ambitions in the name of improving biological life forms.[27] Titles of recent books, such as *Modern Prometheus: Editing the Human Genome with CRISPR-Cas9* by Kozubek (2016) and *A Crack in Creation: Gene Editing and the Unthinkable Power to Control Evolution* by Doudna and Sternberg (2017), reinforce, to some extent, the theme of playing God in the public arena. If scientists are accused of playing God, the logical question that follows is, what is life?

As with disagreements on the species concept, there is no agreement on a single definition of life. The current definition in biology is based on seven characteristics shared by living systems: (1) cellular organization; (2) ordered complexity; (3) sensitivity; (4) growth, development, and reproduction; (5) energy utilization; (6) homeostasis; and (7) evolutionary adaptation.[28] Based on these characteristics, synthetic biologists cannot create life any more than biochemists Stanley Miller and Harold Urey could in 1953.[29] They created the building blocks of life (i.e., amino acids) by simulating the conditions of Earth's early atmosphere in the laboratory, but they did not create life itself. Moreover, synthetic biologists simply assemble existing biological components — "*modifying life to come up with new life forms*."[30] Venter assembled a team of twenty scientists, including microbiologist and biochemist Clyde A. Hutchison III and 1978 Nobel laureate and microbiologist Hamilton O. Smith, who were the first to design a partially synthetic species of bacterium called *Mycoplasma laboratorium* [*M. laboratorium*] (derived from the genome of *Mycoplasma genitalium* [*M. genitalium*]).[31] To do this, they transfected a cell with a synthetic chromosome. Synthetic biologists believe that if they could understand the basic principles of life, then they could be better bioengineers.[32] This was the motivation behind identifying, designing, and synthesizing a minimal cellular genome so simple that the molecular and biological function of every gene could be determined.

Based on previous research indicating that the mycoplasmas were the simplest cells capable of autonomous growth, Venter and colleagues used the mycoplasmas as models for understanding the basic principles of life. They reported the results of their research in a landmark 2016 paper.[33] Venter and colleagues initially sequenced *M. gentialium* in 1995 in the hope of isolating and eliminating the genes that were not essential for cell growth under ideal conditions in the laboratory.[34] This would fulfill their desire to understand the molecular and biological function of every gene that was essential for life. But they found that deciphering the mechanics of the cell was a very difficult task. This difficulty reinforced the fact that we *Homo sapiens* are children in the grand scheme of universe, tinkering with the machinery of life that has a history on this planet going back to the Cambrian period approximately 540 million years ago.

It is a given that bacterial cells must be capable of adapting to numerous environments in order to survive. However, Venter's team knew that some bacteria growing in restricted environments have lost genes that were unnecessary in a stable environment — a process known as gene reduction through evolutionary time.[35] This had implications for the mycoplasmas because, as mentioned above, they had the smallest known genomes of any autonomous replicating cells. And they typically grew in the nutrient-rich (stable) environment of animal hosts. Venter and colleagues compared the genome sequences of *Haemophilus influenza* (1,815 genes) and *M. gentialium* (525 genes) and found to their surprise a common core of only 256 genes.[36] They designated these 256 genes as the minimal gene set for life.

To understand which genes were essential and which were nonessential, Venter and his team used a biological process called global transposon mutagenesis.[37] This process allowed genes to be transferred to a host organism's chromosomes, interrupting the function of a gene on the chromosome by mutating it. If the mutation led to detrimental effects, then the gene was essential for physiological processes. If the organism continued to function normally, then the gene was nonessential. Using this method, Venter and colleagues cataloged 150 nonessential genes and 375 essential genes in *M. genitalium*.[38] Following these results, they believed that it was possible to "produce a minimal genome that was smaller than any found in nature but larger than the common set of 256 genes."[39] The scientists were now ready to attempt the synthesis of a hypothetical minimal genome.

They abandoned their initial model *M. genitalium* and decided to use *Mycoplasma mycoides* (*M. mycoides*) for minimization because it grew faster.[40] They transplanted *M. mycoides* genomes, as isolated DNA molecules, into cells of *Mycoplasma capricolum* (*M. capricolum*). In the process, the *M. capricolum* genome was sacrificed, resulting in a cell with the transplanted genome designated as *M. mycoides* JCV1-syn1.0 (1,078,809 base pairs).[41] While this painstaking process resulted in an almost exact copy of the wild-type *M. mycoides* genome, there were critical problems: *M. mycoides* JCV1-syn1.0 for the most part was not viable.[42] Before the next series of experiments, Venter and his team made substantial changes in all aspects of their research: They improved their global transposon mutagenesis method, which was important not only to reliably classify essential and nonessential genes but also to identify quasi-essential genes needed for robust growth but not detrimental to survival.[43] More importantly, they established several rules for "removing genes from our genome design without disturbing the expression of the remaining genes."[44] The result after these methodological changes was a working approximation to a minimal cell labeled JCV1-syn3.0 and controlled by a 531-kilobase synthetic genome.[45] It was smaller than *M. genitalium*, grew faster, and encoded 438 proteins.

The information gleaned from this research had serious implications for genetic engineering: (1) identifying essential from nonessential genes was critical in deciphering the cell's operation; (2) retaining quasi-essential genes yielded viable segments of the hypothetical minimal genome but no complete genome; (3) deleting pairs of redundant genes for essential functions resulted in a nonviable genome, so had to be avoided; and (4) most importantly, there were genes that could not be assigned to any specific function.[46] These factors were not applied when *M. mycoides* JCV1-syn1.0 was synthesized with its design flaws. This was rectified when the researchers used Tn5 mutagenesis (a section of chromosome that can move from one place to another) to get a more accurate discrimination of essential versus nonessential genes.[47]

The team initially assigned genes catalogued by global transposon mutagenesis to the essential and nonessential categories.[48] After using Tn5 mutagenesis, they added the third category of quasi-essential genes. By the end the Tn5 analysis, genes that were not hit by insertions in the terminal 20 percent of the 3′ end or the first few bases of the 5′ end indicated to the scientists that any interruption to the function of these genes would be detrimental to survival and therefore they classified the genes as essential.[49]

In a new round of Tn5 mutagenesis on *M. mycoides* JCV1-syn2.0, Venter and his team classified 90 genes as nonessential.[50] On further analysis, 53 of these genes were frequently classified as essential or quasi-essential. The remaining 37 genes were classified as nonessential and selected for deletion from *M. mycoides* JCV1-syn2.0 (along with two vector sequences and the ribosomal operon in segment 6). They subsequently synthesized eight new reduced genome design (RGD) 3.0 segments, which were reassembled in yeast to obtain several versions of RGD3.0 genomes as yeast plasmids. When the RGD3.0 genomes were transplanted from yeast into *M. capricolum*, several were viable; the team chose one RGD3.0 clone that they called *M. mycoides* JCV1-syn3.0 for detailed analysis.[51]

The purpose of discussing the research performed by Venter and his team in such detail is because it is instructive for a wide variety of research related to genetic engineering. In discussing their complex experiments, Venter's team made loud statements such as "be cautious in your modifications of the genome, the basic principles of life are extremely complicated."[52] It follows that scientists engaged in genetic engineering should slow down, despite their positive goals of preventing disease or being the first to create a therapy that works. Careless enthusiasm could do more harm than good. For instance, Venter and his team could assign no biological function for 31 percent (149) of the apparently essential genes in *M. mycoides* JCV1-syn3.0; they had no choice but to classify them as *generic* (genes encoding identifiable proteins but with no clue of their function) and *unknown* (genes that could not be categorized with any activity).[53] Additionally, the ambiguous nature of quasi-essential genes and the danger of deleting pairs of redundant genes for essential functions warrants treading lightly and carefully in the minefield of the cellular system.

In conclusion, the charge by critics that Venter and his colleagues were playing God is wrong in this case. These researchers acknowledged, in detail, the difficulties in synthesizing a minimal bacterial genome. Replying to the charge that he was playing God, Nobel laureate Smith simply replied, "We don't play."[54] It is ironic that the critics of synthetic biology have the audacity to think that humans can create life. This very charge *is* the greater hubris.

## Notes

1. Mary W. Shelley, *The Complete Frankenstein* (1818; repr., Washington, DC: Literary Folio, 2017), 100.
2. Antonio Regalado, "Biologist Venter Aims to Create Life from Scratch," *Wall Street Journal*, June 29, 2005, http://www.post-gazette.com/pg/05180/530330.stm.
3. Ibid.
4. Ibid.
5. Henk van den Belt, "Playing God in Frankenstein's Footsteps: Synthetic Biology and the Meaning of Life," *Nanoethics* 3, no. 3 (November 29, 2009): 262.

6. Ibid.
7. Seth Borenstein, "Scientists Struggle to Define Life," *USA Today*, August 19, 2007, http://www.usatoday.com/tech/science/2007-8-19-life_N.html.
8. Henk van den Belt, "Playing God in Frankenstein's Footsteps," 262; John Brockman, "Constructive Biology: George Church," *Edge*: *The Third Culture*, June 26, 2006, https://www.edge.org/3rd_culture/church06/church06_index.html.
9. Henk van den Belt, "Playing God in Frankenstein's Footsteps," 262; Alexander Reed, "Designing Life: A Look at Synthetic Biology," Retrieved October 24, 2018, from http://scienceinsociety.northwestern.edu/content/articles/2008/medill-reports/jan/endy/designing-life-a-look-at-synthetic-biology.
10. Borenstein, "Scientists Struggle to Define Life," http://www.usatoday.com/tech/science/2007-8-19-life_N.html.
11. Rainer Breitling, Eriko Takano, and Timothy S. Gardner, "Judging Synthetic Biology Risks," *Science* 347, no. 6218 (January 9, 2015): 107.
12. Ibid.
13. Ibid.
14. Ibid.
15. Edmund T. Whittaker, *A History of the Theories of Aether and Electricity: From the Age of Descartes to The Close of the Nineteenth Century* (1910; repr., Missoula, MT: Kessinger Publishing LLC, 2007), 67.
16. Ibid., 67–70.
17. Ibid., 70, 72–74.
18. Jon Turney, *Frankenstein's Footsteps: Science, Genetics, and Popular Culture* (New Haven: Yale University Press), 22; Samuel H. Vasbinder, *Scientific Attitudes in Mary Shelley's "Frankenstein,"* (Ann Arbor, MI: UMI Research Press, 1984), 79.
19. Whittaker, *A History of the Theories of Aether and Electricity*, 76–78.
20. Henk van den Belt, "Frankenstein, or Beauty and Terror of Science," *Journal of Geek Studies* 4, no. 1 (2017), 5.; Whittaker, *A History of the Theories of Aether and Electricity*, 76.
21. Mary W. Shelley, *The Complete Frankenstein* (1818; repr., Washington, DC: The Literary Folio, 2017), 51–52.
22. Ibid., 71.
23. Shelley, *The Complete Frankenstein* (1831; repr., Washington, DC: The Literary Folio, 2017), 244.
24. Ibid., 485.

25. James A. Hijiya, "The Gita of Robert Oppenheimer," *Proceedings of the American Philosophical Society* 144, no. 2 (June 2000): 123.
26. Ibid.
27. Jim Kozubek, *Modern Prometheus: Editing the Human Genome with CRISPR-CAS9* (New York, NY: Cambridge University Press, 2016), 196–199.
28. Peter Raven, George B. Johnson, Kenneth A. Mason, Jonathan B. Losos, Susan R. Singer, *Biology*, 9th ed. (New York: McGraw-Hill, 2008), 508.
29. Stanley L. Miller and Harold C. Urey, "Organic Compound Synthesis on the Primitive Earth," *Science* 130, no. 3370 (July 31, 1959): 245–251; Raven *et al.*, *Biology*, 509–510. In 1953, biochemists Stanley L. Miller and Harold C. Urey conducted an experiment to find out if life began in the water. Specifically, they replicated the atmosphere (a reducing atmosphere rich in hydrogen and excluding gaseous oxygen) and oceans (heated water) of early Earth. Within a week, some of the carbon originally present in methane gas had converted into other simple carbon compounds (i.e., formaldehyde and hydrogen cyanide). These compounds combined to form simple molecules but, more importantly, complex molecules such as the amino acids alanine, glycine, glutamic acid, valine, proline, and aspartic acid. Miller and Urey did not create life; they simply created the building blocks of life.
30. Borenstein, "Scientists Struggle to Define Life," http://www.usatoday.com/tech/science/2007-8-19-life_N.html.
31. Clyde A. Hutchison III, Ray-Yuan Chuang, Vladimir N. Noskov, Nacyra Assad-Garcia, Thomas J. Deerinck, Mark H. Ellisman, John Gill *et al.*, "Design and Synthesis of a Minimal Bacterial Genome," *Science*, Research Article Summary, 351, no. 6280 (March 25, 2016): 1414.
32. Ibid.
33. Clyde A. Hutchison III, Ray-Yuan Chuang, Vladimir N. Noskov, Nacyra Assad-Garcia, Thomas J. Deerinck, Mark H. Ellisman, John Gill *et al.*, "Design and Synthesis of a Minimal Bacterial Genome," *Science* 351, no. 6280 (March 25, 2016): 1–11; aad6253, http://dx.doi.org/10.1126/science.aad6253.
34. Ibid., aad6253-1.
35. Ibid.
36. Ibid.
37. Ibid.
38. Ibid.
39. Ibid.

40. Ibid.
41. Ibid.
42. Ibid.
43. Ibid., aad6253-3.
44. Ibid.
45. Ibid., aad6253-1.
46. Ibid., aad6253-8–10.
47. Ibid., aad6253-2.
48. Ibid.
49. Ibid.
50. Ibid.
51. Ibid., aad6253-5.
52. Ibid., aad6253-8–10.
53. Ibid., aad6253-6.
54. Henk van den Belt, "Playing God in Frankenstein's Footsteps," 262; Borenstein, "Scientists Struggle to Define Life," http://www.usatoday.com/tech/science/2007-8-19-life_N.html.

# Chapter 9

# Bioengineering and the Emergence of Dual Use

The prospect of human gene editing inevitably recalls past abuses of human rights involving the biological sciences, and especially the history of eugenics in the first half of the 20th century.

— Daniel Kevles, science historian and participant in the First International Summit on Human Gene Editing at New York University, 2015[1]

During the 1800s, US government agents were alleged to have deliberately infected the Plains Indians by giving them trading blankets infected with the deadly disease [smallpox], decimating the population.

— Committee on Research Standards and Practices to Prevent the Destructive Application of Biotechnology, 2004[2]

Bioengineering, as stated in the beginning of this book, is the manipulation of DNA (including that of nonhuman animals and plants) to improve human health, environment, and safety by means of gene therapy; iPSCs; cloning; xenobiology; in vivo and ex vivo genetic editing to cure genetic illnesses; and genetic editing applied to ART (IVF and PGD). These biotechnologies have the potential to unintentionally resurrect the old eugenics by stigmatizing individuals racially, physically, mentally, and economically. As discussed elsewhere in the book, editing the MSTN (Myostatin) and SCN9A (Sodium ion channels expressed in pain neurons) genes in healthy individuals, who could afford the therapy, would increase their muscle mass and make them insensitive to pain. Athletes, soldiers, and others would gain an unfair advantage. Adding a memory-enhancing drug (originally created for Alzheimer patients) or smart pill would produce a subgroup of humans who might assume a sense of superiority — probably reinforced by their modified brain structure — and in turn use their wealth to gain political power. The same goes for using Easy PGD to select for sex, intelligence, hair color, eye color, height, and absence of disease. Essentially, such a baby would be designed with traits that have a high value in society. At the first international summit on human gene editing, Kevles warned the audience about the dangers of the forces that are fueling the new eugenics: "Though [old] eugenics is no longer a powerful movement several of the forces that animated the eugenics movement a century ago remain vital."[3] The National Academies of Sciences, Engineering, and Medicine's Committee on Science, Technology, and Law published a brief report about this summit, expanding on Kevles's warning above:

> Economic forces to reduce health care costs could put pressure on people to change genetic sequences associated with disease. The association of racial, ethnic, and other groups with particular diseases could lead to new forms of stigmatization. The belief that genes influence particular behaviors or other complex traits could lead to pressures to change those genes in future generations. And consumer demand for particular attributes in the offspring could lead people to pursue private sector options for human gene editing that are difficult to regulate.[4]

The warning regarding the new eugenics and the related dual-use aspect of genetic engineering (i.e., examples presented above) deserves elaboration. Recall the hypothetical memory-enhancing drug for Alzheimer patients that would help healthy individuals gain an advantage. Editing the MSTN and SCN9A genes would also give healthy people an advantage. Using these techniques, national governments — allies and adversaries alike — could produce soldiers who are stronger, more intelligent, and possess a higher pain threshold than at present. A rogue country could exploit the traits in these troops for nefarious purposes. Or, in another example, gene drives could be used to spread genes for male or female sterility in an insect or plant species, leading to its extinction. Sterility in plants could lead to famine, and a country with the gene drive technology could blackmail another country with the threat of famine. How countries decide to use GM humans (or other modified life forms) to help or hurt non-GM humans and/or countries goes to the heart of dual-use dilemma.

In 2004, the NRC's Committee on Research Standards and Practices to Prevent the Destructive Application of Biotechnology said, "In the life sciences, for example, the same techniques used to gain insight and understanding regarding the fundamental life processes for the benefit of human health and welfare may also be used to create a new generation of BW [biological warfare] agents by hostile governments and individuals."[5] The committee echoed the concern of the US intelligence community, which is that technologies intended to benefit humankind can be modified for military purposes.[6] While the term *dual use* and its meaning were clearly understood in the late 20th century and are commonly found in the 21st-century literature, the offensive use of biological agents has been documented and alleged for centuries, according to the committee.[7] The committee also stated that it was not until the 19th century that infectious disease agents were used on a regular basis as weapons of war, but they added that this early form of eugenics occurred long before Galton coined the term.[8] For instance, French and British soldiers and civilians were alleged to have deliberately infected North American Indian populations with smallpox during the 17th and 18th centuries. The committee also discussed an incident in June 1763 when a Captain Ecuyer of the Royal Americans met with two Indian chiefs under the pretense of friendship and gave them blankets infected with smallpox. And it is alleged that the US government adopted this same strategy against the Plains Indians in the 19th century.[9] These events were frequently dramatized in the television westerns of the 1960s and 1970s.

During World War I, Germany was accused of using highly infectious diseases like glanders and anthrax to infect horses, mules, cattle, and reindeer.[10] During World War II, the Japanese targeted cities in China with plague-causing bacteria.[11] Essentially all major combatants had some type of biological weapons program by the end of World War II. Unfortunately, these programs are top secret and still in operation today. They are a large part of the cause of the dual-use dilemma in connection with biotechnology. If governments ended biotechnology research aimed at engineering dangerous pathogens, there would be no dual-use dilemma. But the overwhelming theme of the committee's report was that the benefits of dangerous biotechnology research outweigh the risks.[12] The first sentence of the preface of the committee's report read, "The charge to our Committee was to consider ways to minimize threats from biological warfare and bioterrorism *without hindering the progress of biotechnology, which is essential for the health of the nation*"[13] (italics added). The committee elaborated on this in the conclusion of the executive summary:

> Throughout the Committee's deliberations there was a concern that policies to counter biological threats should not be so broad as to impinge upon the ability of the life sciences community to continue its role of contributing to the betterment of life and improving defenses against biological threats. Caution must be exercised in adopting policy measures to respond to this threat so that the intended ends will be achieved without creating "unintended consequences."[14] (italics added)

This governmental position, moreover, has encouraged scientists to come close to the "red line" and sometimes cross it in the course of their research. In early 2000, several researchers crossed the red line, prompting the committee to propose a set of recommendations. One key recommendation was to review each stage of research to make sure that the end product would not be weaponized.[15] One could speculate as to whether the impetus for proposing recommendations was solely due to the high-risk experiments or whether residual effects of the 9/11 terrorist attacks on the United States also played a part. Regardless, it is interesting that the application of new genetic technologies to improve human health, environment, and safety have produced not only designer babies but also designer diseases that could potentially be used eugenically.[16]

## 9.1. Questionable Viral Studies in the Early 2000s

In 2001, researchers Ronald Jackson of the Commonwealth Scientific and Industrial Research Organization and Ian Ramshaw of the Division of Immunology and Cell Biology at the Australian National University in

Canberra, Australia, and their colleagues inserted into a strain of *ectromelia* (mousepox) virus an ovary-specific antigen and an immunomodulator (cytokine).[17] The aim of the experiment was to create an infectious immunocontraceptive (contraceptive vaccine) for wild mice to boost production of antibodies in the infected mice that targeted their own fertilized eggs and caused permanent infertility. The scientists incorporated the gene for zona pellucida glycoprotein 3 (*ZP3*), encoding the ovary specific antigen, and the gene for interleukin 4 (special proteins that activate T cells and B cells) encoding the immunomodulator.[18] But the unintended effect was increased virulence of the mousepox, which normally causes only mild symptoms in the species of mice used in the study.[19] The virus killed all of the mice within a week, even mice from a genetically resistant strain. According to the NRC's Committee on Research Standards and Practices to Prevent the Destructive Application of Biotechnology, this experiment was not completely original, because Jackson and his colleagues drew upon a previously published experiment in which researchers incorporated interleukin 4 into the genome of a recombinant *vaccinia* virus, which enhanced the virulence of *vaccinia* in mice.[20] It is believed that the increased virulence of *vaccinia* was due to suppression of the antiviral immune response (the immunomodulator).

Even though Jackson and his team published their research six months before the 9/11 terrorist attacks in the United States, some in the scientific and intelligence communities initially questioned the wisdom of publishing this research — particularly the methods section of the manuscript, which detailed how the mousepox virus was modified: "The publication of this paper provides a blueprint or road map for terrorists to engineer a more virulent strain of smallpox that could overwhelm the human immune system in even well-vaccinated individuals."[21] In the end, the *Journal of Virology* editors and reviewers recommended publication of this research because Jackson's work built upon published research showing the effects of decreased or increased levels of interleukin 4 on the virulence of viruses and microorganisms.[22] The NRC committee agreed with the decision to publish this work on mousepox and interleukin 4 because, in its view, the scientific community should know that the increased virulence attained by interleukin 4-gene-expressing viruses could overwhelm the immune system of well-vaccinated individuals.[23]

It follows that scientists could then synthesize a countermeasure if an interleukin 4-gene-expressing virus were engineered for bioterrorism.

In 2002, the German American virologist Eckard Wimmer, professor of microbiology at the State University of New York, Stony Brook, and fellow virologists Aniko Paul and Jeronimo Cello at the same university published the results of their research in which they had reconstructed a full-length poliovirus complementary DNA by chemically synthesizing oligonucleotides of plus and minus strand polarity.[24] They subsequently transcribed the synthetic poliovirus complementary DNA with RNA polymerase into viral RNA, generating de novo the infectious poliovirus. When they conducted experiments in tissue culture using neurovirulence tests in CD155 transgenic mice, they found that the synthetic virus had both biochemical and pathogenic characteristics of the poliovirus.[25] Like the mousepox interleukin 4 research, de novo synthesis of infectious poliovirus provoked the imaginations of lawmakers, because terrorists could read the publication and manufacture the virus. In the 107th Congress, Rep. Dave Weldon (R-FL) introduced House Resolution 514, which criticized the editors of *Science* for publishing the research and compromising the security interests of the nation.[26] The resolution also called upon the executive branch to be proactive in this matter:

> To examine all policies, including national security directives, relevant to the classification or publication of federally-funded research to ensure that, although the free exchange of information is encouraged, information that could be useful in the development of chemical, biological, or nuclear weapons is not made accessible to terrorists or countries of proliferation concern.[27]

But, as in the mousepox interleukin 4 experiment, the information for producing and manipulating the genome of RNA viruses was already in the public arena and provided a foundation on which Wimmer and colleagues built their research and added to available knowledge. The NRC committee added that the Wimmer experiment would not have been a great bounty of bioinformation for terrorists because the synthesized poliovirus was much less virulent compared to wild-type strains of poliovirus.[28]

However, two months before the synthesized poliovirus experiment was publicized, a team of researchers led by pathologist Ariella Rosengard in the Department of Pathology and Laboratory Medicine at the University of Pennsylvania and colleagues Yu Liu, Zhiping Nie, and Robert Jimenez, in the same department, investigated the difference in virulence between the virus that caused smallpox (*variola*) in humans and the weakened form of the virus (*vaccinia*) used to vaccinate against it.[29] The scientists believed that the key to the difference in virulence between the homologues was their inhibition of immune-response enzymes designated virus complement control protein (VCP) for *vaccinia*. Since live *variola* proteins were not available for study, the scientists engineered the smallpox inhibitor of complement enzymes (SPICE). In their analysis, they learned that "SPICE is nearly 100-fold more potent than VCP at inactivating human C3b and 6-fold more potent at inactivating C4b. SPICE is also more human complement-specific than is VCP."[30] This is why the *vaccinia* virus is used as a vaccine against smallpox: it is associated with low mortality.

As mentioned above, Rosengard's experiment was reported two months before the synthesized poliovirus experiment was publicized. But it was addressed last in the "Contentious Research in Life Science" section of the NRC committee report. Why? The speculation is that some officials believed that the Rosengard study posed a greater risk to US national security than the other experiments and therefore should appear last in the section to promote mental recall. One can propose a single scenario for all these viral experiments: A bad actor could use the information to increase the virulence of, for example, the smallpox vaccine (from the *vaccinia* virus). The human complement system (i.e., C3b and C4b) is an important part of the immune system, and if it were inhibited, then many people would die.[31] In the end, the NRC committee believed that the benefits of publishing this research outweighed the risks of bioterrorism:

> A commentary written on this paper pointed out that it is very unlikely *vaccinia* virus carrying SPICE in place of VCP would approach the pathogenicity of *variola*. Furthermore, publication of the article alerted

> the community of scientists to this mechanism for virulence. *This information should stimulate scientists in both the public and private sectors to identify compounds or immunization procedures that disable SPICE. These could form the basis for new treatments or vaccines both to immunize against the naturally occurring smallpox virus and to counteract the genetically engineered variety.*[32] (italics added)

The committee's perspective, particularly the italicized part, was later adopted by two scientists in 2012 to justify the publication of their high-profile and provocative research.

As mentioned in the previous section, these three potentially troublesome experiments prompted the NRC's Committee on Research Standards and Practices to Prevent the Destructive Application of Biotechnology to propose new recommendations. One recommendation has already been discussed. The others included educating scientists through national and international professional societies and related organizations about the nature of the dual-use dilemma in biotechnology and their responsibility to reduce the risk; augmenting resources of the Department of Health and Human Services (HHS) and NIH to enhance the already established review system for experiments involving microbial agents that can be misused; self-policing by scientific organizations and scientific journals to review manuscripts for experiments that can be misused; HHS creating a National Science Advisory Board for Biodefense (NSABB) to provide advice and guidance when reviewing experiments of concern; national security developing sustained communication with life sciences and advising scientists on how to reduce the risks of bioterrorism; and the scientific community creating an International Forum on Biosecurity.[33] Fortunately, aided by the backdrop of post-9/11 America, these recommendations were adopted, and the new NSABB asserted its influence by asking the editors of *Nature* and *Science* to delay publication of the details of two risky studies in 2012.[34] These scientists had used targeted mutagenesis on the highly pathogenic avian H5N1 virus, which became airborne and transmissible by respiratory droplets in ferrets. Linked to the national security concern of "the risk of accidental or malicious release" was the crucial fact that ferrets, like humans, are mammals.[35]

## 9.2. Dangerous Experiments On Avian H5N1 Influenza Virus Leading to Temporary Halt in Publication of Results in 2012

Virologist Yoshihiro Kawaoka, professor of virology in the Department of Pathobiological Sciences at the University of Wisconsin–Madison, and colleagues had monitored avian H5N1 viruses closely and knew that H5N1 continued to cause outbreaks in poultry and in some humans in North Africa and Southeast Asia without acquiring the ability to transmit human to human.[36] Regardless, they were concerned that through natural processes or malevolence, a transmissible H5-hemagglutinin-possessing virus could emerge and cause a pandemic, because humans had no immunity to influenza viruses possessing an H5-hemagglutinin (H5-HA).[37] To prepare for this outcome, Kawaoka and colleagues believed that it was vital for them to conduct gain-of-function studies in order to understand the mutations that an H5-HA-possessing virus may acquire to make it transmissible among mammals.[38] An understanding of the structure of influenza A viruses enabled them to identify the molecular features that they could deliberately modify to produce a change in the virus.

It has been known for some time now that influenza A viruses have an RNA genome consisting of eight segments.[39] For instance, segment one encodes the polymerase proteins polymerase basic protein 2 (PB2), polymerase basic protein 1 (PB1), and polymerase acidic protein (PA). Segment two is identical to segment one except for the second small protein (PB1-F2) that segment two encodes. Segment three is identical to segment one. Segments four and six encode the viral surface glycoproteins HA and neuraminidase (NA). HA is responsible for binding to the viral receptors known as sialic acids (SAs) on host cells and the fusion of the viral membrane with host cells, called endocytosis. NA is a sialidase that cleaves SAs, enabling the virus to be released from host cells and virus particles. Segment five encodes the nucleocapsid protein (NP) and forms ribonucleoprotein complexes (RNPs) with the polymerase proteins. Segment seven encodes two proteins that constitute the viral membrane: the viral matrix structural protein (M1) and the ion-channel protein (M2). Finally, segment eight encodes the nonstructural protein (NS1) and the nucleic-export protein (NEP). Collectively, these segments determine the efficacy of respiratory droplet or airborne transmission of influenza A viruses, and "host restriction of replication and transmission of influenza A viruses is partly determined by specific SA receptors on the surface of susceptible cells."[40] In particular animal species, influenza viruses either have a strong or weak affinity for these receptors. For instance, avian influenza viruses overwhelming bind to $\alpha$-2,3-linked SA receptors, and human influenza viruses bind to $\alpha$-2,6-linked SA receptors.[41] Ron Fouchier, professor of molecular virology in the Department of Viroscience at Erasmus Medical Center in Rotterdam in the Netherlands, and his colleagues learned from their experiment that the receptor distributions in ferrets are similar to what is found in humans, in that the $\alpha$-2,6-linked SA receptors are mostly present in the upper respiratory tract (i.e., nasal passages and throat), and the $\alpha$-2,3-linked SA receptors are mainly in the lower respiratory tract (i.e., bronchi and lungs).[42] Fouchier used the example of chickens and other birds, in which both $\alpha$-2,3-linked (that predominate in birds) and $\alpha$-2,6-linked SA receptors are present throughout the respiratory and intestinal tracts. *The implications of these SA receptors on the surface of susceptible cells is that a change in receptor specificity from avian $\alpha$-2,3-SA to human $\alpha$-2,6-SA by mutagenesis in the receptor binding site of HA can make an avian virus transmissible in humans and possibly cause a pandemic.*[43]

Kawaoka and colleagues began their study by introducing random mutations into the globular head (amino acids 120–259) and receptor-binding pocket of avian H5N1, isolated from humans in Vietnam in 2004 (H5N1; VN1230/2004) in order to identify new mutations that confer human-type receptor-binding preference.[44] The scientists subsequently generated a complete genome of the H5N1 virus by pasting the mutated HA gene, the unmodified NA gene of VN1230/2004, and viral genes of avian Puerto Rico 8/34 (H1N1; PR8) together into plasmids and then transfected these plasmids into human embryonic kidney (293T) cells.[45] Kawaoka, in short, engineered a hybrid virus.

To test the affinity of the hybrid virus for human $\alpha$-2,6-SA receptors, turkey red blood cells (which possessed both $\alpha$-2,6-SA receptors and $\alpha$-2,3-SA receptors on their surfaces) were treated with sialidase to remove $\alpha$-2,3-SA receptors, leaving only the human $\alpha$-2,6-SA receptors.[46]

The hybrid virus was then absorbed to the $\alpha$-2,6-SA turkey red blood cells. At the end of the experiment, Kawaoka and colleagues isolated 370 viruses that bound to $\alpha$-2,6-SA-turkey red blood cells but only nine agglutinations with varying efficiencies.[47] For the parental control virus (VN1230/PR8) with untreated turkey red blood cells (possessing both $\alpha$-2,3-SA and $\alpha$-2,6-SA receptors), the scientists found agglutinated avian $\alpha$-2,3-SA turkey red blood cells only.[48] In the nine viruses in which agglutinations occurred, mutations occurred in regions that were targeted for random mutagenesis, clustering around the receptor-binding pocket. Other important findings from this experiment were that several of the viruses possessed mutations that were known to have strong affinity to the human-type receptors.

Another key step in the Kawaoka study was to identify the amino acid changes responsible for conversion of avian $\alpha$-2,3-SA to human $\alpha$-2,6-SA in the mutant that were critical for the shift from $\alpha$-2,3-SA to $\alpha$-2,6-SA receptors. Additionally, this shift in specificity had profound effects on virus attachment to human respiratory epithelium when sections of trachea and lung tissues were exposed to the mutant viruses.[49]

Throughout their study, Kawaoka and colleagues used past studies as a guide.[50] For instance, they knew that H1N1 and H5N1 viruses had both been isolated in pigs and read published studies showing genetic compatibility of these viruses. The concern among virologists was that within the pig, reassortment of avian, swine, and human strains could occur.[51] Therefore, "the coexistence of H5N1 and 2009 pandemic H1N1 viruses could provide an opportunity for the generation of transmissible H5 avian-human reassortants in mammals."[52] Based on this concern, the scientists engineered reassortant viruses that possessed the mutant VN1203 HAs and seven gene segments from a 2009 pandemic H1N1 virus (A/California/04/2009; CA04). They reasoned that pandemic influenza viruses are most effective because they spread from human to human. Consequently, to study growth and transmissibility of reassortant viruses, the animal model used must be able to transmit the virus by aerosol via sneezing.[53] The scientists chose ferrets for this transmissibility experiment. Subsequently, the scientists inoculated six ferrets intranasally to determine whether the introduced HA mutations affected replication of the H5 reassortant viruses, and they found that these mutations did not affect the replication of H5 reassortant virus in infected animals.[54]

Next, Kawaoka and colleagues needed to assess whether the H5 reassortant viruses with human-type receptor specificity could transmit between ferrets.[55] Uninoculated ferrets were placed in cages next to ferrets inoculated with several mutants (i.e., rgCA04, rgVN1203/CA04, etc.). The scientists reported recovering rgCA04 from three ferrets via contact.[56]

The last experiment in the study assessed whether current control measures would be effective against H5-transmissible reassortant mutant virus. This experiment is more worrying than the others. The researchers examined the reactivity of sera from individuals vaccinated with a prototype H5N1 vaccine synthesized to stop any virus possessing specific respective mutations in HA.[57] The reaction (with this virus) was at a higher titer than with a wild-type H5 HA virus. And the H5-transmissible reassortant mutant virus was susceptible to the prescription antiviral drug oseltamivir. The primary message Kawaoka and colleagues wanted to emphasize was that "appropriate control measures would be available to combat the transmissible virus" described in their study.[58]

Finally, Kawaoka and colleagues acknowledged that they were unsure whether the mutations identified in their study that allowed the virus to be transmissible between ferrets would also sustain human-to-human transmission.[59] Nevertheless, the human-virus-characteristic amino acids in the seven segments of the RNA genome "may have critically contributed to the respiratory droplet transmission of the HA(N158D/N224K/Q226L/T318I)/CA04 virus in ferrets."[60]

Fouchier and his colleagues, in a similar study, began with an actual H5N1 virus isolated from a human victim in Indonesia (A/Indonesia5/2005), because they feared that this virus could acquire mutations that would allow it to become more transmissible between humans and cause a pandemic.[61] Fouchier's goal, like Kawaoka's, was to understand the biological mechanisms that would make avian H5N1 transmissible in mammals. Unlike Kawaoka, Fouchier's emphasis was airborne transmission.[62] Based on previous studies, Fouchier and colleagues knew that amino acid substitutions in the receptor binding site of the HA glycoprotein of A/Indonesia5/2005 would change the binding preference from the avian

$\alpha$-2,3-linked SA receptors to human $\alpha$-2,6-linked SA receptors.[63] Consequently, their experimental rationale was to obtain transmissible avian H5N1 viruses with high titers from the upper respiratory tracts of ferrets after infecting them with a mutant avian H5N1 virus with receptor specificity for $\alpha$-2,6-linked SA receptors.[64] As in Kawaoka's study, Fouchier and colleagues, introduced random mutations affecting amino acids 120–259. Specifically, the amino acid substitutions were in the HA of A/Indonesia/5/2005.[65] Some of these amino acid substitutions were similar to the ones generated by Kawaoka.

In their first experiment, Fouchier's team inoculated groups of six ferrets with A/Indonesia5/2005 and the mutant viruses intranasally.[66] Subsequently, they collected throat and nasal swabs to determine viral titer due to shedding from the upper respiratory tract. They learned that ferrets inoculated intranasally with A/Indonesia/5/2005 virus produced high virus titers in the ferret upper respiratory tract during a seven-day period.[67]

Continuing, the team introduced amino acid substitution E627K in PB2, because they knew from previous research that this particular substitution had been associated with increased virus replication in mammalian cells.[68] But in this experiment, they found no significant viral transmission.

Since the initial mutations alone were not effective in causing transmission, Fouchier and colleagues designed an experiment to encourage the virus to adapt to replication in the mammalian respiratory tract by passing other mutant viruses in addition to A/Indonesia/5/2005 from ferret to ferret by directly inoculating uninoculated animals with nasal samples from infected ones and repeating the procedure ten times.[69] In this third experiment as in the second, throat and nasal swabs were collected daily from live ferrets until day four after inoculation, at which time the ferrets were euthanized to collect samples from nasal turbinates and lungs. These samples were used to inoculate the next ferret, regarded as passage 2. This procedure was repeated until passage 6.

In the next procedure, the scientists induced ferrets to sneeze, collected the aerosol spray for intranasal inoculation, and then repeated this for passages 7–10.[70] In nasal turbinates and nasal swabs of ferrets inoculated with the other mutants, as opposed to A/Indonesia/5/2005, the scientists found higher and higher virus titers as the passage numbers increased. Fouchier wrote, "These data indicate that A/H5N1$_{\text{HA Q222L,G224S,PB2,E627K}}$ was developing greater capacity to replicate in the ferret URT [upper respiratory tract] after repeated passage, with evidence for such adaptation becoming apparent by passage number 4."[71] As the virus titers increased with each passage of the virus, amino acid substitutions were detected overwhelmingly in A/Indonesia/5/2005 and the other mutants.[72] Despite the fact that these substitutions could affect receptor binding of avian H1 and the virulence of avian H5 virus, Fouchier wanted to know whether binding to $\alpha$-2,6-linked SA would be enhanced (with some affinity for $\alpha$-2,3-linked SA) by introducing the new substitutions.[73]

At this point, Fouchier and colleagues began their fourth experiment, in which they investigated whether airborne transmissible viruses were present in A/Indonesia/5/2005 and the other mutants during passage.[74] Ferrets were inoculated intranasally with passage 10 nasal wash samples. Uninoculated recipient ferrets were placed in cages near inoculated ferrets (not in direct contact with one another but close enough to allow airflow between them). The following day, the scientists collected nasal and throat swabs and confirmed the presence of one of the mutant viruses (passage 10) in uninoculated recipient ferrets paired with inoculated ferrets.[75] They then used a subsequent throat-swab sample taken from the recipient ferret with the highest virus titer to inoculate two other uninoculated ferrets. These recipient ferrets transmitted the virus to other ferrets. The scientists continued this procedure several times, and transmission occurred efficiently. Additionally, they consistently detected specific amino acid substitutions in the consensus genome sequences of viruses recovered from six ferrets.[76] Based on these results, Fouchier made an important statement: "As few as five amino acid substitutions (four in HA and one in PB2) may be sufficient to confer airborne transmission of HPAI A/H5N1 virus between mammals."[77]

In their fifth and sixth experiments, Fouchier and colleagues, as in Kawaoka's study, tested whether the mutations in HA in the airborne-transmissible virus were sensitive to antiviral drugs; they also tested whether this virus increased reactivity with antisera (indicating recognition of HA by antibodies) in individuals vaccinated with a potential H5N1 vaccine.[78] First, they found that the airborne-transmissible virus (with nine amino acid substitutions) was sensitive to the drug oseltamivir. Second, chimeric viruses created for this experiment increased reactivity with ferret antisera. When this chimeric virus was used to evaluate the presence of existing immunity against the airborne-transmissible virus in sera obtained from human volunteers aged 70 years or older, there was no reaction.[79]

Fouchier and colleagues concluded that avian H5N1 viruses can acquire the ability to transmit by aerosol or respiratory droplets between mammals without genetic mixing in any intermediate host.[80] They warned that the specific amino acid substitutions identified in the experiments that were associated with airborne transmission represent biological traits that could be determined by other amino acid substitutions. They added:

> Given the large numbers of HPAI A/H5N1 virus-infected hosts globally, the high viral mutation rate, and the apparent lack of detrimental effects on fitness of the mutation that confer airborne transmission, it may simply be a matter of chance and time before a human-to-human transmissible A/H5N1 virus emerges.[81]

On January 20, 2012, a small group of researchers announced that there would be a temporary halt to certain experiments on the avian H5N1 influenza virus, a halt that stemmed from the two studies discussed above.[82] This fragile coalition of researchers agreed to an indefinite moratorium to give influenza researchers and the public time to discuss safety and rationale for conducting potentially risky experiments that could modify H5N1 to make it more dangerous to humans. Microbiologist Adolfo Garcia-Satre, at the Icahn School of Medicine at Mount Sinai in New York City, a leading proponent of the moratorium, advised that "laboratory safety officials and scientists will need to at least have a consensus on the level of biocontainment required."[83] To their credit, both Kawaoka and Fouchier followed strict protocols in conducting their experiments using infectious materials. For instance, both studies were performed in enhanced biosafety level 3 (BSL3+) containment laboratories approved for such use by the CDC in the United States and by the Dutch government (the NIH funded both studies through the National Institute of Allergy and Infectious Diseases).[84] And all personnel were trained extensively for working in the BSL3+ facility. Therefore, any worries about the virus getting out of the laboratory because of carelessness or accidents may have been unfounded. Additionally, fears of terrorists reading the published papers and then duplicating the experiments to create a human-to-human transmissible H5N1 virus were fanciful; the complexity of the experiments in both studies precluded the possibility that a terrorist cell could make avian H5N1 more transmissible between humans.

It is more likely that these H5N1 experiments could aid a nation state's bioweapons program or be used to implement a eugenic goal, such as ethnic cleansing. Hypothetically, for example, rogue elements of the US government obsessed with immigration restriction at the southern border (an obsession analogous to the CIA's singular goal in the 1950s and 1960s to stop communism, including rogue operations) could access the stockpile of human-to-human transmissible H5N1 virus (or any infectious disease in the bioweapons arsenal) and surreptitiously infect an aircraft, train, or bus used to transport immigrants back to Mexico, Guatemala, Honduras, and points farther south.[85] Of course, this would be absolutely foolish because such irresponsible behavior would create a human pandemic beginning in the Americas.

Returning to the moratorium, initially many scientists welcomed what was to be a 60-day moratorium on risky avian H5N1 studies.[86] A few days later, the 60-day ban was changed to a moratorium of indefinite duration. Biosecurity experts renewed their calls to fund agencies to do a better job at screening problematic experiments before they begin.[87] These calls are even louder in the current COVID-19 pandemic. In a controversial decision, the NIH terminated a $3.4 million grant, first awarded in 2014 and renewed in 2019, to the nonprofit EcoHealth Alliance that supported research into bat coronaviruses in China.[88] The impetus for killing the grant was the Trump administration's belief, based on circumstantial evidence, that the coronavirus causing the pandemic escaped from a laboratory in Wuhan, China, that employs Chinese virologist Shi Zhengli, who had received funding from the grant.

Returning to the avian H5N1 studies, immunologist Anthony Fauci, the head of the US National Institute of Allergy and Infectious Diseases, held that "We've still got a lot of homework to do … and some boxes to check before the moratorium should be lifted."[89] But the biggest concern among scientists was the proposed new NIH criteria for funding in which a scientist had to show "that a gain-of-function experiment has *high significance to public health* … and no feasible alternative methods for doing it"[90] (italics added). Additionally, researchers had to provide evidence that the H5N1 virus they wanted to produce *could evolve naturally* and be encountered in the ecology.[91] A study that failed to meet the criteria (i.e., a study where there was a potential risk of creating an exotic or synthetic pathogen) would face an extended review by HHS. Kawaoka and Fouchier stated emphatically that their research met the criteria. Summarizing their findings, the Kawaoka team attempted to justify their research:

> Our study highlights the pandemic potential of viruses possessing an H5 HA. Although current vaccines may protect against a virus similar to that tested here, the continued evolution of H5N1 viruses reinforces the need to prepare and update candidate vaccines to H5 viruses. The amino acid changes identified here will help individuals conducting surveillance in regions with circulating H5N1 viruses (for example, Egypt, Indonesia, Vietnam) to recognize key residues that predict the pandemic potential of isolates. … Moreover, our findings have suggested that different mechanisms (that is, receptor-binding specificity and HA stability) may act in concert for efficient transmissibility in mammals.[92]

And the Fouchier team did the same:

> *Our findings indicate that HPAI A/H5N1 viruses have the potential to evolve directly to transmit by aerosol or respiratory droplets between mammals, without reassortment in any intermediate host, and thus pose a risk of becoming pandemic in humans. Identification of the minimal requirements for virus transmission between mammals may have prognostic and diagnostic value for improving pandemic preparedness.*[93] (italics added)

In March 2012, the NSABB voted in favor of full publication of the two H5N1 studies. Despite all of this, in January 2018, virologist David Evans in the Department of Medical Microbiology & Immunology and Li Ka Shing Institute of Virology at the University of Alberta and colleagues used DNA fragments of horsepox along with sequences from modern *vaccinia* virus to synthesize a live synthetic chimeric horsepox virus.[94] They explained that synthetic horsepox virus exhibited less virulence in mice than *vaccinia* but still provided protection against *vaccinia*. They added that given the negative side effects of *vaccinia* (serious rash, infection, and inflammation of the heart muscle), the synthetic horsepox virus could lead to a safer alternative. The vaccines on the market at the time were Modified Vaccinia Ankara, produced by a German-Danish company, and LC16m8, produced by the Japanese: these vaccines are currently considered to be safe.[95] Therefore, many scientists are unclear as to why this study was done. Kai Kupferschmidt, writing for *Science*, quoted Thomas Inglesby, director of the Center for Health Security at the Johns Hopkins University Bloomberg School of Public Health: "The world is now more vulnerable to smallpox."[96] According to Kupferschmidt, another scientist, microbiologist David Relman of Stanford University, went as far as to ask the editor in chief of *PLOS ONE* not to publish Evans's paper.[97] But unlike the journals *Science* and *Nature*, which temporarily delayed publication of the Kawaoka and Fouchier papers, *PLOS ONE*'s committee on dual-use research unanimously agreed to publish the study because it believed that "the benefits of publication, including the potential improvements in vaccine development, outweighed the risks."[98] This familiar theme is heard (regardless of the temporary delays in research or international scientific summits) among scientific journals, committees, and organizations. Nevertheless, Associate Professor Gregory Koblentz, a biodefense expert at George Mason University in Arlington, Virginia, and virologist Andreas Nitsche at the Robert Koch Institute in Berlin believed that, with time, if someone wanted to recreate smallpox or another poxvirus, they could find the instructions in the Evans publication.[99]

Again, the use of genetic engineering comes down to benefits and risks. If the benefits to human health, environment, and safety are greater than the risks, or if the risks can be minimized, the powers that be will err on the side of technology. Regardless, scientists walk a fine line between benefits and risks in which these two worlds hold many unknown variables.

# Notes

1. The National Academies of Sciences, Engineering, and Medicine, *International Summit on Human Gene Editing: A Global Discussion* (Washington, DC: National Academies Press, 2015), 4.
2. National Research Council, *Biotechnology Research in an Age of Terrorism* (Washington, DC: National Academies Press, 2004), 34, https://nap.national acadeies.org/catalog/10827/biotechnology-research-in-the-age-of-terrorism.
3. The National Academies of Sciences, Engineering, and Medicine, *Human Gene Editing*, 4.
4. Ibid., 4.
5. National Research Council, *Biotechnology Research*, 19.
6. James R. Clapper, "Statement for the Record, 'Worldwide Threat Assessment of the U.S. Intelligence Community,'" Office of Director of National Intelligence, February 9, 2016, 9, https://www.dni.gov>files; Kelly Servick, "CRISPR — A Weapon of Mass Destruction?" February 11, 2016, https://sciencemag.org.
7. National Research Council, *Biotechnology Research*, 19.
8. Ibid.
9. Ibid., 34.
10. Ibid., 20.
11. Ibid.
12. Ibid., vii–ix.
13. Ibid., vii.
14. Ibid., 13–14.
15. Ibid., 11–12.
16. David Hearst, "Smart Bio-weapons Are Now Possible," *Guardian*, May 20, 2003, http://www.guardian.co.uk./uk_news/story/0,3604,959473,00.html.
17. National Research Council, *Biotechnology Research*, 25; Ronald J. Jackson, Allstair J. Ramsay, Carina D. Christensen, Sandra Beaton, Diana F. Hall, Ian A. Ramshaw, "Expression of Mouse Interleukin-4 by a Recombinant Ectromelia Virus Suppresses Cytolytic Lymphocyte Responses and Overcomes Genetic Resistance to Mousepox," *Journal of Virology* 75, no. 3 (February 2001): 1205.
18. National Research Council, *Biotechnology Research*, 2004 (Washington, DC: The National Academies Press), 19. https://doi.org/10.17226/10827; Ronald J. Jackson, Allstair J. Ramsay, Carina D. Christensen, Sandra Beaton, Diana F. Hall, Ian A. Ramshaw, "Expression of Mouse Interleukin-4 by a Recombinant Ectromelia Virus Suppresses Cytolytic Lymphocyte

Responses and Overcomes Genetic Resistance to Mousepox," *Journal of Virology* 75, no. 3 (February 2001): 1205–1210; Rachel Nowak, "Killer Mousepox Virus Raises Bioterror Fears," *NewScientist* (January 10, 2001), https://newscientist.com/article/dn311-killer-mousepox-virus-raises-bioterror-fears/.

19. National Research Council, *Biotechnology Research*, 25–26.
20. Ibid., 26.
21. Ibid.
22. Ibid.
23. Ibid., 26–27.
24. Jeronimo Cello, Aniko V. Paul, Eckard Wimmer, "Chemical Synthesis of Poliovirus cDNA: Generation of Infectious Virus in the Absence of Natural Template," *Science Online* 297, no. 5583 (August 9, 2002): 1016–1018, http://www.sciencemag.org/cgi/ontent/full/297/5583/1016; National Research Council, *Biotechnology Research*, 27–28.
25. Ibid.
26. National Research Council, *Biotechnology Research*, 27–28.
27. Ibid., 28.
28. Ibid.
29. Ariella M. Rosengard, Yu Liu, Zhiping, Robert Jimenez, "Variola Virus Immune Evasion Design: Expression of a Highly Efficient Inhibitor of Human Complement," *Proceedings of the National Academy of Sciences USA* 99, no. 13 (June 25, 2002): 8808.
30. Ibid.
31. Raven *et al.*, *Biology*, 1060, 1080.
32. National Research Council, *Biotechnology Research*, 29.
33. Ibid., 4–12.
34. Martin Enserink, "Public at Last, H5N1 Study Offers Insight into Virus's Possible Path to Pandemic," *Science* 336, no. 6088 (June 22, 2012): 1494.
35. Marc Lipsitch, Joshua B. Plotkin, Lone Simonsen, Barry Bloom, "Evolution, Safety, and Highly Pathogenic Influenza Viruses," *Science* 336, no. 6088 (June 22, 2012): 1529.
36. Masaki Imai, Tokiko Watanabe, Masato Hatta, Subash C. Das, Makoto Ozawa, Kyoko Shinya, Gongxun Zhong *et al.*, "Experimental Adaptation of an Influenza H5 HA Confers Respiratory Droplet Transmission to Reassortant H5 HA/H1N1 Virus in Ferrets," *Nature* 486, no. 4703 (June 21, 2012): 420.
37. Ibid.
38. Ibid.

39. Sander Herfst, Eefje J. Schrauwen, Martin Linster, Salin Chutinimitkul, Emmie de Wit, Vincent J. Munster, Erin M. Sorrell *et al.*, "Airborne Transmission of Influenza A/H5N1 Virus between Ferrets," *Science* 336, no. 6088 (June 22, 2012): 1534.
40. Ibid., 1537.
41. Ibid.
42. Ibid.
43. Ibid.
44. Imai *et al.*, "Experimental Adaptation of an Influenza H5 HA," 420.
45. Ibid.
46. Ibid.
47. Ibid., 421.
48. Ibid.
49. Ibid., 421–422.
50. Ibid., 420.
51. Ibid., 423.
52. Ibid.
53. Ibid.
54. Ibid.
55. Ibid.
56. Ibid.
57. Ibid., 427.
58. Ibid.
59. Ibid.
60. Ibid.
61. Sander Herfst, Eefje J. Schrauwen, Martin Linster, Salin Chutinimitkul, Emmie de Wit, Vincent J. Munster, Erin M. Sorrell *et al.*, "Airborne Transmission of Influenza A/H5N1 Virus between Ferrets," *Science* 336, no. 6088 (June 22, 2012): 1536.
62. Ibid.
63. Ibid., 1536–1537.
64. Ibid., 1537.
65. Ibid., 1538.
66. Ibid.
67. Ibid.
68. Ibid.
69. Ibid., 1538–1539; Enserink, "Public at Last," 1494–1497.
70. Herfst *et al.*, "Airborne Transmission of Influenza," 1539.
71. Ibid.

72. Ibid.
73. Ibid.
74. Ibid.
75. Ibid.
76. Ibid.
77. Ibid.
78. Ibid., 1540.
79. Ibid.
80. Ibid.
81. Ibid.
82. David Malakoff, "How Much Longer Will Moratorium Last?" *Science* 338, no. 6012 (December 7, 2012): 1497.
83. Ibid.
84. Ibid., 1496.
85. Patrick J. McGarvey, *CIA: The Myth and the Madness* (New York, NY: Penguin Books, 1972), 1–240.
86. Malakoff, "How Much Longer Will Moratorium Last?" 1496.
87. Ibid.; Jocelyn Kaiser, "Controversial Flu Studies Can Resume, U.S. Panel Says," *Science* 363, no. 6428 (February 15, 2019): 676–677.
88. Science Magazine, "Nobel Awardees Decry Grant Halt," *Science* 368, no. 6494 (May 29, 2020): 921.
89. Malakoff, "How Much Longer Will Moratorium Last?" 1496.
90. David Malakoff, "Proposed H5N1 Research Reviews Raise Concerns," *Science* 338, no. 6012 (December 7, 2012): 1271.
91. Ibid.
92. Imai *et al.*, "Experimental Adaptation of an Influenza H5 HA," 427.
93. Herfst *et al.*, "Airborne Transmission of Influenza," 1541.
94. Ryan S. Noyce, Seth Lederman, David H. Evans, "Construction of an Infectious Horsepox Virus Vaccine from Chemically Synthesized DNA Fragments," *PLoS ONE* 13, no. 1; e0188453; https://doi.org/10.1371/journal.pone.0188453.
95. Ibid.
96. Kai Kupferschmidt, "Critics See Only Risks, No Benefits on Horsepox Paper: Scientists Say Their Labmade Virus Could Make a New Smallpox Vaccine," *Science* 359, no. 6374 (January 26, 2018): 375.
97. Ibid.
98. Ibid.
99. Ibid.

# Chapter 10

# The Murky Waters of Regulation in the Age of Genetic Engineering

> James Watson was thus wrong when, in 1971, he foresaw that, once they were technically accessible, women's eggs would become a *readily available commodity*. On the contrary, with growing biotechnological demand, oocytes have literally become the *golden eggs*. The need in biotechnological research and infertility treatment for fresh, good quality egg cells is high. Human egg cells are, indeed, fast becoming a *commodity* the price of which is set by supply and demand in more or less illegal egg trafficking on a globalizing market.
>
> — Verena Stolcke, 2012[1]

The FDA is a research-friendly regulatory body because its rationale is that innovative research leads to discoveries that help human health.[2] A few examples are as follows: Immunotherapy is used for immunogenic diseases, such as leukemia, SARS corona virus, and HIV/AIDS. GM rice or Golden Rice provides provitamin A in the diet that can be converted by enzymes in the body to vitamin A. Editing ES cells from the bone marrow permits repairs to mutated β-globin genes responsible for sickle cell anemia or genes responsible for SCID. And intentional genomic alterations in animals and humans, such as knocking out the porcine gene *CD163* that allows viruses to invade pig cells or shutting down pig genes, generates virus-resistant pigs or pigs with humanized organs suitable for human transplant. Additionally, the FDA has recently created a program called the FDA's Center for Veterinary Medicine's Veterinary Innovation Program for certain intentional genomic alterations in animals and animal cells, tissues, and cell- or tissue-based products seeking FDA approval.[3] Several independent research teams have joined the program. The goal is to facilitate advancement of innovative products from animals by simplifying the regulatory process, thereby reducing the overall time to approval. One might argue that safety could be sacrificed in favor of innovation. But the FDA would counter that the pathway to approval is faster not by sacrificing safety but by advising the researchers at each step of the active research, lowering the number of review cycles, and shortening the approval process (as opposed to the research team going it alone without the direct guidance of the FDA).[4] Despite all of this, a reading of the FDA's regulations revealed weaknesses that could be exploited by bad actors. According to the Federal Food, Drug, and Cosmetic Act:

> The FDA ensures the safety of some foods; all drugs and medical devices (an instrument, apparatus, implement, machine, contrivance, implant, in vitro reagent, or any similar or related articles) intended for use in the diagnosis, cure, mitigation, treatment, or prevention of disease in man or other animals and to affect the structure or any function of the body of man or other animals; biological products (including blood, vaccines, and tissues for transplantation); and animal drugs and feed.[5]

Within the copious language of the act, however, one does not find guidance on how the medical devices or drugs are used after they have been approved. With the momentum of science, this fact increases the danger that forbidden experimental boundaries will be crossed. Greely wrote:

> If the FDA decides a drug or biological is safe and effective for that use, it approves the product for that use, which goes on the label. *But generally any physician is free, subject only to his or her own conscience (and malpractice coverage) to use an approved drug, device, or biological product in different ways, for different patients, and for different diseases. The drug maker cannot promote the drug for any unapproved purpose, but the FDA cannot stop the doctor from so using it.*[6] (italics added)

And these medical devices and drugs could easily be ES cells (human ES cells or iPSCs) used in PGD and in human-animal hybrid and chimeric animal experiments; animal organs used in xenotransplantation; SCNT used in cloning; iRNA gene therapy; immunotherapy; or ZFN, CRISPR/Cas therapeutics, and nanobiotechnology.

In showing how the FDA should demand proof of safety and efficacy of the potential technique Easy PGD, Greely proposed three different types of preclinical research requirements: nonhuman animal trials, human gamete studies, and human embryo studies.[7] He noted that one of the nonhuman species included should be a primate species (preferably chimpanzees, the closest relative of *Homo sapiens*), because it would be in the same taxonomic order as humans and, therefore, much more reproductively similar to humans than are mice, rats, rabbits, pigs, cats, and dogs.[8] He added that old-world or new-world monkeys would be alternatives if serious ethical or financial issues were not raised by the public. Nevertheless, researchers would have to create stem cell-derived gametes and learn how to do in vitro fertilization in these species. Greely made it clear that in addition to assessment of efficacy and safety of the process for generating offspring in different species, researchers must study the stem cell-derived gametes and resulting embryos within each species so as to learn what features are normal and abnormal in each species studied.[9]

Human gametes derived from iPSCs (or human ES cells or SCNT) would first be created in the laboratory — not in humans — and tested to see how similar they were to normal human gametes.[10] This variation should be assessed relative to what is seen in the research animal's iPSC-derived gametes when compared to their naturally produced eggs and sperm. If the human gametes derived from iPSCs are sufficiently similar to normal human gametes (and the animal iPSC-derived gametes are similar to the naturally produced animal eggs and sperm) and show no anomalies in preclinical trials, Greely argued that the third kind of research should follow, the creation of embryos.[11] Finally, Greely claimed that the FDA should insist on iPSC-derived gametes so that researchers can analyze how normal early embryo development proceeds compared to the development of normal human embryos. Greely explained:

> These three stages of preclinical research would tell us how safe and effective reproduction using stem cell-derived gametes was in two nonhuman species, including a nonhuman primate. And they would tell us

> how similar human stem cell-derived gametes, and embryos made from them, are to normal human gametes and embryos or to abnormal but safe nonhuman gametes and embryos. That should allow the sponsor, and the FDA, to determine whether the results are sufficiently promising to move to the required human trials. … *The decision to move an intervention for the first time from nonhuman animals to people should always be frightening.*[12] (italics added)

Indeed, Greely's preclinical research protocol could be applied to any field within biotechnology. It is slow and methodical and emphasizes safety and efficacy at every step. More importantly, his approach is humble, as evidenced by the italicized sentence in the above quote, restated here for emphasis: "The decision to move an intervention for the first time from nonhuman animals to people should always be frightening."[13]

## 10.1. A Survey of Different Countries and their Laws On the Creation of Human-Animal Hybrid for Research Purposes

In the last 32 years, biotechnology has revolutionized molecular biology. The exponential advances in genetic engineering have spawned many biotechnology companies, and the competition between these companies to be the first to discover a technique or mechanism collides with medical ethics, environmental concerns, and governmental regulation. To allay some of the bioethical concerns, proponents of genetic engineering have resorted to semantics.[14] The unique characteristics of human language, compared to animal vocalization, are openness and productivity, by which new expressions are produced using the rules of language. These new expressions or semantics now permeate the abortion debate. For example, when discussing ES-cell research and cloning as they relate to abortion, terms such as *personhood* have been created to recognize human beings in all stages of life as human persons having full human rights.[15] At the same time, terms such as *pre-embryo*, *pseudoembryo*, *preimplantation embryo,* and *preperson* have also been created and used, according to some, to deny the early human embryo the status as a person with human rights.[16] To government regulators (and some pro-life activists), these new expressions created an acceptable environment for the manipulation of embryos in cloning ES-cell research. Subsequently, terms such as *reproductive cloning* and, even better, *therapeutic cloning* were added to justify producing human embryos to harvest their stem cells.[17] But these new words did not placate some officials (pro-lifers and opponents of cloning) on President George W. Bush's Council on Bioethics. They simply believed that cloning to produce an embryo only to harvest ES cells was unethical.[18] Embryologists had also been thinking about the possibilities of applying reproductive cloning to solve the problem of infertility; this would give infertile couples genetic offspring (genes of at least one parent).[19] But President Bill Clinton strongly stated that "attempting to clone a human being is unacceptably dangerous to the child and morally unacceptable to our society."[20]

Related to this is animal chimera research, in which human iPSCs are injected into the early embryos of animal species in order to grow human organs.[21] Chimera research has a long history in biology, from using transgenic livestock as drug factories to genetically engineering cattle and mice with human DNA to produce human antibodies when vaccinated against the Ebola virus.[22] Since these technologies posed significant risks to health and safety of citizens, as well as to their values and morality, some form of regulation was needed to protect the people. In the 1980s, the UK's HFEA granted researchers permission to create hybrids using hamster eggs and human sperm as a diagnostic test for the quality of human sperm.[23] However, the HFEA Act of 1990 prohibited any embryos from these intergeneric fusions to develop beyond the two-cell cleavage stage. An extension of the act legalized embryo research for the purposes of finding treatments for serious diseases.

In January 2007, the HFEA considered the question of creating human-animal embryos for research.[24] Before they made their decision, the HFEA consulted the British public and the scientific community. The public's initial reaction was disgust and distrust of scientists; however, with further information, the people who supported the use of embryos in research tentatively favored the creation of human-animal embryos as long as there was a rationale for the research (e.g., finding a cure for a disease) and regulation.[25] In contrast, the scientific community was extremely confident that the creation of human-animal embryos was a viable alternative to using human eggs for studying the mechanisms behind generating patient-specific ES cells. On September 5, 2007, the HFEA permitted individual research teams to conduct research projects involving the creation of *cytoplasmic hybrid* embryos — the term used by the authority — provided that projects were necessary and desirable.[26]

While the British legislation on reproductive technologies is somewhat flexible, Australia's and Canada's regulations are comprehensive and rigid.[27] For instance, Australia's Prohibition of Human Cloning Act of 2002 and the Regulation of Human Embryo Research Amendment Act of 2006 prohibited a host of experiments: (1) both therapeutic and reproductive cloning; (2) ART techniques that might assist older women to achieve pregnancy; (3) creation of embryos to harvest ES cells; (4) in vitro embryonic development for more than 14 days; (5) the creation of human-nonhuman chimeric embryos; and (6) genetic editing of gametes or embryos. At present, heritable effects of genetic editing are unknown.

Canada's Assisted Human Reproduction Act of 2004, which was in full force by 2007, prohibited using reproductive technologies for sex selection; banned creation of human-nonhuman chimera and hybrid embryos; banned the creation of embryos to harvest ES cells; and banned the commercialization of human reproduction (i.e., paying for sperm, eggs, and in vitro embryos from donors or paying surrogate mothers for services).[28]

The United States had also passed legislation concerning reproductive technologies and related research (i.e., human-animal embryos).[29] But support for this particular legislation by the president or Congress depended on where the public stood in their values and morality (a complex mix of the abortion debate, the health and safety of the public, and lobbyists for religious groups and the biotechnology industry). In other words, politicians took their cues from the public and people's public feelings on these social issues.

When it was first suggested in the early 1990s that a human-animal hybrid be created for research purposes, there was no specific legislation to deal with such an entity.[30] It was not until 1997 when Stuart Newman, a professor of cell biology and anatomy at New York Medical College in Valhalla, made his case public to encourage public discussion of emerging technologies.[31] Newman, an outspoken critic of using developmental biology to modify the human species, applied for a patent on a human-animal chimera to challenge the US Patent and Trademark Office and the US Congress.[32] Congress had established by law that genetically engineered plants and animals could be patented. Although Newman's patent application was eventually rejected, he won in the sense that his moral opposition to patenting living organisms was upheld, and he publicized constitutional and moral questions.[33] The *Washington Post* ran an article on the controversy, saying that "it had raised profound questions about the differences and similarities between humans and other animals, and the limits of treating animals as property."[34]

President Bush subsequently signed into law an appropriations bill passed by Congress in 2005 that banned patents on humans or embryos.[35] Additionally, federal funding was allocated only to human ES-cell research using specified, preexisting stem cell lines rather than creating new human ES-cell lines. In that same year, on July 11, Senator Sam Brownback, a Republican from Kansas, proposed the Human Chimera Prohibition Act of 2005 to the 109th Congress.[36] This bill died in Congress the next year. It was based on several findings: (1) advances in research and technology have made it possible to merge human and nonhuman tissues to create new forms of life (i.e., chimeras); (2) serious ethical objections are raised because chimeras blur the line between human and animal; (3) respect for human dignity may be threatened by chimeras; (4) the uniqueness of the human species manifested through the brain and reproductive organs may be threatened by chimeras; and (5) chimeras present optimal means for genetic transfer of zoonotic infections that could increase the efficiency or virulence of diseases and cause pandemics.[37] Incidentally, the transfer of zoonotic infection is more real and less hypothetical than the other four findings. The moral implications of this issue struck such a chord with the core elements of President Bush's religious beliefs that he touched on the issue in his 2006 State of the Union address. He called for the banning of the following: "human cloning in all its forms; creating or implanting embryos for experiments; creating human-animal hybrids; and buying, selling, or patenting human embryos."[38] He concluded by stating that "a hopeful society has institutions of science and medicine that do not cut ethical corners and that recognize the matchless value of every life."[39] Most scientists accept and try to adhere to the meaning of these words every day, but some fail to do so (for many reasons addressed in this book, including lack of legal restrictions in using private funds to conduct controversial experiments that government agencies refuse to fund).

In 2007, Senator Brownback repackaged his previous proposal as the Human-Animal Hybrid Prohibition Act of 2007 and introduced it in the Senate. It failed. Subsequently, Representative Chris Smith, a Republican from New Jersey, took charge of the measure and proposed it to the 110th Congress in 2008.[40] The text of this act was similar to the previous versions, although it emphasized that "human dignity and the integrity of the human species are compromised" if human-animal hybrids exist.[41] Although this act, unlike like the others, managed to attract many cosponsors, it failed to pass Congress. It is difficult to understand why politicians have failed to pass these acts, especially with the language used by President Bush during his State of the Union address. Were scientific advancement or the biotechnology lobby contributing factors to the failures of these bills?

Unlike Congress, where chimera-prohibition bills failed, giving hope to developmental biologists, the NIH in September 2015 announced that it would not fund "controversial experiments that add human stem cells to animal embryos."[42] Today, according to Jocelyn Kaiser, staff writer for *Science,* the NIH had tentatively reversed its decision with a caveat: the experiments must pass an ethics review before funding.[43] Kaiser noted that the NIH's decision was based on the results of consultations with the research committee and with a bioethicist who deemed using chimeras to produce human organs "valuable studies."[44] Developmental biologists were thrilled about the decision, because "some were using non-NIH research fund to inject stem cells into pig or sheep embryos, in an effort to grow human pancreases or other organs inside animals."[45] These experiments frighten the public and have caused some countries to ban them. The NIH review process would be strict for experiments that add human stem cells to very early vertebrate embryos through the gastrulation stage and experiments that introduce human cells into the brains of postgastrulation animals, wrote Kaiser.[46] Another key concern is how human cells might change the animal's behavior and appearance.

As of this writing, countries such as France and Germany have enacted laws forbidding the creation of a chimeric human embryos, but these laws are ambiguous about whether adding human cells to animal embryos is allowed.[47] Some countries with more permissive policies, such as China, Japan, and South Korea, have allowed the creation of embryos for research through SCNT. The majority of these countries do not completely ban human-animal embryos.[48] In fact, Japanese law limits research on human-animal chimeric embryos by not allowing development beyond the appearance of the primitive streak (early gastrulation).[49]

Like the regulations governing the creation of chimeric human embryos, the regulation of gene-editing research varies in strength and ambiguity around the world. Kathleen Vogel, associate professor in the School of Public Policy at the University of Maryland, and her students from PLCY 306 discussed these policies in 2018.[50] Beginning with France, Vogel noted that on April 4, 1997, the French ratified the Oviedo Convention for the Protection of Human Rights and Dignity of the Human Being with Regard to the Application of Biology and Medicine. Article 13 of the convention is pertinent: "An intervention seeking to modify the human genome may only be undertaken for prevention, diagnostic, or therapeutic purposes and only if its aim is not to introduce any modification in the genome of any descendants."[51] This statement was a crucial factor in the restriction of genome editing in France. Additionally, Vogel wrote about a campaign called Stop GMO Babies, which demanded that scientists not use genome editing to modify human embryos. Members of this campaign believed that it was unethical and unnatural and that the embryo, as a person, could not give consent to the procedure.[52] Regardless, the French government, wrote Vogel, developed a plan called *Médicine France génomique 2025* with the goal of creating a nationwide system in which genome information and technology, such as CRISPR/Cas9, could be shared to help patients with various diseases.[53] Funding would come from government and the private sector. Moreover, there was pressure from trade unions and other public interest groups to strengthen regulations on organisms created through genome editing.

Reviewing the policies in Germany, Vogel wrote that research on human germ line editing using CRISPR technology was currently limited due to the *Embryonenschutzgesetz — EschG* (Embryo Protection Act) of 1990.[54] This act stipulated that no more than three embryos can be created per cycle of IVF, and all three, regardless of quality (or viability), must be transferred to the patient's womb at one time and cannot be genetically altered, harvested, and then frozen or discarded.[55] In agriculture, the federal environment ministry (disagreeing with the agricultural ministry) favored a strict approval process for gene editing in crops due to the unknown long-term impact on the ecology.[56] Despite some restrictions, German pharmaceutical giants Bayer, Merck, and Curevac have invested close to $500,000 in CRISPR research to generate innovative treatments for human diseases.[57] In fact, according to Vogel, Merck is partnering with an Israeli biotechnology company to edit the genomes of plants to produce therapeutic treatments.

Canada, compared to Germany and France, has very tough policies on genetic editing research. According to Vogel, human germ line editing is banned, and researchers who ignore the ban will be charged for violating the Canadian 2004 Assisted Human Reproduction Act and punished with a hefty fine or prison time.[58] In theory, the fear that gripped the Canadian public after the cloned sheep Dolly was announced to the world might have provided the rationale for a ban on any type of research involving genes and embryos. Echoing the tug-of-war between ministries in Germany, the Canadian Institutes of Health Research challenged the law, hoping for future decriminalization and legalization of human germ line editing.[59]

In the United Kingdom, Brexit has shaped gene-editing research in addition to religion and politics. In 2016, for instance, the HFEA approved an application to use CRISPR technology on human embryos against the backdrop of Brexit and freedom from European Union bioengineering restrictions.[60] However, Brexit has caused uncertainty about collaborations with European Union scientists in terms of immigration (many European Union scientists need to work in the United Kingdom) and sources of funding. These problems born of Brexit will, as Vogel correctly asserted, affect the United Kingdom's progress in genetic editing.

Some members of the clergy (both Catholic and Protestant) have raised the concern (already addressed in this book) that a researcher might play God and endow an embryo, through gene editing, with greater intelligence or strength. Any attempt at genetic enhancement is fraught with danger because of the complex interactions of genes with the cultural and natural environment (most traits will be polygenic).

Overall, the biggest headache emphasized by Vogel is how European countries could accommodate CRISPR and other genetic-editing technologies within the current regulatory policies governing GMOs established to protect humans, animals, and the environment.[61] Recall from Chapter 2 that GMOs contain foreign genes, and gene-edited organisms contain alterations to existing genes without adding foreign DNA. Is a genetically edited organism the same as a GMO? Scientists like Doudna would say no![62] Vogel stated that the *process* that produces the organism makes it a GMO, but if the *result* of the process is not different from what occurs naturally, then the organism is not a GMO; Doudna would accept the latter part of the sentence. Fortunately for scientists using genome-editing technologies, the European Court of Justice in early 2018 ruled that there were no grounds to change current regulatory laws governing GMOs to include CRISPR-modified plants.[63] Consequently, these gene-edited plants (if naturally identical) would be exempt from European Union regulations. But European Union countries can still prevent the planting of gene-edited crops in the name of public safety.

Outside of Europe and the Americas, there have also been conflicts with governments concerning genetic engineering in general, and genetic editing in particular. For example, Vogel wrote about the National Association for Sustainable Agriculture Australia (an organization promoting organic practices) protesting strongly against gene-edited crops.[64] But at the same time, environmental and conservation groups are advocating deregulation so that CRISPR technology can be used to help restore Australia's dying coral reefs.[65] South Korea and China are at roughly opposite ends of the spectrum on altered genomes. In South Korea, the Bioethics and Biosafety Act prohibited any form of genetic experimentation on human embryos, while its Notification Act banned any genetic engineering technology that altered genes.[66] In China, meanwhile, approvals for use of genetic engineering in clinical trials are relegated to the review boards of institutions such as universities or biotechnology companies instead of a government agency.[67]

The danger of this type of laissez-faire policy was highlighted in November 2018, when Jiankui He allegedly claimed that he created two

CRISPR-edited babies resistant to HIV (see Chapter 6). China was one of the first countries to take advantage of the new CRISPR technology. In 2016, Chinese scientists were the first to launch clinical trials using CRISPR, based on results of previous research in which they created gene-edited monkeys and human embryos.[68] Dennis Normile, writing for *Science*, noted that China was recruiting patients for ten planned CRISPR-related clinical trials — more than the rest of the world combined — to treat diseases such as cancer and AIDS.[69] But in Jiankui He's case, Chinese scientists were unanimous in their condemnation of his alleged work.[70] Embryo editing and implantation violated the Ethical Guiding Principles on Human Embryonic Stem Cell Research ratified jointly by the Chinese health and science ministries in December 2003. Unfortunately, this law was weak, according to bioethicist Zhai Xiaomei of the Chinese Academy of Medical Sciences, who stated, "In some countries, such activity would lead to imprisonment, but China still lacks a relevant legal and policy framework."[71] Surprisingly, other scientists from the very same academy also called for tighter regulation and oversight. Essentially, the consensus among Chinese scientists is the same as that among other scientists around the world: gene editing is in its infancy and altering the germ line should be banned. The seriousness of this issue in China can be gauged from the fact that the *Global Times*, a daily tabloid and website controlled by the Chinese Communist Party, was allowed to print a quote from former vice minister of health Huang Jiefu advocating an urgent need for a national ethics review committee (instead of institutional review committees).[72] In conclusion, Jiankui He is similar in some respects to most scientists, who desire recognition, fame, and to outdo their predecessors, all within the framework of helping the sick. They emphasize the short-term benefits and deemphasize the risks, which is an important theme mentioned frequently throughout this book. Norwegian philosopher Reidar Lie, professor of philosophy at the University of Bergen in Norway, would agree with this assessment of the behavior of scientists when it comes to evaluating the benefits and risks relating to their research. Lie stated that "He [Jiankui] shares the same characteristic as many other scientists at the frontiers of knowledge: overestimating the benefits of their own research and underestimating the risks."[73] This is one general reason why the new eugenics persists and why continued vigilance by scientists globally is warranted to prevent additional unintended consequences.

## 10.2. Regulation: Putting the Brakes On Runaway Science

One issue that emerged during the early commercialization of biotechnology, and which still is a problem today, is private companies and universities encouraging their researchers to secure patents for new biotechnological discoveries. For example, Stanford University and the University of California at San Francisco jointly filed a US patent application in 1974 on behalf of their respective faculty members as the sole inventors of the recombinant DNA technology; Dolly's inventors applied for patents in 1997 covering cloning technology for animals and humans; and the Broad Institute of the Massachusetts Institute of Technology and Harvard University became involved in a patent dispute with the University of California, Berkeley, and the Helmholtz Center for Infection Research in Germany over which scientific team originally invented the CRISPR/Cas9 technology (see Chapter 6).[74] Several companies were created beginning in the fall of 2013 to develop CRISPR-based therapies.[75] According to Doudna, Editas Medicine was founded with $43 million in financing from three venture capital firms and subsequently formed a collaboration with Juno Therapeutics in the form of an exclusive multimillion-dollar license to develop T cell therapies. In November 2018, the FDA approved Editas's application for its in vivo CRISPR genome-editing treatment EDIT-101, which treats Leber Congenital Amaurosis type 10 (LCA10), a disease caused by a mutation in the centrosomal protein 290 gene (*CEP290*) that results in blindness.[76] Two other companies were also founded: one called CRISPR Therapeutics, with $25 million in financing, and another called Intellia Therapeutics, with $15 million in funding, which partnered with the health care firm Novartis to pursue cancer immunotherapy. Some refer to all of this as a type of colonialism, a concept sometimes captured with the colorful word *biocolonialism*.

It seems reasonable to accept the late French philosopher Jean Baudrilland's comparison of the genetic manipulation of nature (or the human instinct to control nature) to colonialism, which historically involved annexation and domination.[77] In his 2000 book *The Vital Illusion*, he viewed genetic engineering as a new kind of colonialism — a biocolonialism tied to human instincts to control life. He stated, "The impulse to annex nature, animals, other races and cultures — to put them universally under jurisdiction — is in effect everywhere. ... Does a species have rights to its own genome and to its own eventual genetic transformation?"[78]

Social anthropologist Verena Stolcke, in her 1997 article "Homo Clonicus," wrote that the welfare and rights of women who provide eggs are largely ignored in arguments on the morality of cloning and harvesting ES cells.[79] She presented examples of clinics manipulating women within legal boundaries. First, in 1998, a private fertility clinic in the United States offered twenty times more than the market value in compensation for egg extractions. Second, a National Health Service clinic in the United Kingdom offered women a reduced price for in vitro fertilization in exchange for donating some of their eggs. Third, Wilmut (whose team cloned Dolly the sheep) proposed free fertility treatments for women in exchange for donating eggs for science. When South Korean Hwang Woo-suk's claim that he had obtained eleven human ES-cell lines in the laboratory could not be verified, the debate over stronger regulation in obtaining oocytes resumed.[80]

Another area where stronger regulation is needed is reproductive medicine, especially for PGD and sperm sorting used in nonmedical sex selection. According to Hvistendahl, "The United States has become a mecca for PGD patients because of its remarkably permissive policies. Even when it comes to cutting-edge technologies whose effects are barely understood, America is driven by the notion of choice."[81] In November 2003, the HFEA in the United Kingdom — one among 36 countries that had outlawed sex selection — published a review from a commission that considered the ethical implications and societal impact of new reproductive technologies yet to be applied in the real world.[82] The commission interviewed 2,165 UK adults (aged 16 years or older) face-to-face during January 9–14, 2003. The respondents' rejection of sex selection in general and sex-selection techniques in particular was based, to some extent, on religious beliefs and evolutionary principles such as long-term unknown effects:

> A great many respondents felt that sex selection was unqualifiedly wrong because it involved interference with divine will or with what they saw as the intrinsically virtuous course of Nature. Many of those who used these arguments used them to express a profound concern that human intervention in reproduction to achieve specific goals might result in unintended and undesirable side-effects. In effect they were saying that because human beings are fallible they are not in a position to take into account the full consequences of their actions. Similarly, interfering with a self-regulating natural process might have unanticipated effects that would be ultimately disadvantageous for those involved. (Many developed this argument by alluding to what they saw as the pernicious consequences of such interferences in the past, citing GM crops and antibiotic-resistant diseases as examples.) Considerations like these, relating to the limits of human wisdom, were generally used to support the precautionary view that sex selection should not be used at all, at least at the present.[83]

The caution expressed by respondents from the UK public, which resulted in the ban on sex selection in the United Kingdom, puts to shame some of the US regulatory organizations responsible for ARTs. Despite the fact that PGD and sperm sorting have some success, the long-term consequences are unknown. In fact, one study recommended a long-term follow-up after PGD, because the technique increased the prenatal death rate in multiple pregnancies.[84]

Furthermore, the commission's review warned of the potential harm PGD and sperm sorting might cause if legalized. For PGD testing, risks include (1) damage to the embryo during biopsy procedures, (2) lower live birth rate for women of advanced maternal age, (3) embryo death after thawing due to cryopreservation (storage in a frozen state before the procedure), and (4) increased prenatal death rate.[85] In sperm sorting, the major risk is damage, possibly heritable damage, to sperm DNA caused by a procedure used called flow cytometry sorting.[86] According to Ettore Caroppo, reproductive endocrinologist at ASL Bari in Conversano, Italy, and Sofia Riberio and colleagues at the Clinic of Gynecological Endocrinology in Basel, Switzerland, flow cytometry sorting is a technique used to find out the sex of future offspring by analyzing the DNA content of individual sperm cells and distinguishing the larger X chromosome resulting in females from the smaller Y chromosome resulting in males and then allowing the separation based on these differences.[87] Prior to flow cytometry sorting, according to Caroppo, a fluorescent dye is applied to semen that binds to the DNA of each spermatozoan. The X chromosome spermatozoan absorbs more dye than the Y chromosome spermatozoan because it has more DNA. It isn't hard to imagine possible damage (e.g., mutations in the DNA caused by chemicals within the dye). As a result of the differential dye absorption, X chromosome spermatozoa glow brighter than Y chromosome spermatozoa when exposed to UV light. Here is another opportunity for DNA damage, this time by radiation. As the spermatozoa pass through the flow cytometer in single file, each spermatozoan is captured by a drop of fluid and assigned an electric charge (positive for X and negative for Y), and the charge is used to separate the cells.

The United States has equivalent organizations to the UK HFEA, which are the FDA and, interestingly, the CDC. But these organizations are not as strong on this issue of reproductive technology. According to some bioethicists, their mandate for evaluating new technologies in reproductive medicine is different from similar organizations in other countries.[88] Their focus might be on other more familiar terrain, such as pharmaceuticals and the perpetual fight against infectious and noninfectious disease epidemics and pandemics. In this environment of absent regulation, there are no guidelines for fertility clinics to follow and no requirements to report to the CDC or FDA regarding why patients are undergoing PGD or sperm sorting.[89]

***

Science does not operate in a vacuum; it is affected by the greater culture. This greater culture includes government support and funding of research that leads to monetary benefits in the commercial applications of biotechnology in medicine and agriculture. To be first in the race of discovery means patents and multimillion-dollar contracts in the free market of genetic engineering. Sometimes, unfortunately, this means exaggerating an inconclusive result or questionable benefits. How can society corral these wild stallions called scientists and create guidelines for experimentation? Where should the line be drawn between therapy and elective genetic enhancement? How can a balance be found between public safety and scientific advancement? Or, as stated elsewhere in this book, are these concepts (public safety and scientific advancement) two sides of the same coin or a false dichotomy?

Government regulation works best when the penalties for violating regulations include a substantial fine to the institution or the banning of the scientist from conducting research. Biotechnology is a fairly young field, and governments are just realizing, at least in most Western countries, the hazards posed by genetic engineering. Moratoriums or regulations without penalties are not a sufficient deterrent. Equally worrisome is risky research done in plain sight when scientists find loopholes within the regulations. Sinsheimer recalled that at the Asilomar Conference in 1975, some researchers were continuing recombinant DNA experiments despite the moratorium.[90] Based on his conversations with other participants at the conference, "people were using the technique [recombinant DNA] and getting scientific results. I remember Dave Hogness from Stanford talking about shotgun experiments of the type I mentioned with *Drosophila* genes and so forth. So, it was clear that whatever had been called for, there had not really been any moratorium."[91] This mid-1970s behavior is still prevalent today. How can society slow down or stop the momentum in genetic engineering? The same question can be asked of other emerging scientific fields, such as artificial intelligence (AI), that pose unknown dangers to *Homo sapiens*. Generally, science has momentum and this fact goes to the heart of public safety versus scientific advancement. The US government views scientific advancement as a way to protect the public.[92] Sinsheimer saw no simple solution. Even more depressing is Sinsheimer's discussion of the NIH. Apparently, the NIH formed a committee called the Recombinant DNA Molecule Program Advisory Committee prior to the Asilomar Conference to create guidelines.[93] Sinsheimer, who was invited to by this committee to comment on the guidelines, emphasized the difficulties:

> I think once this NIH committee got into it, they found it was much more difficult to do than they had anticipated. NIH was quite unfamiliar, if you want, with being a regulatory agency. Obviously, when you get into regulation, you find yourself bogged down in a whole mess of things. You have to have definitions of terms. There are so many possible kinds of experiments and so on, that you had to think of.[94]

That, however, was 43 years ago, and the NIH is now a different organization that has educated itself and is arming itself for the biotechnology age. The organization has many weapons to use in situations of noncompliance with guidelines, one being to stop funding for any research that is headed in the wrong direction.

## 10.3. Jennifer Doudna's Fear Realized

In January 2015, 40 years later, another Asilomar-like conference was held in Napa, California, to discuss the path forward for genomic engineering.[95] The questions raised at this conference on the use of gene-editing tools required a global discussion. This led to an international summit on human gene editing, which was held in Washington, DC, in December of the same year. Three years later, in November 2018, a second international conference was held in Hong Kong to address the claim that a scientist used genome editing to make genetic alterations in early embryos that were subsequently implanted.[96] But before continuing, consider Doudna's experience, beginning in mid-2012, of two pivotal moments in the application of the brand-new CRISPR/Cas9 technology that alarmed her sufficiently to call for summits on genome editing.

After Doudna and colleagues published their landmark paper in June 2012, it did not take long for researchers "all around the world, far too many scientists to fit into any one lab," to exploit the biochemical properties of CRISPR/Cas9 to engineer the DNA in plant, animal, and human cells.[97] While most of this research was used to develop treatments for genetic diseases, the pace of change from the revelation that CRISPR regions in bacteria provided immunity from viral infection to their development as a diagnostic tool made Doudna anxious.[98] Her anxiety increased in early 2014 when a Chinese research team used CRISPR/Cas9 to edit the embryos of Southeast Asian monkey species *Macaca fascicularis* (referred to in laboratories as cynomolgus monkeys) and then implanted the edited embryos into surrogate cynomolgus mothers.[99] Since the alterations would be inherited by future generations, Doudna was concerned that scientists might take the next dangerous step and apply CRISPR to human embryos. Her concern was well founded, because in 2015, another Chinese research team edited human tripronuclear zygotes.[100] Although these zygotes are nonviable and would never be used for implantation, a sacred line was crossed, which triggered Doudna's call for a public conversation about genetic engineering. The one-day conference in the Napa Valley was a start, but the fact that CRISPR/Cas9 technology was being used globally, as prominently demonstrated by these Chinese scientists, resulted in an urgent international meeting held late in 2015 and a subsequent meeting in late 2018.

At the December 2015 international summit, hosted by the US National Academy of Sciences, the US National Academy of Medicine, the Royal Society, and the Chinese Academy of Sciences, the participants discussed the wide range of applications of gene-editing technology; the ethical, legal, and social issues; and governance of human gene editing.[101] One concern raised was that scientists were altering gene variants at a time when they were still learning about genetic variation and the complex interaction of genes, populations, and environment. Related to this, Eric Lander noted that "human genetic disease is complex; we still have a lot to learn. … Before we make permanent changes to the human gene pool, we should exercise considerable caution."[102] Another concern was the strong commercial motivation to play on the vulnerability of individuals afflicted by illness with the promise of a cure and on the vanity of others with the promise of enhancing their traits.

John Harris, director of the Institute for Science, Ethics, and Innovation in the School of Law at the University of Manchester, made the entirely reasonable statement that "gene editing will be acceptable *when its benefits, both to individuals and to the broader society, exceeds its risks*"[103] (italics added). In starker terms, German theologian Hille Haker, chair of moral theology at Loyola University, Chicago, proposed a universal ban on germ line gene editing for reproductive purposes beginning with a two-year moratorium until an international ban with binding regulations can be secured through the United Nations. In her assessment of risks and benefits, Haker stated:

> The goal of society should be to promote a better life for all, and to ensure that everybody can live a life in dignity and freedom. … Can this be achieved by germline gene editing? My view is no. *The future risks of gene editing are unpredictable … which means that the long-term harms may well outweigh the benefits.*[104] (italics added)

Haker also addressed the question of personhood: "Researchers and future parents have an obligation to respect the morally relevant status of the human embryo … but germline editing does not meet this obligation because it either renders the embryo morally neutral or diminishes it to the status of property or goods."[105]

Addressing the question of discrimination and equalities, some summit presenters believed that human gene editing and other biotechnologies have the potential to exacerbate existing inequalities. Catherine Bliss, associate professor of sociology at the University of California, San Francisco, bolstered this belief: "Well-meaning science that intends to produce benefits for society can unintentionally reproduce social injustices — for example, *in the way that genomics has inadvertently reinforced certain racial categories*"[106] (italics added). Bliss made an important point, because in the 1960s and 1970s, implementation of mandatory genetic disease screening laws for Tay-Sachs disease and sickle cell anemia brought back memories of mainline eugenics (see Chapter 1). These diseases were forever attached to race: Tay-Sachs disease was a "Jewish" disease — reinvigorating the argument or whether being Jewish is a race — and sickle cell anemia was a "black" disease.

In the end, the organizing committee's statement recognized the fact that gene-editing technologies, like it or not, are already in broad use in biomedical research around the world.[107] However, there needs to be more oversight in terms of regulation by national governments. The members of the committee agreed that human gene editing can occur but must proceed in a more responsible and transparent manner.[108] Most importantly, continual international dialogue among scientists and the public is necessary. Some of the organizing committee's conclusions were that

(1) in preclinical research, edited human embryos should never be implanted; (2) somatic-cell gene editing can occur in clinical trials as long as there are risk-benefit assessments before the trials; (3) clinical trials of germ line gene editing cannot occur because "the safety issues have not yet been adequately explored," such as off-target mutations, mosaicism, and long-terms effects of the alterations to the human genome, including adaptation to changing environments; (4) once introduced into one human population, genetic alterations will eventually spread throughout the species; and (5) genetic enhancements to subsets of the population "could exacerbate social inequities or be used coercively."[109]

Three years later, the second international summit met to continue the dialogue on human gene editing and find out more about Jiankui He's claim of implanted edited human embryos. Unfortunately, Jiankui He did not provide key details of his research, which resulted in both skepticism and concern from participants regarding his claim.[110] Consequently, the organizing committee called for an independent assessment of Jiankui He's research. The committee said:

> Even if the modifications are verified, the procedure was irresponsible and failed to conform with international norms. … Its flaws included an inadequate medical indication, a poorly designed study protocol, a failure to meet ethical standards for protecting the welfare of research subjects, and a lack of transparency in the development, review, and conduct of the clinical procedures.[111]

At this summit, developmental biologist Kathy Niakan of the Francis Crick Institute in London made a case for possible applications of genome editing in human embryos to address the problem of low success rates of IVF by exploring how embryos form, implant, and develop.[112] Likewise, obstetrician-gynecologist Paula Amato of the Oregon Health & Science University considered cardiomyopathy and argued that germ line editing of human embryos could prevent transmission of this disease and others to future generations.[113] But developmental biologist Robin Lovell-Badge, head of the Division of Stem Cell Biology at the Francis Crick Institute, agreed with the consensus, noting that before germ line editing (regardless of method) can be used safely and effectively, "many scientific questions need to be addressed, including questions about the unique features of the first embryonic cell cycle and the complexities of double-strand break repair."[114] Additional presenters argued that other evils, such as PGD or somatic cell editing after birth, are preferable to germ line editing.

Interestingly, the idea of universal, shared governance with United Nations backing — raised at the 2015 summit — emerged again. South African bioethicist Ames Dhai, head of the Department of Bioethics at the University of Witwatersrand in Johannesburg, believed that a global standard for genetic engineering should be established. She raised "the possibility of a United Nations treaty or covenant on the human genome that would globally harmonize the governance of genome editing."[115] Similarly, professor of medical humanities Ock-Joo Kim in the College of Medicine at Seoul National University, encouraged participants to adopt universal values in their application of genetic engineering, given the strong influence of market forces.[116]

In the end, the organizing committee basically reiterated the conclusions established at the 2015 summit, stating that it would be irresponsible to proceed with any clinical use of human germ line editing without continued international discussions and oversight. The committee added:

> The variability of effects produced by genetic changes makes it difficult to conduct a thorough evaluation of benefits and risks. Nevertheless, germline editing could become acceptable in the future if these risks are addressed and if a number of additional criteria are met. These criteria include strict independent oversight, a compelling need, an absence of reasonable alternatives, a plan for long-term follow-up, and attention to societal effects. Even so, public acceptability will likely vary among jurisdictions, leading to differing policy responses.[117]

Hopefully, biomedical scientists will continue to be proactive in organizing these international summits (and inviting social scientists, ethicists, health care providers, policy makers, regulators, research funders, industry representatives, and members of public interest groups) to share the potential benefits and dangers of their research to living organisms and the environment. This type of continued transparency is akin to guardrails that prevent the biological sciences wandering down the dark paths of the old eugenics.

## Notes

1. Verena Stolcke, "Homo Clonicus," in *Clones, Fakes, and Posthumans*, edited by Philomena Essed and Gabrielle Schwad, Thamyris/Intersecting: Place, Sex, and Race (Amsterdam: Rodopi, 2012), 35.
2. Federal Drug Administration, "The History of FDA's Fight for Consumer Protection and Public Health," https://www.fda.gov.
3. Scott Gottlieb and Anna Abram, "FDA Announces Controversial 'Risk-based' Rules for CRISPR-edited Animal Technology," April 4, 2019, https://www.geneticliteracyproject.org/2019/04/08/fda-announces-cntroversial-risk-based-rules-for-crispr-edited-animal-technology/.
4. Federal Drug Administration, "The History of FDA's Fight," https://www.fda.gov.
5. US Food, Drug, and Cosmetic Act, Section 201(g)(1), codified at 21 U.S.C.§321(g)(1). As amended through P.L. 116-118, enacted March 8, 2019.
6. Henry T. Greely, *The End of Sex and the Future of Human Reproduction* (Cambridge, MA: Harvard University Press), 156–157.
7. Ibid., 217.
8. Ibid.
9. Ibid., 217–218.
10. Ibid., 218.
11. Ibid.
12. Ibid., 218–219.
13. Ibid., 219.
14. Stolcke, "Homo Clonicus," 30.
15. Ibid.
16. Ibid.
17. Ibid.
18. George W. Bush Executive Order 13237, "Creation of the President's Council on Bioethics," November 30, 2001, 66FR 59851, https://federalregister.gov/a/01-29948.
19. Stolcke, "Homo Clonicus," 30.
20. Meredith Wadman, "White House Bill Would Ban Human Cloning," *Nature* 387, 6634 (June 12, 1997): 644.
21. Sarah Taddeo and Jason S. Robert, "Hybrids and Chimeras: A Report on the Findings of the Consultation," by the Human Fertilisation Embryology Authority 2007 Report, October 2007, http://embryo.asu.edu/handle/10776/8240.

22. William H. Velander, Henryk Lubon, and William N. Drohan, "Transgenic Livestock as Drug Factories," *Scientific American*, no. 1 (January 1997): 71–74; David Robson, "The Birth of Half-Human, Half-Animal: The Quest to Create Animals with Human Organs Has a Long History — And It Is Now Becoming a Reality. Has Science Taken a Step Too Far?" January 5, 2017, http://www.bbc.com/earth/story/20170104-the-birth-of-the-human-animal-chimeras.
23. Taddeo and Robert, "Hybrids and Chimeras," http://embryo.asu.edu/handle/10776/8240.
24. Ibid.
25. Ibid.
26. Ibid.
27. Kirraley Bowles, "Prohibition of Human Cloning Act 2002 — Summary of Provisions," accessed September 17, 2018, http://www.findlaw.com.au/articles/1269/prohibition-of-human-cloning-act-2002-8211-su.
28. Assisted Human Reproduction Act, accessed September 17, 2018, http://laws-lois.justice.gc.ca/eng/acts/A-13.4/FullText.html.
29. Bowles, "Prohibition of Human Cloning Act 2002," accessed September 17, 2018, http://www.findlaw.com.au/articles/1269/prohibition-of-human-cloning-act-2002-8211-su; Mark Dowie, "Gods and Monsters," *Mother Jones* (January–February 2004), http://www.motherjones.com; Rick Weiss, "U.S. Denies Patent for a Too-Human Hybrid," *Washington Post* (Sunday, February 13, 2005), A03, https://www.washingtonpost.com/wp-dyn/articles/A19781-2005Feb12.html; US Patent Application no. 08/933,564, "Chimeric Embryos and Animals Containing Human Cells."
30. Weiss, "U.S. Denies Patent," A03, https://www.washingtonpost.com/wp-dyn/articles/A19781-2005Feb12.html.
31. Ibid.
32. Dowie, "Gods and Monsters," http://www.motherjones.com; Weiss, "U.S. Denies Patent," A03, https://www.washingtonpost.com/wp-dyn/articles/A19781-2005Feb12.html; US Patent Application no. 08/933,564, "Chimeric Embryos and Animals Containing Human Cells."
33. Weiss, "U.S. Denies Patent," A03, https://www.washingtonpost.com/wp-dyn/articles/A19781-2005Feb12.html.
34. Ibid.
35. Ibid.
36. "S.659 (109th): Human Chimera Prohibition Act of 2005," *GovTrack*. https://www.congress.gov/bill/109th-congress/senate-bill/659/text.

37. Weiss, "U.S. Denies Patent," A03, https://www.washingtonpost.com/wp-dyn/articles/A19781-2005Feb12.html.
38. "President Bush's State of the Union Address — CQ Transcripts Wire," *Washington Post*, accessed August 28, 2018, https://www.washingtonpost.com/wp-dyn/content/article/2006/01/31/AR2006013101468.html.
39. Ibid.
40. "H.R. 5910 (110th): Human-Animal Hybrid Prohibition Act of 2008," *GovTrack*. https://www.govtrack.us/congress/bills/110/hr5910.
41. Ibid.
42. Jocelyn Kaiser, "NIH Plans to Fund Human-Animal Chimera Research: Agency Would Subject Experiments — Such as Growing Human Organs in Pigs — To Extra Ethical Review," *Science* 353, no. 6300 (August 12, 2016): 634.
43. Ibid., 634–635.
44. Ibid., 635.
45. Ibid., 634.
46. Ibid., 635.
47. Ibid., 634–635.
48. Taddeo and Robert, "Hybrids and Chimeras," http://embryo.asu.edu/handle/10776/8240; Kaiser, "NIH Plans to Fund Human-Animal Chimera Research," 634.
49. Kaiser, "NIH Plans to Fund Human-Animal Chimera Research," 634–635.
50. Kathleen M. Vogel and Students from PLCY 306, "Crispr Goes Global: A Snapshot of Rules, Policies, and Attitudes," *Bulletin of Atomic Scientists* (June 5, 2018), https://thebulletin.org/2018/06/crispr-goes-global-a-snapshot-of-rules-policies-and-attitudes/.
51. Roberto Andorno, "The Oviedo Convention: A European Legal Framework at the Intersection of Human Rights and Health Law," *Journal of International BioTechnology Law* 2, no. 4 (January 2005): 140; Vogel and Students from PLCY 306, "Crispr Goes Global," https://thebulletin.org/2018/06/crispr-goes-global-a-snapshot-of-rules-policies-and-attitudes/.
52. Vogel and Students from PLCY 306, "Crispr Goes Global," https://thebulletin.org/2018/06/crispr-goes-global-a-snapshot-of-rules-policies-and-attitudes/.
53. Ibid.
54. Ibid.
55. The Federal Minister for Research and Technology, Riesenhuber, "Act for the Protection of Embryos (The Embryo Protection Act) — Gesetz zum

schutz von Embryonen (Embryonenschutz gesetz-EschG)," *Federal Law Gazette*, no. 69, Part 1 (December 19, 1990): 2–3.
56. Ibid.
57. Ibid.
58. Ibid.
59. Ibid.
60. Ibid.
61. Ibid.
62. Jennifer A. Doudna and Samuel H. Sternberg, *A Crack in Creation: Gene Editing and the Unthinkable Power to Control Evolution* (New York: Houghton Mifflin Harcourt, 2017), 125.
63. Ibid., 127; Vogel and Students from PLCY 306, "Crispr Goes Global," https://thebulletin.org/2018/06/crispr-goes-global-a-snapshot-of-rules-policies-and-attitudes/.
64. Vogel and Students from PLCY 306, "Crispr Goes Global," https://thebulletin.org/2018/06/crispr-goes-global-a-snapshot-of-rules-policies-and-attitudes/.
65. Ibid.
66. Ibid.
67. Ibid.
68. Dennis Normile, "For China, a CRISPR First Goes Too Far," *Science* 362, no. 6419 (December 7, 2018): 1091.
69. Ibid.
70. Ibid.
71. Ibid.
72. Ibid.
73. Ibid.
74. Paul Berg and Janet E. Mertz, "Personal Reflections on the Origins and Emergence of Recombinant DNA Technology," *Genetics* 184, no. 1 (January 2010): 9–17; "Dolly Goes to Market: World Patents on Sheep Clones Include Humans," May 5, 1977, http://users.westnet.gr/cgian/clonepat.htm; Jim Kozubek, *Modern Prometheus: Editing the Human Genome with CRISPR-CAS9* (New York, NY: Cambridge University Press, 2016), 24–30, 306–308, 323–333.
75. Doudna and Sternberg, *A Crack in Creation*, 89, 177.
76. Ibid.
77. Baudrillard, *The Vital Illusion*, 22–23.
78. Jean Baudrillard, *The Vital Illusion* (New York: Columbia University Press, 2000), 22–23; Gabrielle Schwad, "Replacement Humans," in

*Clones, Fakes, and Posthumans*, edited by Philomena Essed and Gabrielle Schwad, Thamyris/Intersecting: Place, Sex, and Race (Amsterdam: Rodopi, 2012), 90.

79. Verena Stolcke, "Homo Clonicus" in Philomena Essed and Gabrielle Schwad, eds., *Clones, Fakes, and Posthumans* (Thamyris/Intersecting: Place, Sex, and Race) (Amsterdam: Rodopi Press, 2012), 35.
80. Woo suk Hwang, Young June Ryu, Jong Hyuk Park, Eul Soon Park, Eu Gene Lee, Ja Min Koo, Hyun Yong Jeon *et al.*, "Evidence of a Pluripotent Human Embryonic Stem Cell Line Derived from a Cloned Blastocyst," *Science* 303, no. 5664 (March 12, 2004): 1669–1674; Woo suk Hwang, Sung II Roh, Byeong Chun Lee, Sung Keun Kang, Dae Kee Kwon, Sue Kun, Sun Jong Kim *et al.*, "Patient-Specific Embryonic Stem Cells Derived from Human SCNT Blastocysts," *Science* 308, no. 5729 (June 17, 2005): 1777–1783.
81. Mara Hvistendahl, *Unnatural Selection: Choosing Boys Over Girls, and the Consequences of a World Full of Men* (New York: PublicAffairs, 2011), 258.
82. Hvistendahl, *Unnatural Selection*, 249–262; Human Fertilisation Embryology Authority, "Sex Selection: Options for Regulation," (2002–2003 Report, November 12, 2003), 17.
83. I. Liebaers, S. Demyttere, W. Verpoest, M. De Rycke, C. Staessen, K. Sermon, P. Devroey, P. Haentjens, M. Bonduelle, "Report on a Consecutive Series of 581 Children Born after Blastomere Biopsy for Preimplantation Genetic Diagnosis," *Human Reproduction* 25, no. 1 (January 2010): 275–282.
84. H. Joris, E. Van den Abbeel, A. De Vos, A. Van Steirteghem, "Reduced Survival after Human Embryo Biopsy and Subsequent Cryopreservation," *Human Reproduction* 14, no. 11 (July 1999): 2833–2837; I. Liebaers, S. Demyttere, W. Verpoest, M. De Rycke, C. Staessen, K. Sermon, P. Devroey, P. Haentjens, M. Bonduelle, "Report on a Consecutive Series of 581 Children Born after Blastomere Biopsy for Preimplantation Genetic Diagnosis," *Human Reproduction* 25, no. 1 (January 2010): 275–282; S. Mastenbroek, M. Twisk, F. van der Veen, and S. Repping, "Preimplantation Genetic Screening: A Systematic Review and Meta-Analysis of RCTs," *Human Reproduction* 17, no. 4 (July–August 2011): 454–466.
85. Ettore Caroppo, "Sperm Sorting for Selection of Healthy Sperm: Is It Safe and Useful," *Fertility and Sterility* 100, no. 3 (June 28, 2013): 695–696; Sofia Ribeiro, Gideon Sartorius, Christian de Geyter, "Sorting of

Spermatozoa with Flow Cytometry," *Fertility and Sterility* 100, no. 3 (July 22, 2013): e15, https://doi.org/10.1016/j.fertnstert.2013.06.006.

86. Ibid.
87. Hvistendahl, *Unnatural Selection*, 258–259; Katie Moisse, "Boy or Girl? Clinic Offers Choice," September 24, 2012, https://abcnews.go.com.
88. Ibid.
89. Sinsheimer, interview by Shelley Erwin, 50.
90. Ibid., 54.
91. Ibid.
92. National Research Council, *Biotechnology Research in an Age of Terrorism* (Washington, DC: The National Academies Press, 2004), vii, https://doi.org/10.17226/10827; The National Academies of Sciences, Engineering, and Medicine, *Review of the Federal Strategy for Nanotechnology-Related Environmental, Health, and Safety Research* (Washington, DC: The National Academies Press, 2009), 19.
93. Ibid., 56.
94. Ibid.
95. The National Academies of Sciences, Engineering, and Medicine, *International Summit on Human Gene Editing: A Global Discussion*, December 1–3, 2015 (Washington, DC: The National Academies Press), 1–8.
96. The National Academies of Sciences, Engineering, and Medicine, *Second International Summit on Human Gene Editing: Continuing the Global Discussion: Proceedings of a Workshop in Brief*, November 27–29, 2018 (Washington, DC: The National Academies Press), 1–10.
97. Jennifer A. Doudna and Samuel H. Sternberg, *A Crack in Creation: Gene Editing and the Unthinkable Power to Control Evolution* (New York: Houghton Mifflin Harcourt, 2017), 86.
98. Ibid., 87.
99. Ibid., 138–139.
100. Ibid., 214–222; Puping Liang, Yanwen Xu, Xiya Zhang, Chenhui Ding, Rui Huang, Zhen Zhang, Jie Lv *et al.*, "CRISPR/Cas9-mediated Gene Editing in Human Tripronuclear Zygotes," *Protein Cell* 6, no. 5 (April 1, 2015): 363–372.
101. The National Academies of Sciences, Engineering, and Medicine, *International Summit on Human Gene Editing*, 1.
102. Ibid., 3.
103. Ibid.

104. Ibid., 4.
105. Ibid.
106. Ibid.
107. Ibid., 6.
108. Ibid.
109. Ibid., 7.
110. The National Academies of Sciences, Engineering, and Medicine, *Second International Summit on Human Gene Editing*, 2.
111. Ibid., 3.
112. Ibid., 4.
113. Ibid.
114. Ibid.
115. Ibid., 5.
116. Ibid.
117. Ibid., 7.

# Chapter 11

# Benefits and Risks: The Eternal Struggle

We are here today to talk — dare I say it? — a brave new world. As a growing number of performance enhancing technologies become a practical reality, we confront a new world in which human beings wield increasing power over their own destinies. But to what end? And what are the assumptions that guide this quest for perfection? And at the end of the day, will the benefits outweigh the costs?

— Luis Lugo, 2004[1]

I come down on the side that says if you can make great gains by making embryo hybrids in preventing premature death and understanding disease then a limited amount of such research is morally justifiable.

— Arthur Caplan, professor of bioethics at New York University's Langone Medical Center's Division of Medical Ethics, 2008[2]

In the 1970s, geneticists were trying to insert elements of simian virus 40 (SV40: cancer-causing virus in monkeys and humans) into eukaryotic cells to investigate signals that control expression of genes.[3] The Mallinckrodt Professor of Genetics at Harvard Medical School, Richard C. Mulligan — an undergraduate in the 1970s — was interested in this gene transfer technology: "I was struck at the time as to how sophisticated SV40 appeared to be with regard to its interactions with different kinds of cells and its strategies for achieving viral expression, and wondered whether viruses such as SV40 might be able to be engineered to transfer and express foreign genes in cells."[4] At the Massachusetts Institute of

Technology at this time, geneticists engineered SV40 to carry genes into a cell, creating a recombinant virus in the process. The benefit, noted by Kozubek, was getting the virus to code for human β-globin to correct genetic blood disorders such as sickle cell anemia and beta thalassemia in which a small amount of the edited gene resulted in the expression of the normal protein.[5] After these experiments, Mulligan was convinced that viral vectors would be critical in gene therapy, and retroviruses would be the optimal vectors to use.

In the late 1970s, Mulligan collaborated with Andrew Chess (now professor of genetics and regenerative biology at Mount Sinai Hospital, New York) and David Altshuler (chief scientific officer at Vertex Pharmaceuticals in Boston) to create an artificial version of horizontal gene transfer (HGT) — a phenomenon once believed to be rare in nature.[6] Artificial HGT allowed geneticists to insert fragments of genes in chromosomes with subsequent expression of proteins.

In the early 1980s, Mulligan and colleagues engineered retroviruses, "turning them into stripped-down delivery vehicles that he could throw any cargo into — and that included genes that he wanted to deliver to the human genome. This was the recombinant virus, a virus construct carrying the cargo of a human gene."[7] In his book, Kozubek does an excellent job of discussing the anatomy and function of a retrovirus and Mulligan's engineered retrovirus.[8] A retrovirus can exist as a free virus or as a provirus once it binds to a cell and injects its genetic material into the genome of the cell. At each end of the provirus is an important multifunctional DNA element known as the long terminal repeat (LTR). Kozubek defined this LTR as controlling the expression of genes on the provirus and their insertion into the chromosomes of the host cell. Equally important are the genes between the LTRs that encode the proteins of the viral coat and an enzyme called reverse transcriptase. Kozubek discussed in detail the important retroviral genes and their specific function:

> In fact, retrovirus carries three major genes: *gag*, which builds basic infrastructure, *pol*, or polymerase, which builds an "RNA-dependent DNA polymerase," most often just called "reverse transcriptase." In effect, the reverse transcription enzyme copies the RNA virus into DNA that can incorporate into a genome; pol also includes the code to build

> an enzyme called integrase which makes "double-strand breaks" to hack into a host genome. It is machinery to turn RNA back into DNA and to incorporate it into the genome of a cell, where it becomes a "provirus." After it is integrated in a chromosome, it can be expressed and packaged, turning into free virus, or more baby viruses, once again. A gene called *psi* is important in its ability to repackage into more copies of free, infectious viral particles, and a gene called *env*, or envelope, enables the packaged virus to make contact with a receptor on new cells to enable uptake of the virus particle.[9]

In a preliminary effect-on-function experiment, Mulligan and colleagues deleted a section of the *psi* gene (responsible for repackaging the free virus and making it infectious) so that the virus particles could insert into the genome of the host cell without spreading to other cells.[10] They subsequently engineered a replication-competent recombinant retrovirus. First, they obtained the provirus of a mouse and removed all the retroviral genes except the two LTRs and the *psi* gene.[11] To test the vector system, they inserted a bacterial antibiotic gene called *neo* into the modified provirus. The result was a recombinant retrovirus — without retroviral genes — that could insert into the genome of a host cell.[12] The scientists could use this engineered viral vector to carry human genes into cells, but they needed the recombinant retrovirus to replicate itself in cells to be effective in gene therapy. Consequently, they obtained another provirus from a mouse retrovirus, and this time, they removed the *psi* gene.[13] In a second effect-on-function experiment, they introduced the provirus without the *psi* gene into a mouse, and it integrated into the mouse's chromosomes. No virus was produced because the RNA copies could not be packaged into viral coats.[14] Mulligan's team then infected the cells with the recombinant retrovirus, and repackaging occurred, with replication of retrovirus containing the *neo* gene. In the mid-1980s, Mulligan and colleagues published papers on their new technique resulting in a flurry of activity in the development of gene therapies for diseases, including hematopoietic stem cells, beta thalassemia, and SCID.[15] However, there were still serious concerns about mutagenesis inadvertently knocking out tumor-suppressor genes and immune reactions. In 1999, for instance, an 18-year-old man named Jesse Gelsinger died from an overwhelming immune response four days after being injected with an adenoviral vector during gene therapy trials at the University of Pennsylvania Children National Medical Center in Philadelphia.[16] This young man suffered from ornithine transcarbamylase (OTC) deficiency, an X-linked genetic disease of the liver.[17] OTC is an enzyme that moves free nitrogen molecules out of the body. In the absence of this enzyme, nitrogen builds up in the body and binds with hydrogen to form ammonia resulting in death. An adenovirus was selected for trial, despite warnings about its immunogenic properties (ironically, a retrovirus was rejected by these researchers because of its link to leukemia-like conditions).[18] These dangers did not deter NIH geneticist William French Anderson, who began filing applications for gene therapy patient trials.[19] In these early days of gene therapy, the short-term benefits outweighed the known risks. Eventually, the NIH Human Gene Therapy Subcommittee for Clinical Trials approved the application.

The first approved gene therapy treatment was given to a four-year-old girl named Ashanti DeSilva, who was born with ADA-SCID.[20] Based on previous studies, in which the SCID mouse was created, scientists knew how to locate the mutant genes and replace them in order to restore the immune system. Anderson extracted white blood cells from the child and mixed them with recombinant retroviruses in a petri dish. Then the cell mixture was returned to the child's body. According to Kozubek, "DeSilva was not cured through the procedure, and *it only worked for a few months. But the treatment appeared safe*, and due this fact alone, it had an effect of boosting support for the fledgling field. The scientific consensus supported experimental gene therapy for children who'd virtually run out of luck. Anderson's team showed it could safely be done"[21] (italics added).

For a short time, the benefits of gene therapy were being recognized overseas, especially in western Europe. In fact, according to Kozubek, by the year 2000, 18 of 20 children in trials in the United States and western Europe had been successfully treated for X-SCID by introducing a new gene absent in their immune systems.[22] But then, unexpectedly, the bill became due, so to speak. Five of those patients developed leukemia-like conditions because the recombinant virus LTRs landed near the *LMO2* cell division gene and triggered the LTRs to switch on permanently, resulting in nonstop cell division (in this case an increase in the number of white blood cells).[23]

This concise history of viral vectors used in gene therapy mirrors the histories of other techniques in biotechnology in which short-term benefits and unknown risks result in the death of a patient. Researchers motivated by fame, funding, patents, prizes, and curing the sick — items mentioned frequently in this book — consciously or unconsciously ignore the warning signs in their research. They prefer to plow straight ahead into tragedy. They are willing to promote the short-term benefits over the notable risks to long-term health.

The pattern of behavior is the same for the newly emerging technology CRISPR/Cas9, although the long-term risks to human health in general and the human genome in particular are still unknown. Nevertheless, like recombinant DNA in the 1970s, CRISPR/Cas9 is a revolutionary technology that has serious implications for life on this planet. In her book, Doudna asked the following question: "If CRISPR and related technologies can eliminate inhumane practices like dehorning, reduce antibiotic usage, and protect livestock from deadly infections, can we afford not to use them?"[24] This is an interesting question, which is more or less addressed in different sections of this book as it relates to other technologies. Here this question will be explored as it relates to CRISPR/Cas9. Scientists are using CRISPR to generate, with great precision, animal models of human disease in animals (for example, monkeys, pigs, mice, ferrets, etc.) that best exhibit particular diseases of interest, such as obesity, SCID, Duchenne muscular dystrophy, Parkinson's disease, Alzheimer's disease, Huntington's disease, and cystic fibrosis.[25] Some of these diseases are not expressed in certain animals. In other words, when the normal gene malfunctions, no pathology occurs.[26] Therefore, scientists must be cognizant of gene expression, because identical genes in different animals are expressed differently. For example, the gene responsible for cystic fibrosis (*CFTR*, coding the for cystic fibrosis transmembrane conductance regulator, which affects a chloride ion channel in the lungs) is found in humans and mice.[27] Defects in the human *CFTR* gene cause serious damage to the lungs, but the same mutations in the mouse *CFTR* gene result in no lung damage.[28] Gene expression has serious implications for genetic editing.

In 2013 and 2014, gene-edited mice and monkeys were created by injecting CRISPR into one-cell embryos (see Chapter 6). However, it is the nonhuman primates, not the mice, that are irreplaceable as models for the study of the diseases listed in the previous paragraph. In fact, Chinese scientists programmed CRISPR in cynomolgus monkeys to simultaneously target genes associated with SCID and obesity in humans.[29] Certainly, this monkey model had significant benefits to human health. This work has led to other research using cynomolgus monkeys and rhesus macaques in which genes implicated in certain diseases (e.g., human cancers, Duchenne muscular dystrophy, and some neurological disorders) have been deliberately altered and corrected.[30]

Furthermore, the pig has reemerged as a popular animal model of human disease because of CRISPR.[31] Transgenic pigs (such as the Enviropig or the pig engineered to express a gene from a jellyfish) were created long before CRISPR. Their anatomical similarity to humans already made pigs a go-to animal model of human disease; with the advent of CRISPR they are even more attractive. Doudna wrote that gene-edited pigs have been used to model deafness, Parkinson's disease, and immunological disorders.[32] Pigs have also been genetically altered to be smaller for ease of handling in the laboratory.[33] CRISPR enabled scientists to replace pig genes with human gene counterparts in so-called humanized pigs so that pig organs might escape the array of immunological defenses in the human body. Doudna continued, "Gene editing is now being harnessed to shut down pig genes that might provoke the human immune response and to eliminate the risk that porcine viruses embedded in the pig genome could hop over and infect humans during transplantation."[34] With thousands of people currently on the waiting list for organ transplants, it is certainly beneficial that animals in general and pigs in particular can offer a source of whole organs for xenotransplantation. However, infection caused by animal diseases (zoonoses) as a result of this type of transplantation is a serious concern (see Chapter 6).

Doudna noted that as of 2015, the use of CRISPR/Cas9 for gene editing of staple crops such as rice, sorghum, wheat, soybeans, tomatoes, oranges, and corn had been successful.[35] She discussed barley and the Mildew Locus O gene (*Mlo*), which is linked to a harmful fungus that causes the disease known as powdery mildew.[36] In the middle of the 20th century, it was found that exposing seeds to radiation produced powdery-mildew-resistant barley. Recently, it was learned that identification and disruption of the *Mlo* gene in other plants also resulted in resistance to powdery mildew. For example, scientists at the Chinese Academy of Sciences used CRISPR to edit copies of the *Mlo* gene in bread wheat (*Triticum aestivum*), which created powdery-mildew-resistant bread wheat without any undesired effects.[37] Doudna expressed the new (and potential) benefits of CRISPR to human health, environment, and safety:

> CRISPR has been harnessed to edit genes in rice that confer protection against bacterial blight; to endow corn, soybeans, and potatoes with natural resistance to herbicides; and to produce mushrooms that are impervious to browning and premature spoiling. Scientists have used CRISPR to edit the genome of sweet oranges, and a team of California researchers is now attempting to apply the technology to save the U.S. citrus industry from a bacterial plant disease called huanglongbing — a Chinese name that translates as "yellow dragon disease" — that has devastated parts of Asia and now threatens orchards in Florida, Texas, and California. In South Korea, scientist Jin-Soo Kim and his colleagues hope gene editing in bananas can help save the Prized Cavendish variety from extinction, an outcome threatened by the spread of a devastating soil fungus. And elsewhere, researchers are even toying with the possibility of inserting the entire bacterial CRISPR system, reprogrammed to slice up plant viruses, into crops, providing them with a completely new antiviral immune system.[38]

Civilization cannot survive without agriculture; therefore, no one can doubt that these genetic modifications to food crops are beneficial. In addition, this technology can make foods healthier by reducing the high levels of trans fat and sugars. According to Doudna, soybean genes have been edited to generate seeds with lower levels of fatty acids, and the gene encoding glucose and fructose sugars has been knocked out in Ranger Russet potatoes.[39] There were no unintended mutations in either case. Do not forget CRISPR gene drives, discussed in Chapters 1 and 14, which spread genes for resistance to *P. falciparum* in several generations of wild mosquitoes so that the mosquitoes were unable to transmit the malaria parasite. These drives could be applied to other mosquito species that carry infectious diseases such as dengue virus, West Nile virus, yellow fever virus, Chikungunya virus, Zika virus, Keystone virus, Ross River fever, Barmah Forest fever, Rift Valley fever, filariasis, tularemia, dirofilariasis, and the various categories of encephalitis that cause death in millions of people in the tropical regions of the world.[40] In short, "CRISPR-based gene drives might be the best weapon we have against this pervasive threat, whether we use them to prevent mosquitoes from harboring specific pathogens or to wipe out the insects altogether."[41]

Furthermore, animal model studies and initial clinical trials using the CRISPR technology to fix Mendelian or monogenic genetic diseases are showing promising results.[42] In most of the clinical trials and laboratory experiments, scientists are using cultured human somatic cells derived from the patient's tissue samples to repair or knock out a particular gene using CRISPR ex vivo, and they are then returning these cells back into the patient.[43]

The first clinical trial to demonstrate this was conducted in 2008 by a team of physicians and researchers led by Pablo Tebas and David Stein, from the Perelman School of Medicine's Division of Infectious Diseases at the University of Pennsylvania and Albert Einstein College of Medicine in the Bronx, who used the first gene-editing technology, ZFN, donated by Sangamo Biosciences.[44] The patients involved in this clinical trial were all infected with HIV. The CCR5 receptor site allows HIV to attach to the T cell, leading to infection.[45] What if there is a mutation at this receptor site? Then HIV is denied an entry point, resulting in no infection. In fact, there is a mutant form of the CCR5, recognized as CCR5Δ32, with 32-base-pair deletion that changes the biochemical properties of the receptor site.[46] Individuals homozygous for this mutation cannot contract HIV/AIDS. Interestingly, the CCR5Δ32 mutation is at 14 percent frequency only in Europe, and coalescence analysis (showing how two copies of an allele share a common ancestor in the past) indicates that the CCR5Δ32 mutation originated in approximately 1300.[47] The selective pressure that caused the increase in frequency of this mutation was smallpox (rather than the competing epidemic disease during the fourteenth century, bubonic plague), because the smallpox virus was a chronic problem for more than 2,000 years.[48] Over time, an exaptation occurred: the CCR5Δ32 mutation initially provided resistance to fourteenth-century Europeans in an environment of smallpox, and then over time, as smallpox was slowly eradicated, the mutation provided resistance to HIV.[49] This mutation (and historical background) was the primary rationale for conducting a clinical trial using ZFN to knock out the gene for CCR5.

The research team enrolled twelve chronic aviremic HIV-infected patients who were receiving highly active antiretroviral therapy in an open-label, nonrandomized, uncontrolled study after injection with Sangamo Biosciences-728-T (SB-728-T), which consisted of autologous (meaning the individual is both donor and recipient) CD4-enriched T cells that had been modified at the *CCR5* gene locus by ZFNs.[50] The ZFN-modified T cells (*CCR5*-modified cells) were manufactured at the University of Pennsylvania. The primary objective, according to the research team, was to assess whether a single dose of the ZFN-modified T cells would cause side effects raising safety concerns.[51] The secondary objective was to assess changes in CD4 T cell count and persistence of ZFN-modified cells, which indicated that HIV could not bind to them.[52]

The results of the clinical trial using ex vivo gene editing therapeutically were encouraging. According to the research team, ZFN-modified cells were generally accepted and persisted after adoptive transfer resulting in increased numbers of CD4 T cells and slower increase in HIV levels than when antiviral therapy was temporarily interrupted.[53] And the longest delay in HIV resurgence occurred in patients who were heterozygote for the *CCR5* mutation. Based on this, the scientists noted that the relative survival advantage of ZFN-modified T cells during treatment interruption "suggests that genome editing at the *CCR5* locus confers a selective advantage to CD4 T-cells in patients infected with HIV."[54] Additionally, they believed that there was a strong correlation between heterozygosity and control of viral load. It follows that continuing clinical application of ZFN-modified CD4 T cells will depend on inactivation of both copies of the *CCR5* gene in homozygote individuals.

But CRISPR has burst upon the scene and has, in most cases, overshadowed ZFNs.[55] According to Doudna, CRISPR will have a greater role in generating therapies aimed at eliminating HIV by editing the virus's infectious genome.[56] Furthermore, as discussed elsewhere in the book, doctors can use CRISPR to repair mutated β-globin genes in stem cells isolated from a patient's bone marrow. Finally, the mutated gene responsible for Duchenne muscular dystrophy may potentially be repaired in the future, based on in vivo studies in mice. Doudna wrote that four independent laboratories have packaged genetic instructions for CRISPR into adeno-assisted virus vectors (which provoke a mild immune response, unlike retroviruses and lentiviruses) and injected them directly into the muscles of adult mice suffering from Duchenne muscular dystrophy.[57] The results were positive: healthy dystrophin genes were generated, resulting in increased muscle strength in the mice.

But what are the long-term risks of all this genetic editing? A frequent argument from scientists is that they are doing the same alterations nature does but at a faster rate. There must be a reason why nature takes its time, despite many mistakes. The changing environment must be considered, and as scientists correct some of nature's so-called mistakes, they must be cognizant of the long-term impact of what they are doing on the ecology; organisms do not live in isolation. For instance, gene drives used on disease-carrying insects or other animals or plants might result in their eventual extinction, particularly gene drives that spread genes for female sterility.[58] And the drives might impact other organisms in the ecosystem. Is this risk acceptable? In her book, Doudna quoted a statement from an unnamed entomologist: "If we eradicated them [mosquitoes] tomorrow, the ecosystems where they are active will *hiccup* and then get on with life"[59] (italics added). The problem here is that the intensity and long-term effects of this "hiccup" on organisms in the ecosystem, including humans, is really unknown. It is instructive to note that during the agricultural and industrial revolutions, humans modified their environment to suit their purposes. Over time, the modified environment impacted humans' own biology, through infectious and noninfectious diseases, for example.

And what about the ZFN-modified autologous CD4 T cells? Will there be a long-term negative impact on individuals? It seems that homozygotes for the mutant gene for CCR5Δ32 have an increased susceptibility to fatality from West Nile virus.[60] Additionally, the long-term persistence of the ZFN-modified T cells could have been because of the compromised immune system in HIV-infected individuals (no immune cells to remove the modified cells) and nothing else. The scientists involved, cognizant of the compromised immunity scenario, made the important point that previous studies had shown that even in the late stages of HIV infection, gene-modified T cells engineered with immunogenic viral vectors were eliminated by immune cells.[61] But once again, how great must the benefit be before the risk is acceptable? To use an example, CRISPR has the problem of off-target cleavage activity or edits at unintended sites not matching the guide RNA (see Chapter 6). Doudna wrote:

> With gene editing … any off-target DNA sequence, once edited, is irreversibly changed. Not only will unintended edits, to the DNA be permanent, they will also be copied into every cell that descends from the first one. And although most random edits are unlikely to damage the cell, *if we've learned anything from certain diseases and cancers, it's that even a single mutation can be enough to wreak havoc on an organism.*[62] (italics added)

Since off-target edits by CRISPR and other gene-editing technologies are fairly predictable (i.e., they only affect the DNA sequences that are most similar but not identical to the matching RNA guide), is the risk of off-target mutations acceptable? The italicized part of Doudna's statement above deserves repetition: "If we've learned anything from certain diseases and cancers, it's that even a single mutation can be enough to wreak havoc on an organism." It seems that we, as members of the global community, have much thinking to do about this and other biotechnologies.

## 11.1. Therapeutic Cloning and Genetic Editing via CRISPR: Benefits and Risks?

Sexual reproduction, from a genetic and evolutionary perspective, transmits the father's and mother's genetic history to the offspring, and it also creates variation each generation. What if a human child could be conceived using the technique of SCNT (used to create Dolly the sheep), in which somatic cells are taken from an adult male, an egg is taken from an adult female, and these donor cells are fused together (after the egg's nucleus is removed). The resulting child would be a clone of the donor male — not a shared descendant of both father and the mother. Although the mother would only contribute her mitochondrial DNA, fatherhood becomes more *artificial*. The definition of fatherhood is sperm, and sperm is cheap. In therapeutic or reproductive cloning, the value of sperm is further reduced: it becomes expendable. Is this risk bearable?

First, what are the benefits of therapeutic cloning? Can nuclear transplantation of a mutant genome, for example that of a cancer cell, into an enucleated egg give rise to a normal cell? In other words, would the subsequent developing embryo be cancer-free? The question of whether a malignant phenotype could be reversed was posed in King's lab in the late 1950s.[63] In 1961, King and McKinnell reported that they had infected adult frogs of the species *R. pipiens* with herpesvirus Lucké renal carcinoma.[64] Subsequently, the nucleus of a cancerous cell was transferred to an enucleated egg. Initially, King and McKinnell needed to make sure that the tumor (carcinoma) developed from the inserted cancer nucleus and not from an accidentally retained egg nucleus or from stromal cells present in the tumor. The results showed that cancer cell nuclei could induce partial and complete cleavage in enucleated eggs, and some blastulae developed into abnormal embryos and larvae.

The researchers grafted tissue fragments of tumor nuclear transplants to normal hosts of a different ploidy to determine whether the tumor genome could direct advanced tissue differentiation.[65] After 40 days, there was no evidence of carcinoma: well-differentiated tissue of all three germ layers, similar to the normal tissue of controls, was observed during histological examination.[66] Certainly, the reprogramming of Lucké cancer cells or any other mutant cell types in the relatively new field of differentiation therapy would provide a benefit and would be preferable to the nasty side effects of chemotherapy.[67] Will such tissue remain benign? And will there be any heritable effects? There are no definitive answers to these questions.

In the 1990s, CAR T cell trials (see Chapter 2) involved the addition of chimeric genes to T cells to focus them better on finding and destroying cancer cells. The benefits were great in the short-term, but the risks were also great. Kozubek recalled the events:

> Doctors saw it wipe out 3.5 to 7.7 pounds of cancer in a matter of days among three different patients. The deployment of CAR T-cells can also stimulate a cascade of "adverse events" including "cytokine release

> syndrome," which is an effect of activating T-cells en masse, in effect, sending the immune system into overdrive with cascades upon cascades of cytokines being released, swirling up fevers, a collapse of blood pressure, very rarely, death. Not only that, but patients can be at risk for "tumor lysis syndrome," which is a constellation of metabolic disturbances that can occur when a glut of cancer cells are decimated all at once, spilling out all the breakdown products of dying cells, leading to skyrocketing potassium and phosphorous levels, and plummeting calcium levels, ions, or electrically charged molecules that maintain cellular homeostasis, and their disruption can lead to organ failure.[68]

The FDA stopped the Perelman School of Medicine CAR T cell trials in 2000 after a young patient died as a result of a powerful immune reaction.[69]

In similar cancer-research activities, Normile reported on CRISPR therapy used by Chinese scientists to treat cancers of the stomach, lung, and esophagus.[70] He wrote about how they extracted T cells from patients and use CRISPR to knock out the gene that encoded programmed cell death protein 1 (PD-1). Programmed cell death protein 1 prevents the immune system from overreacting to viral infections of tumors when it induces T cells to die.[71] The scientists injected the modified cells back into the patient's body to attack the cancer. Cancers can trigger PD-1 activity to protect themselves from the immune response.[72] China's CRISPR therapy is also being applied to HIV-infected individuals and individuals with beta thalassemia. There is great excitement among these scientists about treating the sick. Yet, they realize — like American scientists — that the routine use of CRISPR to treat disease is years away, and they must weigh the benefits and risks regardless of their excitement.

In an alternative strategy, CRISPR could be used to stop cancer before it happens by editing the DNA mutations before they become cancerous.[73] This would involve selecting the correctly mutated cells, which would be difficult because cancer-causing mutations are intermixed with other mutations that are not related to the disease pathology: "In fact, one of the hallmarks of cancer is the increased rate at which DNA mutations creep into the genome, making it difficult to identify the mutations that are actually playing the largest role in causing tumors."[74] Scientists can then tweak this methodology: instead of repairing the mutant cells, akin to a treatment for cancer, CRISPR can be used as a support system for present therapy by introducing mutations so that scientists can understand the genetic factors that cause cells to continue to divide regardless of regulatory signals to stop.

Doudna briefly mentioned an ongoing experiment in which eight candidate genes that might be associated with myeloid leukemia were knocked out in the stem cells of mice.[75] The cells were then injected back into the bloodstream to see if the mice came down with acute leukemia. In the end, scientists have to weigh the potential benefits (e.g., reduction in the growth of the tumor) against the risks of causing illness (e.g., off-target edits that can mutate and turn on oncogenes).

## 11.2. Resurrecting Extinct Animal Species Using Cloning: A Good Idea?

A benefit of cloning, that excited ecologists, is its use for "the expansion of critically endangered [animal] populations."[76] The Species Survival Commission of the International Union for Conservation of Nature (IUCN), in its *2008 Red List,* listed troubling findings concerning the state of mammalian species:

> Nearly one-quarter (22% [1,207/5,488]) of the world's mammal species are known to be globally threatened or extinct, 63% are known to not be threatened, and 15% have insufficient data to determine their threat status. There are 76 mammals considered to have gone extinct since 1500, and two are extinct in the wild. The most diverse country for mammals is Indonesia (670), followed closely by Brazil (648). China (551) and Mexico (523) are the only other two countries with more than 500 species. The country with by far the most threatened species is Indonesia (184). Mexico is the only other country in triple figures with 100 threatened species. Half of the top 20 countries for numbers of threatened species are in Asia; for example, India (96), China (74) and Malaysia (70). However, the highest levels of threat are found in island nations, and in particular the top three are islands or island groups in the Indian Ocean: Mauritius (64%), Réunion (43%) and the Seychelles (39%).

> Habitat loss, affecting over 2,000 mammal species, is the greatest threat globally. The second greatest threat is utilization which is affecting over 900 mammal species, mainly in Asia.[77]

Most scientists would agree that humans have had an enormous impact, in terms of pollution, deforestation, and hunting, on Earth's biotic and abiotic resources. Since all species are connected ecologically, extinction reverberates throughout the ecosystem. The magnitude of this can been seen in these *2008 Red List* findings.

Most biologists, ecologists, and wildlife managers are working tirelessly to preserve Earth's biodiversity. Traditional conservation strategies would be to declare several acres of land protected under an environmental law or to remove a whole animal population to a new geographical region. Some conservation biologists, however, have become interested in using reproductive biotechnologies to improve the reproductive performance of endangered animals.[78] When Dolly the sheep was revealed to the world, conservation biologists believed that cloning technology could be used to increase the population size of endangered animals or even restore extinct species.[79] In collaboration with geneticists, they began establishing DNA banks for mammals listed as endangered by the IUCN.

Pasqualino Loi of the Department of Animal Pathology and Biotechnology at the University of Teramo in Italy and colleagues argued that endangered wild animals could be salvaged by using the eggs and wombs of their domestic counterparts.[80] They allayed fears of reduced genetic variability within species by arguing that cell lines could be collected when population numbers are high, so that the original level of diversity would be present in the cloned animals.[81] Loi and colleagues admitted that attempts to clone endangered mammals such as the argali (a Central Asian wild sheep) and gaur (a Southeast Asian wild ox) resulted in minimal success. In fact, the gaur clone died of dysentery two days after birth.[82]

The worrisome problem with using cloning technology to increase the population size of endangered wild animals and restore extinct species is the unknown impact of cloning on the gene pool within the species and biodiversity more generally. Another equally worrisome, even frightening, problem is single-minded, well-meaning individuals motivated by flawed logic using wealth and genetic engineering to support research into the resurrection of prehistoric extinct animal species, such as the saber tooth tiger or the pterodactyl. This would be an abuse or misuse of science with devastating consequences to the ecosystem that humans inhabit. Unfortunately, there are men (and women) like the fictional industrialist John Hammond from the 1990 film *Jurassic Park*, who was a good-natured, likeable conservationist naive enough to think that he could control the reproduction and behavior of dinosaurs restored, via genetic engineering, to an island sanctuary. Although *Jurassic Park* was fantasy, it did provide a glimpse of the risk of runaway science, shadows of which appear in today's genetic engineering.

CRISPR, unfortunately, provides a means for de-extinction: resurrecting extinct contemporary, historic, and prehistoric animals.[83] Researchers in Church's laboratory have applied CRISPR to convert elephant gene variants to woolly mammoth gene variants for 14 genes (out of the 1,668 genes that differ between the two genomes).[84] What are the benefits of resurrecting the woolly mammoth? Would they (or other species) pay the price for disruption of the ecosystem, similar to the disruption caused by the introduction of invasive species? Would they pay an even higher price in terms of being killed by big game hunters whose only interest is to demonstrate their skill in hunting and killing a woolly mammoth? If the answer is yes, then the purpose in resurrecting them is defeated from the start. The scenario in which Paleo-Indians hunt the woolly mammoth to extinction, then the woolly mammoth is resurrected by contemporary humans, then contemporary humans hunt the woolly mammoth to extinction again amounts to a vicious cycle.

## 11.3. Induced Pluripotent Stem Cells Promise and Their Propensity to Form Tumors

Human iPSC derivation was one of the most significant advances in genetic engineering and was based on a foundation of earlier work by scientists trying to elucidate SCNT.[85] These iPSCs have tremendous potential benefits for medicine. A primary example is solving the problem of immune reaction. The potential to produce pluripotent cells from adult skin cells would be very beneficial for skin grafts, for example. Related to this is an issue that provokes the consciousness of many Americans: abortion. Deriving pluripotent cells from adult cells eliminates the ethical problem of terminating a preimplantation embryo to harvest the stem cells. Another medical benefit of iPSCs is the creation of human organs for transplantation, particularly the use of pig or sheep embryos in an effort to grow human organs to alleviate the shortage of donated organs.[86] But the risks could be a possible increase in new zoonotic diseases, heritable mutations, changes in animal behavior and appearance, and, philosophically, sacrificing the essence of what makes humans human.

Nevertheless, in 2013, Japanese biologist Takanori Takebe and colleagues at the Department of Regenerative Medicine in the Yokohama City reported growing human liver buds from a mixture of three different kinds of stem cells: hepatocytes (for liver function) from iPSCs; endothelial stem cells (to form blood vessel lining) from umbilical cord; and mesenchymal stem cells (to form connective tissue).[87] After growing in vitro for several days, Takebe and colleagues transplanted the liver buds into mice, where the buds connected to blood vessels and continued to grow. Subsequent tests indicated that the liver was functioning normally, producing liver-specific proteins and metabolizing drugs.

Biologist Tea S. Park and a team of collaborators from the Johns Hopkins School of Medicine, the University of Maryland School of Medicine, the Stanford University Institute for Stem Cell Biology and Stanford University's Department of Cardiovascular Medicine induced embryonic cord blood cells to form pluripotent stem cells.[88] For these cells to be useful in vascular tissue repair, the scientists needed to identify multipotent vascular stem cells. They accomplished this by using cell surface endothelial markers $CD_{31}$ and $CD_{146}$.[89] When the iPSCs were injected directly into the damaged retinas of mice, they grew and repaired the vascular vessels.[90]

The ability to study the cellular basis of disease, and subsequent drug development, is another attractive benefit of human iPSCs.[91] The iPSCs are not only pluripotent but also self-renewing; consequently, they potentially represent an unlimited supply of patient-derived cells. Other human-derived cells have limited uses in the laboratory (i.e., they stop growing after a short time), which makes iPSCs the gold standard in in vitro cellular research.[92] For example, iPSCs have been generated for a wide variety of Mendelian and complex inheritance diseases such as Down syndrome or trisomy 21, polycystic kidney disease, ADA-SCID, Shwachman-Bodian-Diamond syndrome, Gaucher disease Type III, Duchenne muscular dystrophy, Becker muscular dystrophy, Parkinson's disease, Huntington's disease, juvenile-onset type 1 diabetes, and Lesch-Nyhan syndrome.[93] Since the patient-derived iPSCs are defective (compared to normal iPSCs), scientists can study the physiology of the disease. In 2012, for instance, 10 pharmaceutical companies and 23 universities pooled funds and resources in an international collaboration project called StemBANCC, managed by Oxford University, to build a collection of iPSC lines for drug screening for a variety of diseases.[94] By 2017, StemBANCC generated a library of 1,500 iPSC lines from 500 people to be used to simulate a human disease environment — generating human organs — for drug testing and toxicity. The raw materials for the project were skin and blood samples taken from healthy and unhealthy patients, such as those with disorders of the peripheral nervous system and central nervous system, those with diabetes, and those displaying adverse drug reactions.[95] But what are the risks of iPSCs? Two of the genes introduced to reprogram the nuclei of cells are oncogenes (see Chapter 2) that might cause cancer: unrestrained cell growth and division.

Recall the following question, posed frequently throughout this book: does scientific advancement outweigh the risk of alterations to human biology? Chapter 2 mentions that iPSCs have the propensity to form tumors far more frequently than ES cells, which should be a serious concern to researchers. In 2011, for example, Satoshi Nori, in a collaborative effort with colleagues from the orthopedic, pathology, and cell biology departments at several Japanese universities, transplanted human iPSCs into mice after spinal cord injury.[96] Nori noted that the cells differentiated into three neural lineages in the spinal cord and stimulated regrowth of the damaged spinal cord. These positive outcomes were observed for approximately four months until a follow-up study found colonies of tumors that were previously iPSCs.[97] Since members of the c-Myc are oncogenes, it was good news when Yamanaka reported that iPSCs could be created without c-Myc.[98] While inefficiency in generating iPSCs dropped dramatically, no cancer in chimeric mice was reported. Scientists in general are motivated to help humankind and, logically, choose efficiency. But will they trade public safety (i.e., avoidance of tumor generation) for efficiency? Inactivation of the tumor suppressor p53 significantly increased nuclear reprogramming efficiency (but generated tumors).[99] Additionally, efficiency was reduced using plasmids, adenoviruses, and transposon vectors instead of retroviruses. As stated earlier in this chapter, the insertion of retroviruses into a target cell's genome could have disastrous effects. Certainly, scientists are working tirelessly to find methods that reduce tumorigenicity. Moreover, in the biotechnological race to unravel the mysteries of the cell, scientists merge scientific advance with public safety, but scientific advance dominates because scientists view it as a way to protect the public. Is this view not similar, to use an analogy, to a dictatorship that controls the content of radio, television, the internet, and newspapers to "protect" the people, when in fact the dictatorship is a danger to the people?

## 11.4. Assisted Reproductive Technology and the Long-term Dangers of Sex Selection

During the Pew Research Center and Brookings Institute cosponsored event the Pursuit of Perfection discussed elsewhere in the book, Sandel discussed the MicroSort technology.[100] This technology, originally developed for cattle breeding, was used in a clinic in Fairfax, Virginia, in 2004 to allow parents to select the sex of their children in advance. The machine sorts Y-bearing sperm that would produce a boy from X-bearing sperm that would produce a girl. The success rate of the machine in sorting was 81 percent for girls and 76 percent for boys. While the technology is new, according to Hvistendahl, the practice of sex selection of children goes back to the 18th century in Asia.[101] She speculated that female infanticide might have been linked to distribution of wealth (so that financially strapped fathers would not have to pay bride price or dowry to the bridegroom of their daughter), and over time, the practice increased thanks to British colonialism and the imposition of taxes on the locals. "One way to safeguard their wealth," noted Hvistendahl, "was to not have daughters in the first place."[102] In more modern times, sex-selective abortion has occurred in the name of population control. For example, in 1975, poor men in India underwent forced sterilization on a scale rivaling early-20th-century eugenics in Germany.[103] The eugenic aspect of this is the preference for males over females because of the establishment of male domination in prehistory and, over time, the subsequent erroneous enculturation of boys and girls into believing in the greater power of males with respect to intelligence, strength, and control over females. Women across China, India, and South Korea are under great pressure to give birth to sons.[104] Hvistendahl quoted Liao Li, one of her informants in China who aborted two female fetuses: "Girls are very good. ... They're soft. And they can take care of you when you're older. ... If you don't have a boy, you lose face."[105]

Amniocentesis and ultrasound have propelled sex-selective abortion into the realm of the new eugenics, which is quite alarming from an evolutionary perspective. For instance, the sex ratio for children under five years of age in China is 163 boys for every 100 girls, and in India 126 boys for every 100 girls (the natural sex ratio, according to demographers, is 105 boys for every 100 girls).[106] In evolutionary biology, an uneven ratio of males to females in a species is not good for its long-term survival. Hvistendahl made an important point about unknown evolutionary effects and tinkering with poorly understood biological mechanisms: "We still don't know the evolutionary effects of fundamentally altering the sex ratio at birth, but a cursory glance back at history suggests *it is not a great idea to mess with something we don't understand*"[107] (italics added).

Sex selection is one of the benefits of Greely's futuristic Easy PGD. Specifically, he stated that there would be no surprises: parents would get the children they want.[108] Easy PGD would be a benefit to parents if they could select their children's traits (e.g., female, good athlete, scientifically inclined, tall, etc.). More importantly, parents could "weed out" disabling and fatal genetic conditions. Greely discussed this potential benefit: "If we set that amount at 2 percent of current births and assume that easy PGD prevents even only half of those births, that would 'prevent' the suffering of about 40,000 children — and their parents and other family members — each year in the United States."[109]

No doubt the benefits of these scientific breakthroughs are tremendous. It is unrealistic to believe that anyone can slow or even control the momentum of scientific advancement, because the products of science belong to many researchers (not just one), past and present, working independently (and at times collaboratively) to solve nature's puzzles. What each scientist can do is take a deep breath and proceed cautiously, giving weight not only to short-term successes, the shiny objects, so to speak, but also to the long-term changes for the species. What happens today, however, is that scientists give more weight to the short-term benefits and are more than willing to accept these benefits *at any cost*. For example, during the gene therapy clinical trials in the 1990s, scientists were ready to continue despite the deaths of Jesse Gelsinger and other patients in the United States and Europe.[110] Kozubek quoted Claudia Mickelson, then chair of NIH's Recombinant DNA Advisory Committee, responding to CNN after Gelsinger's death: "We have no interest in stopping gene therapy. None whatsoever."[111] This attitude of trying to repair all defective genes in humans at any cost imparts momentum to the new eugenics. And this attitude currently dominates the field of biotechnology.

## Notes

1. Luis Lugo, "The Pursuit of Perfection: A Conversation on the Ethics of Genetic Engineering," March 31, 2004, 1, http://www.pewforum.org/2004/03/31/the-pursuit-of-perfection-a-conversation-on-the-ethics-of-genetic-engineering.
2. Maggie Fox, "Human-Cow Hybrid Embryos Made in Lab," *Australian Broadcasting Corporation*, April 3, 2008, http://www.abc.net.au/science/articles/2008/04/03/2206835.htm.
3. Ibid.
4. Jim Kozubek, *Modern Prometheus: Editing the Human Genome with CRISPR-CAS9* (New York, NY: Cambridge University Press, 2016), 105–113.
5. Ibid., 135.
6. Ibid., 135–136.
7. Ibid., 136–137.
8. Ibid., 138.
9. Ibid., 138–142.
10. Ibid., 139–140.
11. Ibid., 141.

12. Ibid.
13. Ibid.
14. Ibid.
15. Ibid
16. Ibid., 144.
17. Ibid., 43, 154–155.
18. Ibid.
19. Ibid., 155.
20. Ibid., 151.
21. Ibid.
22. Ibid., 152.
23. Ibid., 153.
24. Jennifer A. Doudna and Samuel H. Sternberg, *A Crack in Creation: Gene Editing and the Unthinkable Power to Control Evolution* (New York: Houghton Mifflin Harcourt, 2017), 137.
25. Peter Raven, George B. Johnson, Kenneth A. Mason, Jonathan B. Losos, Susan R. Singer, *Biology*, 9th ed. (New York: McGraw-Hill, 2008), 249, 345; Doudna and Sternberg, *A Crack in Creation*, 138.
26. Doudna and Sternberg, *A Crack in Creation*, 138; Raven *et al.*, *Biology*, 484.
27. Raven *et al.*, *Biology*, 484.
28. Ibid.
29. Yuyu Niu, Bin Shen, Yiaqiang Cui, Yongchang Chen, Jianying Wang, Lei Wang, Yu Kang *et al.*, "Generation of Gene-modified Cynomolgus Monkey via Cas 9/RNA-Mediated Gene Targeting in One-cell Embryos," *Cell* 156, no. 4 (February 13, 2014): 836–843; Doudna and Sternberg, *A Crack in Creation*, 138–139.
30. Doudna and Sternberg, *A Crack in Creation*, 139.
31. Ibid.
32. Ibid.
33. Ibid., 142.
34. Ibid., 141.
35. Ibid., 99.
36. Ibid., 120.
37. Ibid., 120–121.
38. Ibid., 122.
39. Ibid., 123.
40. Ibid., 152.

41. Ibid.
42. Ibid., 156–157, 161.
43. Ibid.
44. Pablo Tebas, David Stein, Winson W. Tang, Ian Frank, Shelley Q. Wang, Gary Lee, S. Kaye Spratt *et al*, "Gene Editing of *CCR5* in Autologous CD4 T Cells of Persons Infected with HIV," *New England Journal of Medicine* 370, no. 10 (March 6, 2014): 903; Doudna and Sternberg, *A Crack in Creation*, 164.
45. Ibid., 902; J. C. Stephens, D. E. Reich, D. B. Goldstein, H. D. Shin, M. W. Smith, M. Carrington, C. Winkler *et al.*, "Dating the Origin of the CCR5-Δ32 AIDS-Resistance Allele by the Coalescence Haplotypes," *American Journal of Human Genetics* 62, no. 6 (June 1998): 1507; Alison P. Galvani and Montgomery W. Slatkin, "Evaluating Plague and Smallpox as Historical Selective Pressures for the CCR5-Δ32 HIV-Resistance Allele," *Proceedings of the National Academy of Sciences USA* 100, no. 25 (December 9, 2003): 15276.
46. Stephens *et al.*, "Dating the Origin of the CCR5-Δ32 AIDS-Resistance Allele," 1507; Tebas *et al.*, "Gene Editing of *CCR5* in Autologous CD4 T Cells," 902.
47. Stephens *et al.*, "Dating the Origin of the CCR5-Δ32 AIDS-Resistance Allele," 1507–1515; Galvani and Slatkin, "Evaluating Plague and Smallpox," 15276–15279.
48. Galvani and Slatkin, "Evaluating Plague And Smallpox," 15277–15278.
49. Ibid., 15278.
50. Tebas *et al.*, "Gene Editing of *CCR5* in Autologous CD4 T Cells," 903.
51. Ibid.
52. Ibid.
53. Ibid., 906–907.
54. Ibid., 906.
55. Doudna and Sternberg, *A Crack in Creation*, 33, 166.
56. Ibid., 166.
57. Ibid., 169–170.
58. Ibid., 149.
59. Ibid., 153.
60. Ibid., 165; Paul Knoepfler, *GMO Sapiens: The Life-Changing Science of Designer Babies* (Hackensack, NJ: World Scientific), 155–156.
61. Tebas *et al.*, "Gene Editing of *CCR5* in Autologous CD4 T Cells," 907.
62. Ibid., 179.

63. Robert G. McKinnell and Marie A. DI Berardino, "The Biology of Cloning: History and Rationale," *BioScience* 49, no. 11 (November 1999): 879; Thomas J. King and Robert G. McKinnell, "An Attempt to Determine the Developmental Potentialities of the Cancer Cell Nucleus by Means of Transplantation," in R.W. Cumley and J. McCoy, eds., *Cell Physiology of Neoplasia* (Austin: University of Texas Press 1961): 591.
64. King and McKinnell, "An Attempt to Determine the Developmental Potentialities," 591.
65. Ibid.
66. Ibid.; McKinnell and DI Berardino, "The Biology of Cloning," 879.
67. McKinnell and DI Berardino, "The Biology of Cloning," 879.
68. Jim Kozubek, *Modern Prometheus: Editing the Human Genome with CRISPR-CAS9* (New York, NY: Cambridge University Press, 2016), 244–245.
69. Kozubek, *Modern Prometheus*, 43; 154–156.
70. Dennis Normile, "China Sprints Ahead in CRISPR Therapy Race: Human Trials Are Using the Genome-Editing Technique to Treat Cancers and Other Conditions," *Science* 358, no. 6359 (October 6, 2017): 20.
71. Ibid.
72. Ibid.
73. Ibid., 21.
74. Doudna and Sternberg, *A Crack in Creation*, 172.
75. Ibid., 173.
76. Pasqualino Loi, Grazyna Ptak, Barbara Barboni, Josef Fulka Jr., Pietro Cappai, and Michael Clinton, "Genetic Rescue of an Endangered Mammal by Cross-Species Nuclear Transfer Using Post-Mortem Somatic Cells," *Nature Biotechnology* 19, no. 10 (October 2001): 962.
77. International Union for Conservation of Nature, IUCN Red List of Threatened Species, version 2018.1, accessed July 21, 2018, http://www.iucnredlist.org.
78. Deshun Shi, Fenghua Lu, Yingming Wei, Kuiqing Cui, Sufang Yang, Jingwei Wei, and Qingyou Liu, "Buffalos (*Bubalus bubalis*) Cloned by Nuclear Transfer of Somatic Cells," *Biology of Reproduction* 77, no. 2 (May 2007): 285–291; Ziyi Li, Xingshen Sun, Juan Chen, Xiaoming Liu, Samantha M. Wisely, Qi Zhou, Jean-Paul Renard *et al.*, "Cloned Ferrets Produced by Somatic Cell Nuclear Transfer," *Developmental Biology* 293, no. 2 (May 15, 2006): 439–448; Martha C. Gómez, C. Earle Pope, Angelica Giraldo, Leslie A. Lyons, Rebecca F. Harris, Amy L. King, Alex Cole *et al.*, "Birth of African Wildcat Cloned Kittens Born from Domestic Cats,"

*Cloning and Stem Cells* 6, no. 3 (2004): 247–257; Min Kyu Kim, Goo Jang, Hyun Ju Oh, Fibrianto Yuda, Hye Jin Kim, Woo Suk Hwang, Mohammad Shamin Hossein *et al.*, "Endangered Wolves Cloned from Adult Somatic Cells," *Cloning and Stem Cells* 9, no. 1 (Spring 2007): 130–137; Pasqualino Loi, Grazyna Ptak, Barbara Barboni, Josef Fulka Jr., Pietro Cappai, and Michael Clinton, "Genetic Rescue of an Endangered Mammal by Cross-Species Nuclear Transfer Using Post-Mortem Somatic Cells," *Nature Biotechnology* 19, no. 10 (October 2001): 962–964.

79. Kim *et al.*, "Endangered Wolves Cloned," 130.
80. Loi *et al.*, "Genetic Rescue of an Endangered Mammal," 962.
81. Ibid., 962.
82. Gretchen Vogel, "Cloned Gaur Short-lived Success," *Science* 291, no. 5503 (January 19, 2001): 409.
83. Doudna and Sternberg, *A Crack in Creation*, 144.
84. Ibid., 145–147.
85. Nobel Media AB, "The Nobel Prize in Physiology or Medicine — 2012 Press Release," October 2012, https://www.nobleprize.org/nobel_prizes/medicine/laureates/2012/press.html. Shinya Yamanaka and Sir John Gurdon shared the 2012 Nobel Prize for the discovery that adult cells can be reprogrammed to be pluripotent. John Gurdon proposed reprogramming in his 1962 experiments deriving adult frogs from the nuclei of somatic cells and transplantation of nuclei between *Xenopus* frog species.
86. Susan Scutti, "First Human-Pig-Embryos Made, Then Destroyed," January 30, 2017, https://www.cnn.com/2017/01/26/health/human-pig-embryo/index.html; Colin Fernandez, "A Human-Pig Hybrid Embryo Has Been Created in a World First: Breakthrough Could Open Up the Possibility for 'Designer' Animal Organs to Be Used in People," January 26, 2017, https://www.dailymail.co.uk/sciencetech/article-4161022/Human-animal-hybrid-embryo-created-time.html.
87. Takanori Takebe, Keisuke Sekine, Masahiro Enomura, Hiroyuki Koike, Masaki Kimura, Takumori Ogaeri, Ran-Ran Zhang *et al.*, "Vascularized and Functional Human Liver from an iPSC-derived Organ Bud Transplant," *Nature* 499, no. 2 (July 25, 2013): 481.
88. Tea S. Park, Imran Bhutto, Ludovic Zimmerlin, Jeffrey S. Huo, Pratik Nagaria, Diana Miller, Abdul Jalil Rufaihah *et al.*, "Vascular Progenitors from Cord Blood-Derived Induced Pluripotent Stem Cells Possess Augmented Capacity for Regenerating Ischemic Retinal Vasculature," *Circulation* 129, no. 3 (October 25, 2013): 359.

89. Ibid., 359–361.
90. Ibid., 362–363.
91. M. Grskovic, A. Javaherian, B. Strulovici, G. Q. Daley, "Induced Pluripotent Stem Cells — Opportunities for Disease Modeling and Drug Discovery," *Nature Reviews, Drug Discovery* 10, no. 12 (November 11, 2011): 915; Junying Yu, Maxim A. Vodyanik, Kim Smuga-Otto, Jessica Antosiewicz-Bourget, Jennifer L. Frane, Shulan Tian, Jeff Nie *et al.*, "Induced Pluripotent Stem Cell Lines Derived from Human Somatic Cells," *Science* 318, no. 5858 (December 21, 2007): 1917.
92. Ivan Guiterrez-Aranda, Veronica Ramos-Mejia, Clara Bueno, Martin Munoz-Lopez, Pedro J. Real, Angela Mácia, Laura Sanchez *et al.*, "Human Induced Pluripotent Stem Cells Develop Tetratoma More Efficiently and Faster Than Human Embryonic Stem Cells Regardless the Site of Injection," *Stem Cells* 28, no. 9 (September 2010): 1568.
93. I. H. Park, N. Arora, H. Huo, N. Maherali, T. Ahfeldt, A. Shimamura, M. W. Lensch, C. Cowan, K. Hochedlinger, G. Q. Daley, "Disease-Specific Induced Pluripotent Stem Cells," *Cell* 134, no. 5 (September 5, 2008): 877–886.
94. Grskovic *et al.*, "Induced pluripotent stem cells — Opportunities for," 928.
95. Ibid., 915–929.
96. Satoshi Nori, Yohei Okada, Akimasa Yasuda, Osahiko Tsuji, Yuichiro Takahashi, Yoshiomi Kobayashi, Kanehiro Fujiyoshi *et al.*, "Grafted Human-Induced Pluripotent Stem Cell Derived Neurospheres Promote Motor Functional Recovery After Spinal Cord Injury in Mice," *Proceedings of National Academy of Sciences USA* 108, no. 40 (October 4, 2011): 16825.
97. Ibid., 16828.
98. Nikhil Swaminathan, "Stem Cells — This Time without Cancer," *Scientific American News* (November 30, 2007), http://www.sciam.com/article.cfm?id=stem-cells-without-cancer.
99. R. M. Mario, Katerina Strati, Han Li, Matilde Murga, Raquel Blanco, Sagrario Ortega, Oscar Fernandez-Capetillo, Manuel Serrano, Maria A. Blasco, "A p53-mediated DNA Damage Response Limits Reprogramming to Ensure IPS Cell Genomic Integrity," *Nature* 460, no. 7259 (August 27, 2009): 1149.
100. Verena Stolcke, "Homo Clonicus" in Philomena Essed and Gabriele Schwab, eds., *Clones, Fakes, and Posthumans* (Thamyris/Intersecting: Place, Sex, and Race) (Amsterdam: Rodopi Press, 2012), 26–27.

101. Michael Sandel, "The Pursuit of Perfection: A Conversation on the Ethics of Genetic Engineering," March 31, 2004, http://www.pewforum.org/2004/03/31/the-pursuit-of-perfection-a-conversation-on-the-ethics-of-genetic-engineering.
102. Mara Hvistendahl, *Unnatural Selection: Choosing Boys Over Girls, and the Consequences of a World Full of Men* (New York: PublicAffairs, 2011), 61–71.
103. Ibid., 69.
104. Ibid. On June 25, 1975, Prime Minister Indira Gandhi used a provision in the Indian constitution to suspend democracy and invoke marshal law or "Emergency Powers" in India. For over a year, Gandhi's government clamped down on dissent and civil liberties. According to Hvistendahl, this was "an especially bleak era for reproductive rights" in that health officials saw an opportunity "to enforce drastic measures on Indians who had previously resisted birth control." A massive campaign was instituted to sterilize poor men. "The task of overseeing the gruesome campaign fell to Indira's son Sanjay Gandhi, who held no official title … At first his mother's government rewarded men who consented to vasectomies. Before long, however, Sanjay Gandhi was issuing quotas so high that local officials could meet them only by dragging men to the operating room — typically a makeshift camp that had sprung up practically overnight (Nearly two thousand men died from botched operations). In some areas, police surrounded villages in the middle of the night and apprehended all the men. In others, they combined sterilization with slum clearance, razing whole neighborhoods and robbing men of both their reproductive ability and homes at the same time. … By the time democratic rule was restored, 6.2 million Indian men had been sterilized in just one year — fifteen times the number of people sterilized by the Nazis" (87–88).
105. Hvistendahl, *Unnatural Selection*, 19–28.
106. Ibid., 27.
107. Ibid., 4–5.
108. Ibid., xvi.
109. Henry T. Greely, *The End of Sex and the Future of Human Reproduction* (Cambridge, MA: Harvard University Press), 191–196, 204.
110. Ibid.
111. Jim Kozubek, *Modern Prometheus: Editing the Human Genome with CRISPR-CAS9* (New York, NY: Cambridge University Press, 2016), 151–156.

# Chapter 12

# Making New Mistakes? Bionanotechnology and Nanomedicine

Thus hath the candle singd the moth O, these deliberate fools, when they do choose, They have the wisdom by their wit to lose

— William Shakespeare, *The Merchant of Venice*, 1596[1]

Many Americans may not realize that engineered nanomaterials are widely used in consumer products such as cosmetics, paints, fabrics, and electronics.[2] In fact, by the mid-2000s, the nanotechnology field had garnered increased scientific, political, and commercial attention. For instance, more than 800 manufacturer-identified first-generation and second-generation nanotech products were available, such as titanium dioxide in sunscreen, zinc oxide in sunscreens and cosmetics, medical imaging and drug delivery, food products, carbon allotropes used to produce gecko tape, silver in food packaging, clothing, disinfectants and household appliances, surface coatings, paints and outdoor furniture varnishes, and cerium oxide as a fuel catalyst.[3] According to the members of the Project on Emerging Nanotechnologies, tennis balls last longer, golf balls fly straighter, and bowling balls become more durable when constructed with nanomaterials.[4] Furthermore, trousers and socks last longer, bandages infused with silver nanoparticles heal injuries faster, and computers contain more memory thanks to nanotechnology. In the early 2000s, the US National Nanotechnology Institute envisaged four generations of nanotechnology: the current era, which includes the second generation, is that of passive (designed for one task) and active nanostructures employed, for example, in drug-delivery devices, sensors, and medical imaging.[5] The third and fourth generations are currently in the experimental stage and involve nanosystems with thousands of interacting components and integrated nanosystems functioning like mammalian cellular systems.[6] The main focus of this chapter is nanotechnology applied directly (and to a lesser extent, indirectly) to biological organisms in order to enhance them or otherwise modify them for better or worse.

In the 1980s and 1990s, the first widespread description of nanotechnology emphasized the technological goal, which was to precisely manipulate atoms and molecules for generating macroscale products.[7] The US NNI subsequently established a generalized definition of nanotechnology that emphasized research: "the manipulation of matter at the atomic, molecular, or macromolecular levels on a length scale of about 1–100 nm (nanometer)."[8] Since many technologies in medicine and electronics, for example, deal with the special properties of matter within and below the given length scale, officials of the NNI believed that any research in nanotechnology must elucidate the physical and chemical properties of materials at the level of atomic and molecular clusters with the goal of manipulating these properties.[9] But the novel physiochemical properties of nanomaterials raised concerns about their adverse effects on biological systems. Indeed, in 2004, the Royal Society and Royal Academy of Engineering published what has come to be seen as a seminal report on the development of safe and beneficial nanotechnologies. In this report, they suggest that nanomaterials affect biological systems at the molecular, cellular (i.e., nuclear or mitochondrial), and protein levels and may deposit in tissues and organs: "The fact that nanoparticles are on the same scale as cellular components and larger proteins has led to the suggestion that they might evade the natural defences of humans and other species and damage cells."[10] Moreover, scientists studying the health and environmental impact of nanotechnology are reviewing long-term research on particle toxicology, such as inhalation of ambient ultrafine particles (less than 100 nanometers in diameter) of asbestos, coal, or cement dust, and its relationship to lung damage.[11]

Andre Nel, chief of nanomedicine at the University of California, Los Angeles, School of Medicine, and colleagues deduced (from a combination of previous research and their own studies) that occupational exposure to coal and asbestos fibers induces inflammation, fibrosis, and cytotoxicity.[12] And experimental studies infusing titanium dioxide and carbon black nanoparticles in animal lungs showed similar injuries. From these studies, Andre and colleagues noted that "a small size, a large surface area, and an ability to generate reactive oxygen species (ROS) play a role in the ability of nanoparticles to induce lung injury."[13] In essence, as particles shrank, lung injury increased. The body's immune system immediately comes into play to remove these foreign materials. But fibers longer than 15 micrometers are not easily removed by macrophages.[14] Fibers narrower than three micrometers — as in the case of asbestos — have aerodynamic properties enabling them to reach the alveoli and capillaries — the gas-exchanging parts of the lungs.[15] Subsequent initiation of a widespread immune reaction leads to inflammation and possibly lung cancer. Particles can also be absorbed into the skin or gut where, again, the body's defenses will kick in to neutralize the toxicity. People whose health is already impaired will probably die as a result.

Unfortunately, although it has been approximately 20 years since nanotechnology burst onto the scene, mechanistic understanding remains limited. According to Eugenia Valsami-Jones and Iseult Lynch, professors of environmental nanoscience at the University of Birmingham in the United Kingdom, there is limited evidence for acute toxicity from nanomaterials at realistic doses, and, more importantly, "there is also no simple correlation between toxic responses and nanoparticle size or other predictable pattern of toxicity."[16] Valsami-Jones and Lynch added:

> Upon entering the environment, some engineered nanomaterials, such as metals (silver, copper) and metal oxides (zinc oxide, iron oxide), may dissolve quickly, whereas others are more persistent (for example, titanium dioxide, silicon dioxide, carbon nanotubes, and graphene). From a safety point of view, soluble nanomaterials represent a best-case scenario, because any hazard from exposure is likely to be no different from that of their constituent ions. Given their small size and the limited quantities used in applications, such release is unlikely to amount to a

> major environmental threat. *Direct exposure of humans or organisms would, however, still be a concern, particularly if the nanomaterials were internalized by cells as particles, only to dissolve and deliver toxic metals inside the cell after entry* — a mechanism known as "Trojan horse" in analogy with Greek mythology.[17] (italics added)

Essentially, the Trojan horse effect is the potential for nanomaterials to be recognized by cellular receptors and internalized by cells, causing damage to pulmonary and digestive systems of humans and other organisms.

## 12.1. Nanotechnology and Health Safety in U.S.

In the middle of the 1990s, federal science and technology agencies began holding informal discussions on the future of nanotechnology in the United States.[18] Certainly, the 2004 Royal Society nanotechnology report provided the momentum, direction, and research strategies with which to frame the eventual US report. Later, informal discussions culminated in the establishment in 2000 of the NNI, which "serves strictly as a coordination mechanism for government agencies that support nanoscale research … ensuring continuing leadership by the United States in nanoscale science, engineering, and technology; and contributing to the nation's economic competitiveness."[19] As stated elsewhere in this book, this is a theme that runs through US science as a whole: benefits seem to override risks. Evidence for this theme comes from the 2009 National Academy of Sciences *Review of the Federal Strategy for Nanotechnology-Related Environmental, Health, and Safety Research*, which listed the total budget of NNI-related research and participating agencies:

> Thirteen NNI-participating agencies currently report investments in Nanotechnology: the Department of Agriculture (USDA) (including the Forest Service [FS] and the Cooperative State Research, Education, and Extension Service [CRSEES]), Department of Defense (DOD), Department of Energy (DOE), Department of Homeland Security (DHS), Department of Justice (DOJ), Department of Transportation (DOT), Environmental Protection Agency (EPA), National Institute for

> Occupational Safety and Health (NIOSH), National Institute of Standards and Technology (NIST), National Institutes of Health (NIH), and National Science Foundation (NSF). *In FY 2007, the total investment by those agencies in NNI-related research was about $1.425 billion; DOD, DOE, NIH, NIST, and NSF contributed over 80% of the total NNI budget. The president's research and development (R&D) budget request for the NNI for FY 2009 was $1.527 billion.*[20] (italics added)

The potential dangers of nanotechnology to biological organisms (supported by evidence in the Royal Society 2004 report) prompted the NNI to form the Nanotechnology Environment and Health Implications (NEHI) working group to examine both applications and implications of nanomaterials.[21] This work, according to NEHI, was not easy:

> The combination of the heterogeneity of the enormous variations among nanomaterials and their applications; the potential for novel forms of toxicity created by their unique size and structural and physical characteristics, and the variations in the frequency, magnitude and duration of releases and exposures, introduces considerable complexity into design of research programs necessary to understand their potential toxicity.[22]

Regardless of the difficulties, NEHI created a strategy in the form of categories for nanotechnology-related environmental, health, and safety research. The research categories were (1) instrumentation, metrology, and analytic methods; (2) nanomaterials and human health; (3) nanomaterials and the environment; (4) human and environmental exposure assessment; and (5) risk-management methods.[23]

The first research category was divided into five research needs: (1) developing methods to detect nanomaterials in biological matrices and the environment, because previous research deemed it essential for understanding doses at the molecular, cellular, organ, and tissue levels and the movement of nanomaterials in the body and environment; (2) developing methods for standardizing assessment of particle size, size distribution, shape, structure, and surface area; (3) developing certified reference materials for chemical and physical characterization of nanomaterials; (4) developing methods to characterize a nanomaterial's spatiochemical composition, purity, and heterogeneity; and (5) understanding how chemical and physical modifications affect the properties of nanomaterials.[24]

The second research category was also divided into five research needs: (1) understanding the absorption and transport of nanomaterials throughout the human body; (2) understanding the relationship between the properties of nanomaterials and uptake via the respiratory or digestive tracks or through the eyes and skin; (3) developing appropriate in vitro and in vivo models to predict in vivo human responses to nanomaterials exposure; (4) developing methods to quantify and characterize exposure to nanomaterials; and (5) understanding the mechanisms of interaction between nanomaterials and the body at the molecular, cellular, and tissue levels.[25]

The third research category proposed to (1) understand the effects of engineered nanomaterials in individuals of a species in order to develop methods for measuring the effects at the genomic, molecular, cellular, organismal, and population levels; (2) understand environmental exposures through identification of principal sources of exposure and exposure routes; (3) understand the transformation of nanomaterials under different environmental conditions; and (4) evaluate ecosystem-wide effects.[26]

The research concerns in the fourth category were more focused on consumer products that people use directly or interact with every day (e.g., paint, building construction, gases, dust particles, and cosmetics).[27] Consequently, research must identify population groups and environments exposed to engineered nanoscale materials and identify industrial and consumer products containing nanomaterials that could cause illness in the exposed population and contaminate the surrounding environment.

For risk-management methods, in the fifth research category, the government is biased (as far as funding goes) toward risk assessment and risk management relevant to environmental, occupational, and consumer exposure as opposed to human health and therapeutic applications.[28] The 2009 National Academy of Sciences report noted:

> The small number of projects addressing the research needs in the nanomaterials and human health section and their bias toward therapeutic applications rather than materials relevant to the environmental, occupational, and consumer exposure settings constituted sufficient evidence

> that *the funded research will not support risk-assessment and risk-management needs for these classes of nanomaterials, generate the information needed to support EHS [Environment, Health, & Safety] risk assessment and risk management, or provide critical data for regulatory agencies.*[29] (italics added)

Despite this bias, one of the major themes in this book is weighing benefits and risks to the health of humans and other organisms. As such, the next section will discuss examples of bionanotechnology and nanomedicine and review some of the familiar social and ethical issues (inequality, human enhancement, patents, dual use, and unintended consequences) they create.

## 12.2. Bionanotechnologies: Nanomedicine and Dangers

*Bionanotechnology* is the term for the manipulation and exploitation of the natural nanoengines (i.e., molecules, cells, and proteins) of biological systems, and nanomedicine is the application of newly emerging nanotechnologies to molecular processes at the cellular level.[30] Since these definitions have some overlap, the respective technologies are merged in the following discussion, which begins with nanomaterials for tissue regeneration. Harry Tibbals, research professor of materials science in the School of Engineering at University of Texas, Arlington, wrote a book titled *Medical Nanotechnology and Nanomedicine*, which was published in 2011 and was one of the first comprehensive introductions to nanoscience and nanotechnology for those with medical (primarily) and biological backgrounds.[31] Several edited volumes similar to Tibbals's book soon followed. In one chapter of his book, Tibbals discussed the role of nanotechnology in tissue regeneration. He stated that nanotechnology plays an important role in building materials and structures for tissue regeneration to match biological cells and tissues.[32] He added that this has the potential to lead to clinical trials for a host of ailments, such as arthritis, musculoskeletal diseases and disorders, and disorders of the cartilage, discs, ligaments, menisci, muscles, tendons, and skin. And these nanomaterials could also be used for medical challenges like cancer, diabetes, Parkinson's disease, Alzheimer's disease, cardiovascular problems, and inflammatory and infectious diseases.[33] Furthermore, he noted that nanotechnology is being used to monitor outcomes in tissue regeneration by analyzing cell distribution, proliferation, and differentiation after cell transplantation.

Nanotechnology is also providing tools for in situ tissue regeneration, guidance of tissue growth, stopping and reversing pathological processes, fabrication of scaffolds, and delivery of signaling molecules and stem cells.[34] Signaling biomolecules, for example, control the growth and metabolism of cells such as hormones, growth factors, receptors, cytokines, histamines, and other messenger molecules that trigger regenerative activity.[35] Since regulation of growth, inflammation, and healing are controlled by a complex signaling network, understanding the correct signaling sequences is critical for the controlled delivery of proteins, peptides, and genes to guide tissue regeneration.[36]

Tibbals discussed the ongoing research to use nanotechnology for efficient harvesting, reprogramming, and delivery of adult stem cells and for the development of intelligent bioactive materials to support the growth process.[37] He then proposed the next step in this research:

> An ultimate vision would be the development of implantable, cell-free, intelligent, bioactive materials that would provide signaling to initiate, promote, and guide self-healing by the patients own stem cells. This approach integrates the concepts of biomaterial scaffolding, cell signaling, and stem cell therapy into a single, unified nanotechnology-based therapy.[38]

Tibbals also envisioned the use of bioactive stimuli to active, silence, or regulate genes as a regenerative or preventative therapy, forestalling deterioration due to aging or disease.[39]

Since the human brain is subject to deterioration due to aging or disease, the creation of nanoengineered materials for treating brain lesions is an active area of research being pursued by multidisciplinary teams of neuroscientists, neurosurgeons, and neurologists, according to Tibbals.[40] Nanoengineered materials for brain regeneration immediately bring to mind a brain-machine interface involving AI (which is addressed in the next chapter). This is not totally science fiction; based on new technology and years of research on neural activity of the brain, computer systems are being made in which neural impulses can be translated into electronic controls for communication and to allow mobility for paralyzed people.[41] However, in non-brain-machine interface regeneration, researchers have found that nanomaterials are good platforms for free radical scavengers that can "protect the brain from immediate and secondary cell death caused by superoxides, nitric oxide, and other free radical excitotoxins, which can be produced in ischemia associated with stroke or injury to the brain or spinal cord."[42] Fullerenes (structures of carbon atoms connected by single and double bonds) and their properties have been exploited in the destruction of free radicals in injured brain tissue, reducing edema and infarction.[43] Any nanoengineered scaffolds must mimic the environment present in brain development to promote regeneration of functional tissue, according to Tibbals. Additionally, every aspect of the interaction of the scaffolds with the cells (deterioration rate, release of growth-promoting factors, etc.) must be carefully engineered. Experiments with rats in which porous biomaterial scaffolds were implanted in surgically formed lesions showed decreased local cell death and lesion growth and increased neuron growth.[44]

Interestingly, scientists in the United Kingdom designed a scaffolding system to which neural cells can attach and grow and, subsequently, be injected into the brain through a fine needle to fill lesion cavities of arbitrary shape and size.[45] As the particles fill the cavity, tissue connections are created because the cells on neighboring particles maintain contact. As the particles biodegrade over time, more space is created for axons, blood vessels, and growing tissue.

Scientists have also studied how nanoparticles could be used for imaging and drug delivery. One example is insulin delivery.[46] Presently, injection of insulin into the skin using a hypodermic needle is painful and sometimes fatal if mistakes are made (i.e., insulin shock that occurs if insulin is injected directly into bloodstream). A lot of effort over the years has been devoted to several approaches using nanoparticles to (1) encapsulate insulin in a protective coating that would allow its release after passing through the digestive system (since insulin is rapidly broken down by this system) or encapsulate it for injection in soft tissues with gradual release in bloodstream; (2) encapsulate zinc insulin in polyester and polyanhydride nanospheres for slow release; and (3) use nanodiamond-insulin complexes as drug carriers for pH-dependent insulin delivery.[47]

In another example, nanoparticles could be functionalized with antibodies allowing pathogenic cells (such as cancer cells) to be separated out of tissue samples (such as bone marrow samples) for analysis and early detection of cancerous tumors.[48] Or nanomagnetic particles could be used for magnetically controlled drug *targeting*. According to Tibbals, established anticancer drugs could be bound with nanomagnetic particles that concentrate the drug to a tumor site (by using generated magnetic fields) before separating so that the drug particles can interact with the tumor.[49]

Another way of targeting a cell type involves adding a molecular recognition factor to the nanoparticle so that it binds to selected cells.[50] In adopting this as an experimental tactic, scientists have mimicked several natural processes: immunoglobulins bind to specific antigens with their membrane-bound antibodies; DNA or RNA templates target genetic material in cells; and diseased tissue gives off signals that alert immune cells. But the problems are enormous, according to Tibbals: (1) targeting drugs must avoid the immune system; (2) nanoparticles and drugs with smaller molecular weights (in terms of the ratio of surface area to volume) interact differently with cellular physiology and DNA; and (3) interactions with living cells at extremely small dimensions could have unexpected effects.[51] Applying the nanoparticles directly into tumors or other sites of diseased tissue is another approach to avoiding healthy cells. This way, powerful drugs that would be too toxic to the circulatory system can be merged with nanoparticles and directed onto the tumor.[52]

Finally, nanoparticles are being investigated and tested for use against infectious diseases and for development of advanced vaccines.[53] Compared to all of the applications (and possible applications) of nanotechnology discussed in this section, using nanoparticles to attack specific types of infectious pathogens is the most frightening. In science fiction, the unintended effects of a viral — nanoparticle or a bacterial — nanoparticle mutation might include the generation of an unstoppable new super biological-synthetic microbe. The same might apply to nanovaccines and their complex interactions with the human immune system. Nevertheless, scientists have tested nanoparticles against tuberculosis (and other mycobacteria) and several viruses: hepatitis C, herpes, and HIV.[54] For infectious diseases in general and viruses in particular, nanoparticles can be engineered to overcome drug resistance and prevent viral capsid conjugates from entering the cell membrane by attaching to these capsid conjugates; these are, no doubt, practical benefits.[55] Collectively, these benefits will induce the spending of billions of dollars on nanotechnology research and development globally. As with most emerging technologies, the competition in the marketplace between pharmaceutical companies and technology companies for patents and customers will be fierce, resulting in a bias toward short-term benefits while playing down the risks.

## 12.3. Bionanotechnology: Social and Ethical Issues

Nanotechnology, like biotechnology, raises similar social and ethical concerns to those discussed throughout this book (inequality, heritable modifications, human enhancement, eugenics, and dual use). But unlike biotechnology, nanotechnologies permeate more areas of basic science, engineering, and applications, creating an enormous challenge for the environment, health, and safety. The previous section presents a frightening hypothetical scenario of the successful merging of nanoparticles with a viral or bacterial strain to create an unstoppable strain of nano-microorganism. There are already enough infectious multidrug-resistant strains of microbes in nature, and nature does not need any more help creating more superbugs.

The 2004 Royal Society report stated that the acceptance of nanotechnology depends on political, economic, and social forces:

> Widespread acceptance and use of nanotechnologies will depend upon a range of social factors including: specific technical and investment factors; consumer choice and wider public acceptability; the political and macro-economic decisions that contribute to the development of major technologies and outcomes that are viewed as desirable; and legal and regulatory frameworks.[56]

This is certainly true for any emerging technology. What is equally true is that those individuals in power (both politically and economically) who believe that the technology benefits the country and its people will attempt to coax the public to their way of thinking. And in this era of the internet and social media, this will probably be easier to do than in the past. It may not work for some people, however, because within any population, there are individuals who are contrary in their politics, economics, race, ethnicity, and ideas about how technology should be applied. Those individuals with financial means may be coaxed into using nanotechnologies for health, enhancement, or personal appliances, while those individuals without financial means must do without these technologies. This would create a "nanodivide" within society, and this division could occur on a larger scale between rich and poor countries, exacerbating the existing problems of forced migrations due to inequality, warfare, famine, and climate change — not all mutually exclusive.[57] There is fear that some short-term developments in nanotechnology (such as cosmetics, drug therapy, and prosthetic enhancement) will be available to those who have money and power, to the detriment of the wider society. For example, improved cochlear and retinal implants to *restore* hearing and eyesight could also be used to *improve* or enhance hearing and eyesight in healthy individuals. Or nanotechnology components used to supplement or replace parts of an injured soldier's body could be used on healthy soldiers to generate super soldiers. This could be seen as dual use (as in the hypothetical nanoviral/bacterial strain). The situation would become worse when these enhancements went on the market and, as usual, only a portion of the population could afford them. Eventually, there would be a modified population and an unmodified population. If the modified population believed that they were superior (in strength and intelligence) compared to the unmodified, then this could lead to the old eugenics. Conversely, the government could create a coercive policy under which nanotechnology modification is mandatory. This would be the new eugenics, in which nanotechnology is used to improve all individuals in the society. But the greatest fear of all might be the convergence of nanotechnology with biotechnology, carrying with it the threat of extinction of various biological life-forms on this planet, including humans.

## Notes

1. William Shakespeare, *The Merchant of Venice*, edited by Horace H. Furness (1596; repr., Philadelphia, PA: Lippincott Company, 1916), act 2, scene 9, 120.
2. Andre Nel, Tian Xia, Lutz Mädler, Ning Li, "Toxic Potential of Materials at the Nanolevel," *Science* 311, no. 5761, (February 3, 2006): 622; Eugenia Valsami-Jones and Iseult Lynch, "How Safe Are Nanomaterials? There Is Still No Consensus on the Toxicity of Nanomaterials," *Science* 350, no. 6259 (October 23, 2015): 388.
3. Rudy Baum, "Nanotechnology: Drexler and Smalley Make the Case for and Against 'Molecular Assemblers,'" *Chemical and Engineering News* 81 (December 1, 2003): 31–42; Nanotechnology Information Center, "Properties, Applications, Research, and Safety Guidelines," December 26, 2014, http://www.americanelements.com/nanomaterials-nanoparticles-nanotechnology.html; The Project on Nanotechnology, "Analysis: This Is the First Publicly Available On-Line Inventory of Nanotechnology-Based Consumer Products," May 5, 2011, http://www.nanotechproject.org/inventories/consumer/analysis_draft.
4. "The Project on Nanotechnology," http://www.nanotechproject.org/inventories/consumer/analysis_draft.
5. Nel *et al.*, "Toxic Potential of Materials," 622–627.
6. Ibid., 622.
7. Eric K. Drexler, *Engines of Creation: The Coming Era of Nanotechnology* (New York; Double Day, 1986), 3–320; Eric K. Drexler, *Nanosystems: Molecular Machinery, Manufacturing, and Computation* (New York, NY: John Wiley & Sons), 1–580.
8. The National Academies of Sciences, Engineering, and Medicine, *Review of the Federal Strategy for Nanotechnology-Related Environmental, Health, And Safety Research* (Washington, D.C.: The National Academies Press, 2009), 13.
9. Ibid.
10. Royal Society, *Nanoscience and Nanotechnologies: Opportunities and Uncertainties* (London: The Royal Society & The Royal Academy of Engineering, 2008), 35, http://www.nanotec.org.uk/finalReport.htm.
11. Nel *et al.*, "Toxic Potential of Materials," 622–627; Valsami-Jones and Lynch, "How Safe Are Nanomaterials?" 388.

12. Nel *et al.*, “Toxic Potential of Materials,” 622.
13. Ibid.
14. Royal Society, *Nanoscience and Nanotechnologies*, 36, http://www.nanotec.org.uk/finalReport.htm.
15. Ibid., 1–111.
16. Eugenia Valsami-Jones and Iseult Lynch, “How Safe Are Nanomaterials? There Is Still No Consensus on the Toxicity of Nanomaterials,” *Science* 350, no. 6259 (October 23, 2015): 388.
17. Ibid.
18. The National Academies of Sciences … Review of, 13
19. Ibid., 13, 15.
20. Ibid., 15.
21. Ibid., 18.
22. Ibid., 19–20.
23. Ibid., 54–55.
24. Ibid., 61–62.
25. Ibid., 62.
26. Ibid., 62–63.
27. Ibid., 63.
28. Ibid., 63–64.
29. Ibid., 70.
30. Harry F. Tibbals, *Medical Nanotechnology and Nanomedicine* (Boca Rotan: CRC Press, 2011), 20.
31. Ibid., 3–484.
32. Ibid., 253.
33. Ibid., 123, 253.
34. Ibid., 255–257.
35. Ibid., 257.
36. Ibid.
37. Ibid., 258–259.
38. Ibid., 259.
39. Ibid.
40. Ibid., 262–263.
41. Ibid., 259.
42. Tibbals, *Medical Nanotechnology*, 275; Noreen M. Gervasi, Jessica C. Kwok, James W. Fawcett, “Role of Extracellular Factors in Axon Regeneration in the CNS: Implications for Therapy,” *Regenerative Medicine* 3, no. 6 (November 2008): 907–923.

43. Tibbals, *Medical Nanotechnology*, 275.
44. Ibid., 282–283.
45. Ibid., 283.
46. Ibid., 125–227.
47. Ibid., 125–126; Rafael A. Shimkunas, Erik Robinson, Robert Lam, Steven Lu, Xiaoyang Xu, Xue-Qing Zhang, Houjin Huang, Eiji Osawa, and Dean Ho, "Nano-Diamond-Insulin Complexes as pH-Dependent Protein Delivery Vehicles," *Biomaterials* 30, no. 29 (October, 2009): 5720–5728.
48. Tibbals, *Medical Nanotechnology*, 22.
49. Ibid., 87.
50. Ibid., 88.
51. Ibid., 88–89.
52. Ibid., 88.
53. Ibid., 123.
54. Haruaki Tomioka and Kenji Namba, "Development of Antituberculosis Drugs: Current Status and Future Prospects," *Kekkaku* 81, no. 12 (December 2006): 753–774; Masaji Okada and Kazuo Kobayashi, "Recent Progress in Mycobacteriology," *Kekkaku* 82, no. 10 (October 2007): 783–799; Thea Thomas and Graham Foster, "Nanomedicines in the Treatment of Chronic Hepatitis C — Focus on Pegylated Interferon Alpha-2a," *International Journal of Nanomedicine* 2, no. 1 (March 2007): 19–24.
55. Tibbals, *Medical Nanotechnology*, 123.
56. Royal Society, *Nanoscience and Nanotechnologies: Opportunities and Uncertainties* (London: The Royal Society & The Royal Academy of Engineering, 2008), 51.
57. Ibid., 52.

# Chapter 13

# Unintended Consequences

The road to hell is paved with good intentions (The earliest iteration of this proverb was made in 1640 by an unknown writer: *L'enfer est plein de bonnes volontés ou désirs* (hell is full of good wishes or desires).

— Christine Ammer, 1997[1]

In a 1936 paper, the late sociologist Robert Merton tried to scientifically analyze the problem of unintended consequences. Although early definitions of unintended consequences, first proposed as far back as the 17th and 18th centuries (by influential Enlightenment thinkers John Locke and Adam Smith and possibly others), were specific to ethics, morality, and economics, Merton defined unintended consequences as "outcomes that are not the ones foreseen and intended by a purposeful action."[2] Unintended consequences permeate every facet of human life, and human history is replete with them: the agricultural revolution, which created a sustainable food system feeding millions of people, led eventually to the modern civilization we love and simultaneously caused increased population density, resulting in infectious "crowd" diseases; Alexander Fleming's habit of keeping an untidy laboratory with petri dishes stacked in one corner led to contamination of staphylococci cultures by fungus, resulting in his accidental discovery of penicillin and his warning in the 1940s about the overuse of antibiotics, which could lead to antibiotic-resistant bacteria; the screening programs for sickle cell anemia and Tay-Sachs disease in the 1960s and 1970s have forever associated these diseases in the public mind solely with blacks and Jews; retroviruses used as a tool for inserting genes into humans initiated runaway cell division leading to cancer; gene-drive modified mosquitos generated to stop the spread of malaria provided opportunities for other insect species carrying infectious diseases to thrive; the study of DNA variation and gene families led to the discovery of restriction fragment length polymorphism (RFLP) and its subsequent application in forensic sciences as DNA fingerprinting; ART technologies have led to designer babies; and the discovery of CRISPR within the genomes of Bacteria and Archaea led to the *exploitation* of this mechanism in creating a genetic-editing technology that is used worldwide.[3] The word *exploitation* is emphasized here because Ruha Benjamin, associate professor in the Department of African American Studies at Princeton University — at the 2015 summit on human gene editing — discussed the potential for human gene editing to exploit the desires of those individuals who seek enhancement of their current traits or those of their embryos, thereby exacerbating existing inequalities in society. Benjamin stated, "The use of gene editing techniques is seeded with values and interests, economic as well as social, that without careful examination could easily reproduce existing hierarchies."[4] This would be an undesirable outcome. Nevertheless, the examples above are a compilation of unintended consequences that are both desirable and undesirable.

Moreover, as stated earlier in this book, the unintended consequences of the old eugenics were the beginnings of human genetic research to improve the health, environment, and safety of all humans, which resulted in the emergence of biotechnology and all that it entails. This is the new eugenics. When Edward Lanphier and colleagues wrote a paper in 2015 calling for a moratorium on human embryo editing, Julian Savulescu countered by arguing the following: "This reasoning is, however, inconsistent with widely accepted practices. Nearly all new technologies have *unpredictable effects* on future generations"[5] (italics added). The unintended consequence of the CRISPR/Cas9 discovery was the exploitation of this technology by scientists globally to edit any biological species they desired. For example, genes had been edited in embryonic kidney cells, human leukemia cells, human stem cells, mouse neuroblastoma cells, bacterial cells, and embryos from zebrafish. Doudna explained:

> In the summer of 2013, as I was marveling at the pace of CRISPR's dissemination, I began keeping a list of all the different cell types and organisms whose genomes had been edited using the technology. *The list was manageable at first and included the zebrafish and cultured bacterial, mouse, and human cells from January and February, followed by yeast, mice, fruit flies, and microscopic worms. At the end of that year, my list included rats, frogs, and silkworms. By the end of 2014, I had added rabbits, pigs, goats, sea squirts, and monkeys, after which, as I acknowledged to audiences at seminars where I shared the list, I honestly lost track.*[6] (italics added)

This CRISPR/Cas9 craze did not stop at animal species. Plant biologists using ZFN and TALENS to study pathogens in rice, bell peppers, sorghum, wheat, soybeans, corn, and tomatoes realized the potential of CRISPR/Cas9 and the ease of using it to edit DNA in these crops and other plant species.[7] In fact, another unintended consequence was the confusion between GMO and non-GMO products; ironically, this distinction was only important to scientists and not the public.[8]

For humans, Doudna listed several potential genetic cures that have been developed with CRISPR/Cas9 for HIV, achondroplasia (dwarfism), chronic granulomatous disease, Alzheimer's disease, congenital hearing loss, amyotrophic lateral sclerosis (ALS), high cholesterol, diabetes, Tay-Sachs disease, skin disorders, fragile X syndrome, and infertility.[9] Ten to 15 years ago, no one could have predicted that there would be possible cures for so many diseases. And the best strategy to attack these diseases was to repair the mutations in the embryo. While this strategy has been known since the early 1960s, there was no way to achieve it until now because the genetic-editing technology was not available then. In essence, human embryo editing was the ultimate unintended consequence of CRISPR/Cas9, even though one might argue that it was a logical step in the evolution of genetic-editing research — despite the fact that PGD was a well-known preventative disease measure.

Like the internet, the CRISPR/Cas9 technology has been widely disseminated to users around the globe. To the surprise of its creators, more than 100,000 CRISPR-related tools have been shipped to many countries around the world, where scientists in academic and commercial laboratories have published their work on editing numerous mammalian genomes.[10] More alarmingly, CRISPR-related tools (intended to modify only bacterial and yeast genes) have been sold online to anyone with 100 dollars, according to Doudna. With the CRISPR/Cas9 technique applied so routinely, "it wasn't hard to imagine biohackers messing with more complex genetic systems — up to and including our own."[11]

Interestingly, the internet serves as an example of how to predict the future of CRISPR/Cas9. For instance, the internet is an electronic manifestation of globalism — the increasing interconnectedness of nations in a world system linked economically and through mass media and modern transportation systems — in which regional networking is merged into a universal *internetworking*.[12] And like the optimistic goals of globalization (economic integration resulting in prosperity and maintaining peace), the internet has, to some extent, promoted commerce, cheap and instant communication, and easy universal cultural exchange. Like the internet, the CRISPR/Cas9 technology is being exploited by scientists globally to solve problems in human biology and ecology. And the CRISPR/Cas9 protocols used to create numerous designer mutations in biological lifeforms are analogous to the numerous social networks (e.g., Facebook, Twitter, Instagram, Snapchat, Flickr, Pinterest, Google, YouTube, to name a few) that are potentially pernicious with problems of off-target edits (or — in the case of social networks — fake news). The CRISPR bacterial immune system was discovered by accident and later converted to a technology. In contrast, the internet began voluntarily as a universal merging of computer networks with the goal of making communication efficient and advancing the science of computing for the military, space exploration, and commerce.[13] Essentially, both technologies began with good intentions. It is close to 30 years since the internet became public, and it has transformed life for better and the worse. In fact, the unintended consequences of the internet are more divisiveness and animosity (rather than unity) coupled with a lack of proper surveillance by its creators. These twin problems have transformed it into a dangerous technology. Similarly, the unintended consequences of CRISPR/Cas9 are its dangerous overuse, including the alleged human embryo editing and uterine implantation. Even though CRISPR scientists have begun to self-regulate themselves, will CRISPR/Cas9 follow the same path as the internet? This question can only be answered by future generations.

## 13.1. Scientific Advancement and the Unintended Consequences

Unlike the rapid progress in genetics and biotechnology research and development (discovery of the structure of DNA, discovery of recombinant DNA, the Human Genome Project, recognition of CRISPR adaptive immunity, and CRISPR/Cas9 technology), AI at this point in the 21st century is extremely rudimentary and nothing close to what is portrayed on screen in science fiction like *Battlestar Galactica*, *The Terminator*, or the Borg of the *Star Trek* universe. Nevertheless, CRISPR/Cas9 technology — which is approximately 10 years old — can be used as an analogy to discuss some possible unintended consequences of AI in terms of biological cognition and brain-computer interfaces.

What if biological brains could be enhanced? Earlier chapters introduce and discuss the possibility of a memory-enhancing drug developed for Alzheimer patients that could improve memory, concentration, and mental energy, which is sold to rich, healthy individuals who desire enhanced cognition. Parents could give the drug to their children, and, at least for the next generation, there would be a small group of superintelligent individuals. Even if the drug was a form of gene therapy, the enhancement would be only in the somatic cells and, therefore, not inherited. At the current stage of technology, obtaining any type of therapeutic superintelligence might be more fiction than fact. But it does seem that development of nootropic drugs (potentially improves cognitive functions) in the mid to late 21st century might be more realistic, at least according to Swedish philosopher Nick Bostrom at the University of Oxford.[14] He noted that "it seems implausible, on both neurological and evolutionary grounds, that one could by introducing some chemical into the brain of a healthy person spark a dramatic rise in intelligence."[15] Bostrom believes that the cognitive functioning of the human brain is complex, and it requires constant stimulation from many different environments, including the home, diet and nutrition, and wider social environments, particularly at the critical stages of development.[16] It follows that parents should focus on enhancing their children's social environments instead of just giving them smart pills.

Selective breeding might be another way of attaining human AI.[17] The problem here is that despite the maximum strength of selection, it would still take several generations to achieve desirable results. And if selection were so strong for intelligence, all other traits (e.g., beauty, height, and eye color) would be at a disadvantage. Regardless of which trait is given preference, selective breeding would be a blatant attempt to resurrect the old eugenics; there would be protest against this. A more realistic scenario, however, is selective breeding for human AI by PGD. The use of PGD is already established in IVF procedures in screening embryos for mutations and sex selection. Therefore, using a science fiction scenario, parents could pay their fertility specialists to select embryos with genes that correlate with intelligence; these genes would be heritable. Parents could then select a larger number of embryos for higher gains in desired genetic traits. While this may seem a great way to increase the number of intelligent individuals, Bostrom argued that selection between 100 embryos does not produce a gain anywhere close to 50 times as large "as that which one would get from selection between 2 embryos" because of the law of diminishing returns.[18] Realistically, the selection needs to be spread out over multiple generations, selecting the top one in 10 over 10 generations, with the offspring of the previous generation in every successive generation, according to Bostrom. But this could occur must faster using viable sperm and eggs from ES cells. An unlimited number of gametes could be fused to produce embryos, which would then be screened and implanted.[19] This technology would give couples who could afford it more selective power in IVF. Bostrom explained:

> Stem cell-derived gametes would allow multiple generations of selection to be compressed into less than a human maturation period, by enabling *iterated embryo selection*. This is a procedure that would consist of the following steps: 1. Genotype and select a number of embryos that are higher in desired genetic characteristics. 2. Extract stem cells from those embryos and convert them to sperm and ova, maturing within six months or less. 3. Cross the new sperm and ova to produce embryos. 4. Repeat until large genetic changes have been accumulated.[20]

While the use of ES cells in research is provocative (forcing US presidents to ban new generation of ES-cell lines), the advantage of this procedure is that it does not have to be repeated for each birth. And many enhanced embryos could be generated from the new cell lines with a high level of average intelligence: "A world that had a large population of such individuals might (if it had the culture, education, communications infrastructure, etc., to match) constitute a collective superintelligence."[21] The unintended consequences here would be programs geared toward GM adults and children and exploitation of the non-GM people. This collective super-intelligent group with economic and political power might not have patience for non-GM people, resulting, sooner or later, in a eugenic purge. In an alternative scenario, cognitive enhancement of the population could lead to a faster rate of scientific advancement accompanied by better biomedical techniques in cognitive enhancement and possible development of machine AI. While the unintended consequences of machine AI are beyond the scope of this book, the problems would be similar to those discussed elsewhere in this book in connection with other technologies and would include societal conflict. Machine AIs could, as portrayed in *Battlestar Galactica* or *The Terminator*, destroy human society in an attempt to cause human extinction or manipulate humans (who are slaves to technology) to unwittingly eliminate other humans.

In a more likely scenario, the future might bring universal health care under which all expectant mothers can benefit from enhanced-intelligence embryos. But, coupled with this universal health care, all citizens with families may be required to have this procedure in an effort to remove any mutational load from the gene pool. This situation would be no better than the forced sterilization of early 20th-century America, and the unintended consequences of this coercive rule would be increased divisiveness (e.g., parents of GM children pitted against parents of non-GM children) and resistance. Additional unintended consequences could be the emergence of different forms of mental illnesses in GM individuals (or "super geniuses") operating at a high cognitive level. Are parents or society willing to run the risk of mental instability in their children for superintelligence? Richard Taite, founder and CEO of a drug rehabilitation center called Cliffside Malibu, wrote that although society may benefit from the productivity of geniuses, their health must also be considered, because "bipolar disorder, depression, addiction, and other psychiatric/psychological disorders are debilitating and can be life-threatening conditions."[22] This is an important statement for all to think about, because if it is true, then greater intelligence would be nonadaptive in the long term. *Medical Daily* contributor Matthew Mientka wrote that some studies have linked bipolar disorder with high achievers such as Vincent Van Gogh, Buzz Aldrin, Emily Dickinson, and Jackson Pollock.[23] Note that two widely read writers, American Ernest Hemingway and Briton Virginia Woolf, both committed suicide.[24] In more recent times, the American writer David Foster Wallace hanged himself in his garage after suffering years of depression.[25] "Like the Sword of Damocles," Mientka wrote, "higher intelligence may in some ways curse its beneficiaries. Aside from the usual desire to self-medicate, smarter people tend to drink alcohol and do drugs more than average — perhaps seeking to drench a burning sense of curiosity."[26] Mientka briefly discussed research at University of Toronto and the Samuel Lunenfeld Research Institute of Mount Sinai Hospital identifying a bimolecular link — in the form of the neuronal calcium sensor-1 protein — between curiosity and intelligence in a mouse model.[27] Interestingly, this protein had also been linked to bipolar disorder and schizophrenia. Other collaborative research between psychiatrist James MacCabe and colleagues at the Institute of Psychiatry, King's College, London, and the Department of Medical Epidemiology & Biostatistics of the Karolinska Institute in Stockholm, in which the researchers searched Swedish national school records for a tentative link between intelligence and bipolar disorder, supported the evidence generated in the study from the University of Toronto and Mount Sinai.[28] MacCabe and colleagues found that "achieving an 'A' grade is associated with increased risk for bipolar disorder, particularly in humanities and to a lesser extent in science subjects. … These findings provide support for the hypothesis that exceptional intellectual ability is associated with bipolar disorder."[29] As stated frequently in this book, there will always be both benefits and risks when using biotechnology for human therapeutic or elective enhancement.

In summary, the unintended consequences of human AI (enhancing biological cognition) would be increased stratification between GM and non-GM individuals, which is connected, more or less, to economics exacerbating the conflict. However, before superintelligence could make an impact on the world, it would take some time to achieve, because the children resulting from these types of germ line interventions have to grow into adults.[30] And human breeding time is slow, with one generation occurring every 20 to 25 years. Nevertheless, the necessary intensity of selection, in terms of the number of people choosing cognitive enhancement, could be created through good advertising showing that the technology "is proven to work and to provide substantial benefit."[31] As cellular phones and the internet have become popular and rooted in modern culture, so could the desire for this enhancement; it would become attractive to most individuals, who would voluntarily choose it without coercion.

What if human AI could be achieved through what Bostrom calls *cyborgization*, the implanting of a mechanical device directly into the brain with computer interface capability?[32] Biased by prior observation of popular science fiction movies like the ones mentioned above, one can predict several obvious (and mostly negative) consequences, including brain damage rather than enhanced intelligence in healthy individuals who attempted the procedure. An additional unintended consequence — particularly with the brain-computer interface — would be governmental or private corporate control of a subset of the population. Bostrom did an excellent job of discussing this aspect of human AI and added the caveat that this technology would not be available anytime soon. First, he discussed the obvious problem of medical complications: infections or unintentional damage caused by displacement of implants resulting in bleeding, pain, or cognitive damage.[33] The Parkinson's implant, for example, supplies a stimulating electric current to the brain of the patient with some success. When it goes wrong, as in some deep implants, the patient's cognitive functions are reduced. In the distant future, "whole brain prosthesis" will replace a malfunctioning brain, similar to the way a damaged leg or ear is replaced with a prosthesis in the fictional *Six Million Dollar Man* or *Bionic Woman*.[34] This prosthetic brain will allow for a more advanced form of brain-computer interfacing than is available today. While one study using rats demonstrated that it was possible to use a neural prosthesis — implanted in two areas of the hippocampus — to enhance performance in working-memory tasks, the interpretation of the results may have been problematic.[35] Most technologies that could be used to enhance human intelligence or achieve human AI are still science fiction. But the possibility of a smart pill or smart chip implant attracts a respectable number of people globally because everyone — given the chance — would want to be a genius in some field. However, people do not immediately consider the risks. And, based on the discussion in this section, the risks seem to greatly outweigh the benefits. It follows, then, that the unintended consequences would be devastating if these cognitive enhancement technologies became operational today. This chapter ends with Catherine Bliss's statement at the 2015 summit on human gene editing, where she emphasized that generating human AI has greater risks — to health and social justice — than benefits. Bliss noted that "well-meaning science that intends to produce benefits for society can unintentionally reproduce social injustices — for example, in the way the genomics has inadvertently reinforced certain racial categories."[36]

## Notes

1. Christine Ammer, *The American Heritage Dictionary of Idioms* (1640; repr., Boston: Houghton Mifflin, 1997), 542.
2. Ibid., 894.
3. Howard Markel, "Scientific Advances and Social Risks: Historical Perspectives of Genetic Screening Programs for Sickle Cell Diseases, Tay-Sachs Disease, Neural Tube Defects, and Down Syndrome 1970–1997" in Neil Holtzman and Michael S. Watson, eds., *Promoting Safe and Effective Genetic Testing in the U.S.: Final Report of the Task Force on Genetic Testing* (National Institutes of Health — Department of Energy Working Group on Ethical, Legal, and Social Implications of Human Genome Research, September 1997), appendix 6, 2, https://biotech.law.lsu.edu/research/fed/tfgt/appendix6.htm; Jim Kozubek, *Modern Prometheus: Editing the Human Genome with CRISPR-CAS9* (New York, NY: Cambridge University Press, 2016), 150–158, 322–325; Jennifer A. Doudna and Samuel H. Sternberg, *A Crack in Creation: Gene Editing and the Unthinkable Power to Control Evolution* (New York: Houghton Mifflin Harcourt, 2017), 149–153; Peter Raven, George B. Johnson, Kenneth A. Mason, Jonathan B. Losos, Susan R. Singer, *Biology*, 9th ed. (New York: McGraw-Hill, 2008), 335–336; Francisco Juan Martinez Mojica, César Diez-Villaseñor, Jesús Garcia-Martínez, and Elena Soria, "Intervening Sequences of Regularly Spaced Prokaryotic

Repeats Derive from Foreign Genetic Elements," *Journal of Molecular Evolution* 60, no. 2 (February 2005): 174–175; Martin Jinek, Krzysztof Chylinski, Ines Fonfara, Michael Hauer, Jennifer A. Doudna, Emmanuelle Charpentier, "A Programmable Dual-RNA-Guided DNA Endonuclease in Adaptive Bacterial Immunity," *Science* 337, no. 6096 (August 17, 2012): 820.

4. The National Academies of Sciences, Engineering, and Medicine, *International Summit on Human Gene Editing: A Global Discussion*, December 1–3, 2015 (Washington, DC: The National Academies Press), 4.
5. Julian Savulescu, Jonathan Pugh, Thomas Douglas, Christopher Gyngell, "The Moral Imperative to Continue Gene Editing Research on Human Embryos," *Protein Cell* 6, no. 7 (July 2015): 476.
6. Doudna and Sternberg, *A Crack in Creation*, 99.
7. Ibid.
8. Ibid., 123–127.
9. Ibid., 87–88, 94–100, 155–176.
10. Ibid., 199.
11. Ibid.
12. Conrad P. Kottak, *Window on Humanity: A Concise Introduction to Anthropology*, 3rd ed. (New York, NY: McGraw-Hill, 2008), 201, G–6; "A Brief History of the Internet," accessed August 11, 2018, http://www.walthowe.com/navnet/history.html.
13. Johnny Ryan, *A History of the Internet and the Digital Future* (London, UK: Reaktion Books), 3–248.
14. Nick Bostrom, *Superintelligence: Paths, Dangers, Strategies* (London: Oxford University Press, 2014), 44.
15. Ibid.
16. Ibid.
17. Ibid., 44–50.
18. Ibid., 46.
19. Ibid., 46–47.
20. Ibid., 47.
21. Ibid.
22. Richard Taite, "Is There a Link Between Intelligence and Mental Illness?" March 2015, https://www.psychologytoday.com/us/blog/ending-addiction-good/201503/is-there-link-between-intelligence-and-mental-illness/.
23. Matthew Mientka, "Why Smarter People Are More Likely to Be Mentally Ill," accessed July 6, 2019, http://www.medicaldaily.com/why-smarter-people-are-more-likely-be-mentally-ill-270039.

24. Ibid.
25. Ibid.
26. Ibid.
27. Matthew Mientka, "Why Smarter People," http://www.medicaldaily.com/why-smarter-people-are-more-likely-be-mentally-ill-270039.
28. Ibid.
29. James H, MacCabe, Mats P. Lambe, Sven Cnattingius, Pak C. Sham, Anthony S. David, Abraham Reichenberg, Robin M. Murray, and Christina M. Hultman, "Excellent School Performance at Age 16 and Risk of Adult Bipolar Disorder: National Cohort Study," *British Journal of Psychiatry* 196, no. 2 (February 2010): 113.
30. Bostrom, *Superintelligence*, 52–53.
31. Ibid., 52.
32. Ibid., 55–56.
33. Ibid., 54–55.
34. Ibid., 55–56.
35. Ibid., 57–58.
36. The National Academies of Sciences, Engineering, and Medicine, *International Summit on Human Gene Editing: A Global Discussion*, December 1–3, 2015 (Washington, DC: The National Academies Press), 4.

# Chapter 14

# The Impatience with Natural Selection

What Nature does blindly, slowly, and ruthlessly, man may do providentially, quickly, and kindly.

— Francis Galton, 1865[1]

Natural selection produces systems that function no better than necessary. It results in ad hoc adaptive solutions to immediate problems. Whatever enhances fitness is selected. The product of natural selection is not perfection but adequacy, not final answers but limited, short-term solutions.

— George A. Bartholomew, American biologist at the University of California, Los Angeles, 1986[2]

Galton's eugenics, sensu stricto, rested on discernment of natural improvement or progress. But natural selection was too slow to stop the "rapidly rising [people] from a low estate."[3] However, "eugenics would accelerate the process, would breed out the vestigial barbarism of the human race and manipulate evolution to bring the biological reality of man into consonance with his advanced moral ideals."[4] Unfortunately, these ideals have endured and with the advent of biotechnology have transformed into the new eugenics, in which scientists tinker with biological organisms to rapidly *improve* the health, environment, and safety of the human species. The late evolutionary biologist Stephen J. Gould wrote that "we are the accidental result of an unplanned process ... the fragile result of an enormous concatenation of improbabilities, not the predictable product of any definite process."[5] The arbitrary and nonprogressive nature of evolution has encouraged modern scientists, with the aid of technology, to violate DNA — the very essence of biological life. It is important to note that the goal of *improving* contemporary populations was initially an idea of the early 19th century, when the words *complexity* and *progress* appeared in its explanation.[6] At this point, a new character enters the story: French naturalist Jean-Baptiste Lamarck, the first scientist who tried to explain how evolution worked by viewing evolution as progress in nature.

In 1809, the year Charles Darwin was born, Lamarck published *Philosophie Zoologique*, in which he discussed the two main components of his theory of evolution: complexity/progress and adaptation of organisms.[7] The first component of Lamarck's hypothesis stemmed from his belief in perpetual spontaneous generation of simple living organisms "through action on matter by a material life force."[8] Using the ancient Greek concept of the four elements (earth, air, fire, and water) as a framework, Lamarck argued that the movements of fluids in living organisms naturally "drove them to evolve toward greater levels of complexity."[9]

The second component of Lamarck's hypothesis, adaptation of organisms, was a teleological (needs of the organism) process that combined with the environment to "move organisms 'up' a ladder of progress" into new and distinct forms by the use and disuse of certain characteristics.[10] Lamarck explained his second law, the inheritance of acquired characteristics:

> Second Law: All the acquisitions or losses wrought by nature on individuals, through the influence of the environment in which their race has long been placed, and hence through the influence of the predominant use or permanent disuse of any organ; all these are preserved by reproduction to the new individuals which arise, provided that the acquired modifications are common to both sexes, or at least to the individuals which produce the young.[11]

The phrase "acquired modifications ... are preserved by reproduction to the new individuals," pilfered from the above quote, introduced what is universally known as the inheritance of acquired characteristics: if an organism had a "need" for the fruits on tops of tall trees, the animal would stretch its neck — translated as improving the trait — until it acquired a long neck. Its offspring would then inherit the long neck trait. While the inheritance of acquired characteristics was taken seriously during the early 20th century, it was later rejected based on advances in modern genetics. Today, in the field of epigenetics — where gene expression can be changed without altering the DNA — some scientists believe that epigenetic inheritance makes possible Lamarck's idea of the inheritance of acquired characteristics.[12]

Australian evolutionary biologist David Haig, professor in the Harvard's Department of Organismic and Evolutionary Biology, wrote in 2006 that epigenetic mechanisms can involve purposeful modifications to an organism's own genetic material in response to the environment.[13] This set the stage for Israeli geneticist and theorist Eva Jablonka of Tel Aviv University and evolutionary biologist Marion Lamb, formerly of Birkbeck University in London, to argue in 2009 that heritable epigenetic modification of DNA in the germ line "makes possible a form of Lamarckian inheritance."[14] They added that "genetic assimilation" (a mechanism proposed in the mid-20th century) of an acquired character — initially induced to enhance adaptation of the organism — could be inherited. The technical example of an epigenetic inheritance system is DNA methylation, in which methyl groups are added to the DNA molecule, changing the activity of a DNA segment without changing its sequence.[15] Environmental stimuli can result in activation of DNA methylation, leading to suppression of gene transcription driving change faster than Darwinian evolution.

Biologists Eugene Koonin and Yuri Wolf at the National Center for Biotechnological Information also believed that various evolutionary phenomena that have been researched and reported in the Darwinian era seemed to fit a general "Lamarckian or at least quasi-Lamarckian" pattern.[16] In a 2009 paper, they discussed the CRISPR/Cas system of anti-virus immunity in Archaea and Bacteria (see Chapter 5). It is known from research that archaeal and bacterial genomes contain "cassettes" of multiple CRISPR units, and some of the inserts in CRISPR cassettes are identical to fragments of bacteriophages and plasmid genes.[17] Recall from Chapter 5 that the CRISPR/Cas system used phage-derived sequences as guide molecules to destroy a phage. In their explanation, Koonin and Wolf reached a startling conclusion about the CRISPR/Cas system:

> The presence of an insert precisely complementary to a region of a phage genome is essential for resistance; the guide RNAs form complexes with multiple Cas proteins and is employed to abrogate the infection; and new inserts conferring resistance to cognate phages can be acquired. ... *The mechanism of heredity and genome evolution embodied in the CRISPR-Cas system seems to be bona fide Lamarckian.*[18] (italics added)

In essence, Koonin and Wolf argued that the Lamarckian system is implicated here because adaptation occurs directly in response to an environmental cue (bacteriophage), with the result being specific adaptation (resistance) for the cell and its progeny in a shorter period of time than it could be achieved with Darwinian selection.[19]

Another genetic mechanism that showed a Lamarckian component is HGT. Recall from Chapter 2 that HGT occurs when prokaryotes (or unicellular eukaryotes) acquire DNA from the environment either indirectly, via phages and plasmids, or directly. The absorbed DNA integrates into the prokaryote's genome, and if this acquired genetic material promotes survival in a given environment, it becomes fixed in the population.[20] The evolution of antibiotic resistance is the best example of this phenomenon. According to Koonin and Wolf, a sensitive microbe must acquire resistance genes by HGT (typically via a plasmid) when it enters an environment where an antibiotic is present.[21] This acquired gene, stimulated by the specific environment, is advantageous (it confers antibiotic resistance) and is passed to the progeny, hence the Lamarckian label. Moreover, acquired genes incorporate new capabilities within preexisting frameworks faster than Darwinian selection would allow.[22] In an environment of higher and higher concentrations of antibiotics, for example, bacteria would have to acquire new genes faster than the slow, unplanned, undirected process of natural selection in order to survive. This interesting scenario leads to a discussion of another approach to accelerating selection called assisted (or human-assisted) evolution.

Human activity since the agricultural revolution has accelerated or exacerbated global warming, resulting in loss of biodiversity and species extinction. These twin threats have impacted all regions of the world, both terrestrial and aquatic. As terrestrial animals, we often have a blind spot for the oceans because we perceive them to be uniform and indestructible.[23] But this is not the case, warn marine biologist Madeleine van Oppen and colleagues at the Australian Institute of Marine Science. They complained:

> Humans have driven global climate change through industrialization and the release of increasing amounts of $CO_2$, resulting in shifts in ocean temperature, ocean chemistry, and sea level, as well as increasing frequency of storms, all of which can profoundly impact marine ecosystems. Coral reefs are highly diverse ecosystems that have suffered massive declines in health and abundance as a result of these and other direct anthropogenic disturbances.[24]

The rate of these disturbances, according to van Oppen, is higher than the corals can adapt to survive. And this may result in loss of reef systems on a global scale.[25]

One approach proposed by van Oppen and colleagues to save the coral reefs is to "enhance the ability of key reef organisms to tolerate stressful environments and to *accelerate recovery* after acute impacts"[26] (italics added). In essence, naturally occurring evolutionary processes (natural selection, random mutations, and acclimatization) are not occurring fast enough to keep up with the rapidly changing environment. The scientists acknowledged the risks of introducing genetically superior organisms or artificially enhanced organisms into the wild (see the discussion of GMOs in Chapter 2).[27] These invasive species, with their advantageous traits, might outcompete the native species, infect the native species with lethal pathogens, change the genetic composition of the native species through hybridization, or reduce genetic diversity.[28]

Van Oppen listed several human activities that could accelerate the rate of naturally occurring evolutionary processes: (1) stress exposure of natural stock to induce preconditioning acclimatization within and between generations through epigenetic mechanisms; (2) active modification of the community eukaryotic and prokaryotic microbes associated with the organism; (3) selective breeding to generate adaptive traits for the environment; and (4) genetic modification to improve traits.[29]

In terms of genetic modification, gene drives, discussed in previous chapters might be more effective for rapid modification of a species. The use of gene drives to alter entire populations of wild organisms seems arrogant regardless of the intention to save the organisms. However, some scientists believe that a little arrogance is necessary to improve life (including the ecosystem) for all humans.[30] Scientists Kevin Esvelt, Church, and Andrea Smidler of Harvard Medical School and immunologist Flaminia Catteruccia of the Harvard School of Public Health see tremendous possibilities in using RNA-guided CRISPR/Cas9. Specifically, they believe that CRISPR/Cas9 will enable scientists to construct efficient RNA-guided gene drives for mosquitoes (and many other species), altering them to stop spreading diseases such as malaria, Zika virus, and dengue fever.[31] For instance, the RNA-guided Cas9 can be directed to cleave several target sequences by expressing more guide RNAs. More guides, according to the scientists, improved cutting rates but increased the probability of cutting nontargeted sequences important for function of the organism.[32] The cut sequence is repaired using homologous recombination to copy the drive, but the problems multiply because homologous recombination rates vary across cell types, developmental stages, species, and phases of the cell cycle. Esvelt and colleagues noted that "the endonuclease gene drive in mosquitoes was correctly copied following 97% of cuts, while a similar drive in fruit flies was initially copied only 2% of the time and never rose above 78%."[33] They believed that the rate of homologous recombination in fruit flies is lower than in mosquitoes. In short, gene drives are limited by homologous recombination rate variation in addition to drive-resistant alleles, reproduction rates and gene flow, and adaptive benefits of the gene drive.[34]

Furthermore, Esvelt and colleagues proposed that an RNA-guided sensitizing drive could potentially replace resistant alleles with their ancestral alleles, making agricultural pests vulnerable to pesticides and herbicides (see Chapter 1).[35] The scientists believed that this approach could eventually lead to the development and use of species-specific pesticides and herbicides to control invasive species (eliminating them from islands or entire continents). RNA-guided suppression drive is one genetic approach for population control or extinction.[36] Esvelt and colleagues also acknowledged the risk of undesired spread of a suppression drive to related species due to rare mating events. And the suppression drive might spread from the invasive species back into the native population.

It is comforting that Esvelt and colleagues strongly emphasized transparency, public discussion, and evaluation before releasing RNA-guided gene drives. They noted:

> Because any consequence of releasing RNA-guided gene drives into the environment would be shared by the local if not global community, research involving gene drives capable of spreading through wild-type populations should occur only after a careful and fully transparent review process.
>
> Technologies with the potential to significantly influence the lives of the general public demand societal review and consent. As self-propagating alterations of wild populations, RNA-guided gene drives will be capable of influencing entire ecosystems for good or ill.[37]

Based on these precautions, the scientists advocated safeguards and control strategies such as reversing genome alterations that have already spread through populations.[38] A reversal drive released later could restore the exact wild-type sequence. This might take several drives. Another strategy proposed was using an immunizing drive to prevent a specific unwanted drive from being copied, blocking the spread of this drive.[39]

Humans continue to intervene in nature with the justification of "improving," "helping" or "saving" biological life because evolution is too slow or "has no vision, no foresight, no sight at all … it does not plan for the future," meaning that humans must take action to enhance organisms.[40] But the environment is always changing, so modifications to the genomes of humans and other organisms today might be detrimental to both groups in the future. Kevles, writing in 1985, quoted a statement written by Lionel Penrose in 1967:

> The social and biological values of hereditary differences are continually altering as the environment changes. … At the moment … our knowledge of human genes and their action is still so slight that it is presumptuous and foolish to lay down positive principles for human breeding. Rather, each person can marvel at the prodigious diversity of the hereditary characters in man and respect those who differ from him

> genetically. *We all take part in the same gigantic experiment in natural selection.*[41] (italics added)

The last sentence of this quote is very important. Most of the time we humans forget that we, like other biological life-forms, are part of the ecology. Not only do we impact the ecology, but the ecology impacts us. Nevertheless, the excitement surrounding biological enhancements accompanied by the idea of a world without illness or disease is a combination that resonates with some who have the means to use enhancement technology "liberally."

Strikingly, some defenders of genetic enhancements have adopted a sophisticated philosophical position with political overtones called liberal eugenics, by which parents can have their children genetically enhanced as long as it is not coercive and does not impact society as a whole.[42] For instance, philosopher Allen Buchanan, professor of philosophy at Duke University, and colleagues wrote that "both justice and our obligations to prevent harm make genetic interventions to prevent disabilities not only permissible but also obligatory."[43] Showing more boldness, British bioethicist John Harris wrote that "it is not only feasible to use genetic technology to make people more healthy, intelligent and longer-lived, it is our moral duty to do so."[44]

Sandel wrote that the libertarian philosopher Robert Nozick proposed a " 'genetic supermarket' that would enable parents to order children by design without imposing a single design on the society as a whole."[45] Furthermore, Sandel discussed the influential school of Anglo-American political philosophers who call for liberal eugenics. They believe that governments should not tell parents how to design their children, and parents should choose designs that do not hamper their children's life plans.[46] Sandel rejected this assertion and argued that liberal eugenics' main principle is state neutrality. He used the example of the state requiring parents to use genetic engineering to boost their children's IQs.[47] This capacity, according to Sandel, is all-purpose and — in this scenario — important in all life plans. As such, the state would encourage enhancements of this capacity. Finally, the late American philosopher and jurist Ronald Dworkin, defending liberal eugenics, argued that "if playing God means struggling to improve our species, bringing into our conscious designs a resolution to improve what God deliberately or nature blindly has evolved over eons, then the first principle of ethical individualism commands that struggle."[48]

Adding the adjective *liberal* in front of *eugenics* does not make it any less offensive (much as with putting *starvation* in front of *cannibalism*). And discussing eugenics in a philosophical context does not make it acceptable to the educated majority. The important point here, however, and the main reason for discussing this subject, is the flawed reasoning of liberal eugenicists that genetic enhancements of children within a family will not impact society as a whole. No human individual or family group lives in isolation. They are part of a community or population — a group of interbreeding individuals that exist in time and space; they share a gene pool. Any genetic modification in one family might ultimately spread throughout the population, just as introduced genes in a crop move into to wild relatives. Social scientists, politicians, and philosophers seldom consider the macroevolutionary consequences of genetic engineering, such as the unintentional alteration of potential future adaptive traits.

The late Robert Edwards, British physiologist, cocreator of human IVF (with the late Patrick Steptoe), and 2010 Nobel Prize winner, was also religious in his beliefs that genetic intervention was necessary to prevent children from being born with disabilities.[49] In 1999, he made this provocative statement: "Soon it will be a sin of parents to have a child that carries the heavy burden of genetic disease."[50] He showed his arrogance and eugenic mentality in 1993 when he reflected on the 25th anniversary of the birth (through IVF) of the first test-tube baby, Louise Brown: "Developing IVF was about more than fertility. … I wanted to find out exactly who was in charge, whether it was god himself or whether it was scientists in the laboratory." Edwards's conclusion? "It was us."[51] Edwards's statement is frightening, and it is one of the key motivations for this book. There are a respectable number of well-meaning scientists who have these same underlying beliefs, which they have translated into using advanced technologies to help people or "raise human standards."[52]

Edwards supported the use of ART for predetermining nonmedical traits (i.e., socially favored characteristics) so that these traits would be selected and heritable.[53] Professor of bioethics Osagie Obasogie of the University of California, Berkeley wrote that Edwards embraced eugenics as morally justified.[54] This should be qualified: Edwards embraced the new eugenics or liberal eugenics, in which individuals have access to enhancement technologies in the free market that enable them to control and modify their traits. Essentially, some individuals are able to circumvent natural selection and accelerate change without the natural environment: these engineered traits (i.e., novel traits) are created in the absence of environmental input. Such novel traits — not found in any human populations — may not be adaptive in the short term or in the long term.

## Notes

1. Francis Galton, *Essays in Eugenics* (London: Eugenics Education Society, 1909), 42.
2. George A. Bartholomew, "The Role of Natural History in Contemporary Biology," *BioScience* 36, no. 5 (May 1986): 325, https://doi.org/io.2307/1310237.
3. Karl Pearson, *The Life, Letters and Labours of Francis Galton* (London: Cambridge University Press, 1914), 207.
4. Daniel Kevles, *In the Name of Eugenics*: *Genetics and the Uses of Human Heredity* (Cambridge, MA: Harvard University Press, 1995), 12.
5. Stephen J. Gould, "Extemporaneous Comments of Evolutionary Hopes and Realities," in *Darwin's Legacy Nobel Conference XVIII* (1983), 101–102.
6. Hugh Elliot, *Zoological Philosophy by Jean Baptiste Lamarck* (1809; trans.; London, UK: Macmillan and Co., 1914), xxx, xxxv–xxxvii, 68–80, 291–314.
7. Ibid., 68–106.
8. Stephen J. Gould, *The Lying Stones of Marrakech Penultimate Reflections in Natural History* (New York: Harmony, 2000), 119.
9. Ibid., 119; Elliot, *Zoological Philosophy*, 348.
10. Elliot, *Zoological Philosophy*, xxx–xxxvii, 68–80.
11. Ibid., 113.
12. David Haig, "Weismann Rules! OK? Epigenetics and the Lamarckian Temptation," *Biology and Philosophy* 22, no. 3 (June 2007): 415–428; Eva Jablonka and Marion J. Lamb, "The Inheritance of Acquired Epigenetic Variations," *Journal of Theoretical Biology* 139, no. 1 (July 1989): 69–83; Eugene V. Koonin and Yuri I. Wolf, "Is Evolution Darwinian or/and

Lamarckian?" *Biology Direct* 4, no. 42 (November 11, 2009): 1–14, http://www.biology-direct.com/content/4/1/42.
13. Haig, "Weismann Rules! OK?" 419–420.
14. Ibid., 416; Jablonka and Lamb, "The Inheritance," 69; Koonin and Wolf, "Is Evolution Darwinian," 3, http://www.biology-direct.com/content/4/1/42.
15. Haig, "Weismann Rules! OK?" 420–421.
16. Koonin and Wolf, "Is Evolution Darwinian," 3, http://www.biology-direct.com/content/4/1/42.
17. Ibid.
18. Ibid., 4–5.
19. Koonin and Wolf, "Is Evolution Darwinian," 5, http://www.biology-direct.com/content/4/1/42.
20. Ibid.
21. Ibid., 7.
22. Ibid.
23. Madeleine J. Van Oppen, James K. Oliver, Hollie M. Putnam, Ruth D. Gates, "Building Coral Reef Resilience through Assisted Evolution," *Proceedings of the National Academy of Sciences USA Early Edition* 112, no. 8 (February 24, 2015): 2307, www.pnas.org/cgi/doi/10.1073/pnas.1422301112.
24. Ibid.
25. Ibid.
26. Ibid.
27. Ibid.
28. Ibid.
29. Ibid., 2309–2310.
30. Henk van den Belt, "Playing God in Frankenstein's Footsteps: Synthetic Biology and the Meaning of Life," *Nanoethics* 3, no. 3 (November 29, 2009): 260, 262; John Carey, "Playing God in the Lab: Gene Pioneers Are on the Threshold of Creating Life," *Business Week*, April 26, 1999, http://www.businessweek.com/magazine/content/07_26/b404004.htm.
31. Kevin M. Esvelt, Andrea L. Smidler, Flaminia Catteruccia, and George M. Church, "Concerning RNA-guided Gene Drives for the Alteration of Wild Populations," *Elife* 3, (July 17, 2014): 2, e03401.
32. Ibid., 3, 6.
33. Ibid., 6.
34. Ibid., 9–10.
35. Ibid., 12.
36. Ibid., 12–13.

37. Ibid., 16–17.
38. Ibid.
39. Ibid., 15–16.
40. Richard Dawkins, *The Blind Watchmaker* (New York: W.W. Norton, 1986), 5.
41. Daniel Kevles, *In the Name of Eugenics: Genetics and the Uses of Human Heredity* (Cambridge, MA: Harvard University Press, 1995), 251; Lionel S. Penrose, "The Influence of the English Tradition in Human Genetics," in *Proceedings of the Third International Congress of Human Genetics*, edited by James F. Crow and James Neel (Baltimore: John Hopkins University Press, 1967), 22–23.
42. Judith Daar, *The New Eugenics: Selective Breeding in an Era of Reproductive* (New Haven, CT: Yale University Press, 2017), 188–190; Michael J. Sandel, *The Case Against Perfection: Ethics in the Age of Genetic Engineering* (Cambridge, MA: The Belknap Press of Harvard University Press, 2007), 75–83.
43. Allen E. Buchanan, Dan W. Brock, Norman Daniels, Daniel Wikler, *From Chance to Choice: Genetics and Justice* (New York: Cambridge University Press, 2000), 302.
44. John Harris, *Enhancing Evolution: The Ethical Case for Making People Better* (Princeton: Princeton University Press, 2007), 20.
45. Sandel, *The Case Against Perfection*, 77.
46. Ibid., 75–76.
47. Ibid.
48. Ronald M. Dworkin, "Playing God: Genes, Clones, and Luck," in *Sovereign Virtue,* edited by Ronald Dworkin (Cambridge, MA: Harvard University Press, 2000), 452.
49. Osagie K. Obasagie, "The Eugenics Legacy of the Nobelist Who Fathered IVF," *Scientific American* 10 (October 4, 2013): 1, http://www.scientificamerican.com/article/eugenic-legacy-nobel-ivf/.
50. Ibid.
51. Ibid., 5.
52. Ibid.
53. Ibid., 5.
54. Ibid.

# Chapter 15

# Transhumanists, *Homo evolutis*, and Hubris

> The human species can, if it wishes, transcend itself not just sporadically, an individual here in one way, an individual there in another way, but in its entirely, as humanity. We need a name for this new belief. Perhaps *transhumanism* will serve: man remaining man, but transcending himself, by realizing new possibilities of and for his human nature.
>
> — Julian Huxley, evolutionary biologist and first director-general of UNESCO (1957; italics added)[1]

Biotechnology (gene therapy, iPSCs, cloning, xenobiology, and genetic editing applied to ART, IVF, PGD, diseases, animals, and plants) is the key component that fuels the new eugenics, which is still essentially about "improving the stock," where the "stock" today consists of all *Homo sapiens* (and nonhuman life-forms) and not just the selected few that were beneficiaries of the old eugenics. The genetic manipulation of nonhuman life-forms is included in bioengineering because of the universality of DNA, which enables gain-of-function and effect-on-function experiments in other species that generate results relevant to biological systems in humans.

Could *transhumanism* accurately replace the provocative term *new eugenics* to describe the techniques used in bioengineering? Before answering this question, a definition of transhumanism is necessary. According to philosophers David Pearce, of the Institute for Ethics and Emerging Technologies, and Nick Bostrom, cofounders of the World Transhumanist Association (now rebranded and incorporated as

Humanity+, Inc.), transhumanism requires a comprehensive definition to be accepted as a legitimate subject of scientific inquiry.[2] For this reason, Pearce and Bostrom gave two formal definitions for transhumanism:

> The intellectual and cultural movement that affirms the possibility and desirability of fundamentally *improving the human condition* through applied reason, *especially by developing and making widely available technologies to eliminate aging and to greatly enhance human intellectual, physical, and psychological capacities.*
>
> *The study of the ramifications, promises, and potential dangers of technologies that will enable us to overcome fundamental human limitations*, and the related study of the ethical matters involved in developing and using such technologies.[3] (italics added)

Compare these to the definition of synthetic biology from Chapter 8: "The application of science, technology, and engineering to facilitate and accelerate the design, manufacture and/or modification of genetic materials in living organisms."[4] While the new eugenics and transhumanism have the same final goals, which are improvement and enhancement of all humans, the focuses of some enhancements are different. For instance, transhumanists advocate enhancements such as AI, nanotechnology, antiaging materials, cybernetics, and brain-computer interfaces in addition to biotechnology, and the end product of transhumanism is, possibly, a new species: *Homo evolutis*.[5] Additionally, transhumanists include safety and ethics in their definitions. The new eugenics via bioengineering, however, focuses overwhelmingly on genome manipulation. And safety and ethics are not necessarily written down but must be inferred by individual researchers. So, are the terms *new eugenics* and *transhumanism* interchangeable? The answer is no.

Transhumanists are not rejecting the general idea of the new eugenics; they are simply distancing themselves from the word *eugenics* and all of the negative baggage of the early 20th century that it carries.[6] Transhumanists prefer the term *reprogenetics,* coined by Silver to mean a national, voluntary use of ART and genetic technologies to produce genetically enhanced humans.[7] In 1997, Silver, who is sympathetic to the use of biotechnology to improve humans, described a view of a future society with GM people (the GenRich) and non-GM people (Naturals):

> In a few hundred years the GenRich — who [will] account for 10 percent of the American population — [will] all carry synthetic genes. … All aspects of the economy, the media, the entertainment industry, and the knowledge industry [will be] controlled by members of the GenRich class. … Naturals [will] work as low-paid service providers or as laborers. … [Eventually] the GenRich class and *the Natural class will become … entirely separate species with no ability to cross-breed and with as much romantic interest in each other as a current human would have for a chimpanzee.* … In a society that values individual freedom above all else, it is hard to find and legitimate basis for restricting the use of *reprogenetics.* … The use of reprogenetic technologies is inevitable. … There is no doubt about it … whether we like it or not.[8] (italics added)

Generally, people have accepted transhumanists' (or genetic engineers') views of the future, Silver's view in particular, but what makes the so-called *Homo evolutis* superior? As mentioned in Chapter 4, synthetic biologist Church pointed to 10 genes associated with superior traits in humans and possibly human ancestors.[9] These 10 genes are listed in Table 15.1. The genetic traits in Table 15.1 are all desirable. And there

**Table 15.1.** Synthetic biologist George Church's 10 genes linked to better traits in humans

| Gene name | Associated trait |
|---|---|
| **LRP5** (Low-density lipoprotein receptor-related protein) | Gain-in-function mutation causes drastic **increases in bone mass** |
| **MSTN** (Myostatin or growth differentiation factor 8) | Mutations in both copies of the myostatin gene results in **bigger stronger muscles** |
| **SCN9A** (Sodium ion channel expressed in pain neurons and neurons in the involuntary nervous system) | Mutations in both copies of the SCN9A gene results in **insensitivity to pain** |
| **ABCC11** (ATP-binding cassette transporter sub-family C member 11) | Mutations in both copies of the ABCC11 gene results in **low odor associated with sweat** |
| **CCR5** (C-C Chemokine receptor type 5), **FUT** Galactoside 2-alpha-L-fucosyltransferase 2) | Mutations in both copies of the CCR5 (CCR5Δ32) or FUT results in **virus resistance** |

(*Continued*)

**Table 15.1.** (*Continued*)

| Gene name | Associated trait |
|---|---|
| **PCSK9** (Proprotein convertase subtilisin/kexin type 9) | Mutations in both copies of PCSK9 results in **low coronary disease** (lower blood low density lipoprotein [LDL]-particle concentration) |
| **APP** (Amyloid precursor protein) | Gain-in-function mutation results in **low Alzheimer's risk** |
| **GHR** (Growth hormone receptor), **GH** (Growth hormone 1) | Mutations in both copies of the GHR or the GH gene results in a **low risk for cancer and Type 2 diabetes** |
| **SLC30A8** (Solute carrier family 30 member 8) | Mutations in one copy of the SLC30A8 results in **low risk for Type 2 diabetes** |
| **IFIH1** (Interferon induced w/helicase C domain 1) | Gain-in-function of IFIH1 gene results in **low risk for Type 1 diabetes** |

*Source*: Adapted from Knoepfler (2016)[9]

are other genes connected to desirable traits: (1) the *CETP* (cholesteryl ester transfer protein) and *FOXO3A* (Forkhead box transcription) genes or "longevity genes," which are purported to slow age-related decline — including memory decline — in older adults; (2) the *APOC3* (apolipoprotein C3) gene, mutations in both copies of which are purported to reduce the risk of cardiovascular ischemic disease; and (3) the *HDEC2* (transcription repressor) gene, mutations in both copies of which are purported to reduce sleep time and simultaneously increase alertness.[10] But what are the risks to humans for these supposedly superior traits? While mutations in the *GHR* and *GH* (growth hormone receptor) genes significantly reduce the risk for cancer and type 2 diabetes, they increase the risk for short stature (Laron syndrome) and obesity.[11] Given the choice between an unpopular trait and an unhealthy medical condition, one would probably choose the former. Mutations in the *APOC3* gene are purported to reduce the risk for cardiovascular disease and increase the levels of high-density lipoprotein (HDL) cholesterol (the so-called good cholesterol, contrasted with low-density lipoprotein, or LDL, cholesterol).[12] But the positive benefits of HDL versus LDL cholesterol have been questioned.[13] The *CETP* gene is purported to slow age-related decline in brain function in older adults but is also associated with increased HDL levels and larger-than-average HDL and LDL particles.[14] Mutations in the *MSTN* gene result in large muscles in pigs. However, birthing problems and lack of flavor are the risks.[15] For instance, molecular biologist Jin-Soo Kim at Seoul National University, embryologist Xi-Jun Yin, at Yanbian University in China, and other collaborators in 2015 edited pig fetal cells by using TALEN to knock out both copies of the *MSTN* gene.[16] The edited cells were transferred to an egg cell, and 32 cloned piglets were subsequently created. However, birthing difficulties resulted due to the large size of the piglets, and only 11 lived to the age of eight months. Only two GM pigs from this research survived to adulthood to display their "double-muscled" trait.[17] Finally, an induced mutation of the *CCR5* gene (i.e., *CCR5Δ32*) prevents HIV from binding to T cells, thereby providing resistance to HIV/AIDS, but the mutation also possibly renders the individual more susceptible to fatality from West Nile virus.[18]

It is clear that tinkering or tweaking — which gives the impression of minor changes — for a desired effect can also lead to drastically fatal results. Consequently, government regulators, in consultation with scientists, must decide if the benefits of preventing and alleviating sickness and disease outweigh the risk of inadvertently causing death or heritable mutations. But Church, as noted earlier in this book, argued that there is a lower risk in editing sperm or egg cells (rather than somatic cells) because fewer cells are involved. Additionally, he boldly told journalist Michael Specter, a science and technology staff writer at the *New Yorker*, that "it strikes me as a fake argument to say that something is irreversible. There are tons of technologies that are irreversible. But genetics is not one of them. In my lab, we make mutations all the time and then we change them back. Eleven generations from now, if we alter something and it doesn't work properly, we will simply fix it."[19] Despite Church's confidence, not all mistakes can be fixed by technology.

It is very difficult to regulate elective enhancements that are completed in private clinics and controlled by market forces. This increases the danger, because the technology might not be strongly regulated, resulting in more harm than good. Alternatively, in a futuristic world, the governments of the world could opt for large-scale genome modification (gene drive), undercutting the market forces, including biotechnology companies. The move by any government to force a genetic modification would be reminiscent of state-sponsored sterilization in the early twentieth century. Even though this futuristic worldwide human gene drive example is presently theoretical, the key problem that must be considered when genetic modification is forced on any sexually reproducing wild organisms is the possibility of irreversible changes to the ecology. Some researchers have argued that gene drives offer substantial benefits to humankind (preventing the spread of disease, supporting agriculture by reversing pesticide and herbicide resistance in insects and weeds, and controlling damaging invasive species, for example) and that it is worth the risk to explore the responsible use of this emerging technology.[20] In fast-reproducing species (insects and small mammals, for example) a gene drive could affect them within a year.[21] Specifically, gene drives might be used to genetically modify mosquitoes and other disease vectors so that they cannot transmit diseases such as malaria and dengue fever. In fact, scientists are confident that they could eradicate malaria within a year by applying the technology to one percent of the wild population of mosquitoes.[22] For practical purposes, this would be welcomed by millions of people in the tropical regions of Africa, the Mediterranean, the Middle East, the Near East, and South Asia who suffer from malaria infection.[23] Unfortunately, most people do not appreciate the potential long-term ecological impact of crashing a mosquito population (or any insect population), although a respectable number of scientists do keep the negative impact of gene drives in mind.

In summary, the technical complexity and constraints of altering entire populations of wild organisms show, glaringly, that humans are blindly tinkering with nature with noble intentions to improve life for humankind. Some scientists, blinded by fame, fortune, patents, or hubris, will proceed without seeing the potential dangers they unintentionally risk for the ecosystem, the human population, and other life-forms.

## 15.1. The Uses and Abuses of CRISPR/Cas9 and Other Technologies in Bioengineering

There has been a long-standing belief in human civilization that technology can solve any problem that arises. The techniques in biotechnology used to modify the genome of humans and other living organisms can transform them or make them "better." In the extreme, human genetic enhancements will be heritable and will lead to creation of a new species — *Homo evolutis* — faster than natural selection can on its own.[24] Whether the process is labeled the new eugenics or transhumanism does not really matter. Scientists are modifying DNA (which has been evolving since long before *Homo sapiens* appeared in the fossil record) without having a full understanding of the basic principles of life. They are unaware of the unintended effects of modifying a gene because they have a difficult time deciphering the operations of the cell. Matthew Porteus, professor of pediatrics and stem cell transplantation at Stanford University Medical School, whose research focuses on using genetic editing to better understand diseases that affect children and infants, noted the high frequency of single nucleotide polymorphisms (SNPs) and the difficulty of finding SNPs that cause disease.[25] Some researchers have acknowledged this difficulty but still argued for genetic editing to find mutant genes. Another view is that random mutations could be advantageous in some unknown future environment. Whether an artificial mutation is adaptive long-term is questionable. Consequently, it is hubristic for humans to think that they can modify DNA without consequences or enhance themselves to the point of creating a new species.

In a limited context, it may be possible to prevent famine, epidemics, and congenital diseases and heal the sick in some regions of the world. But to make humans better requires a definition of *better*, because better today may not be better tomorrow. What if, for example, gene drives could reverse the mutation responsible for sickle cell anemia and restore normal hemoglobin for heterozygotes living in malarial environments. Would they be better then? The answer is no, because all the supposedly superior GM individuals would be at high risk of death from malaria; the heterozygotes had the highest reproductive fitness before the genetic alteration (see Chapter 1).

So why not alter the population of *Anopheles* mosquitoes to cause their extinction, or at least prevent them from spreading the malaria parasite? For most biologists, deliberate extinction of a biological species would be unacceptable. There is, however, another strategy: try to change human behavior so that they do not create environments for mosquitoes to persist. There is a never-ending battle in which organisms are trying to keep up with changing circumstances in the ecosystem; any modifications — natural or artificial — are short-lived. In essence, *modified*, *better*, *enhanced*, and *transhuman* are all relative terms when applied to organisms in the context of the changing environment. If every human being is enhanced via gene drive or nanomedicine with the same desirable traits, then diversity is lost. This is not adaptive given current environmental crises such as global warming, species extinctions, emergence of new microorganisms, emergence of multidrug-resistant strains of existing microorganisms, and childhood obesity. Species extinctions should be a wake-up call for us humans because we are part of the ecosystem, and changes in the ecosystem impact us. Tinkering with the genome, for any purpose, could lead to reduced variation in heritable traits. Being specialized with big bones and muscles, less need for sleep, and good odor sounds great. However, the unintentional negative effects of these traits (already discussed) combined with the changing environment would trap *Homo evolutis* in a specific habitat (not good in a changing environment), resulting in extinction.

In summary, the proposed label of the *new eugenics* for the manipulation of DNA (including that of animal and plants) for health and enhancement by means of gene therapy, iPSCs, cloning, xenobiology, and genetic editing applied to ART, IVF, PGD, diseases, animals, plants, and nanomedicine is not meant to frighten readers but to inform them about these technologies and their potential dangers to life if precautions are not taken. Technology *should* be used to help solve problems, as long as humans do not become slaves to it. Scientists are being pulled by the light speed developments in technology to apply their products to living organisms — in the name of curing diseases — before fully understanding the serious biological, genetic, and cultural ramifications of their actions. This probably, and worryingly, will not change because, from a practical perspective, the momentum of science never stops, and scientific organizations will support new technological innovations that are *sufficiently* safe or low risk and highly beneficial to human health and the environment.[26] In my opinion, there are six key events in history that predict the continued application of biotechnology to living organisms while ignoring the long-term risks because benefits to science, human health, or the ecosystem outweigh the risks. The first event is the creation of the first cloned mammal, Dolly the sheep, who was cloned from an adult somatic cell using nuclear transfer. Dolly's birth was announced with great fanfare in February 1997, but she was actually born on July 5, 1996.[27] This meant that experiments specifically aimed at cloning mammals were occurring in private before 1996. The point is that there may be many provocative experiments occurring in secret in private research institutions around the world. The second event is the origin of recombinant DNA and the advantages of patenting this technology. In 2010 biochemists Berg and Mertz wrote about the approval in 1980 by the US patent office of a joint patent application by Stanford University and the University of California, San Francisco, citing Stanley Cohen and Herbert Boyer, their respective faculty members, as sole inventors of recombinant DNA.[28] The two universities collectively received "nearly $300 million in revenues during the life of this and two other related patents."[29] Berg and Mertz added that shares of the income from the Cohen-Boyer patents were used by their respective departments and institutions to support research and education. While Berg and Mertz wrote that "their claims to commercial ownership of the techniques for cloning all possible DNAs, in all possible vectors, joined in all possible ways, in all possible organisms were dubious, presumptuous, and hubristic," the main point here is that the possibility of fame, lasting name recognition, increased money for salaries and research (the biotechnology industry encourages "academic scientists to patent their research discoveries and to explore their newly discovered entrepreneurial instincts"), and even winning a Nobel Prize motivates scientists to continue their experiments.[30] In the third event, a moratorium was called for in 1974 to assess the risks of recombinant DNA technology and find ways of reducing these risks.[31] Scientists had been engineering SV40 to infect *E. coli* in effect-on-function studies, knowing that *E. coli* was a natural microbe in the human intestinal tract.[32] The concern was that an unsecured lab could inadvertently cause the spread of a cancer-causing gene within the human population. As a result, the Asilomar Conference was convened in February 1975. Regardless of the Asilomar Conference (originating from the moratorium), some scientists were continuing recombinant DNA experiments, according to Sinsheimer.[33] The guidelines proposed as a result of the Asilomar Conference prohibited (1) cloning or recombinant DNA derived from highly pathogenic organisms, (2) deliberate creation of plant pathogens, and (3) deliberate formation of recombinant DNA containing genes from the biosynthesis of toxins.[34] Once the guidelines had been proposed, the moratorium was lifted. But to add insult to injury, these guidelines became more relaxed by 1979 due to lobbying by researchers in the field and financial incentives from the pharmaceutical industry.[35] Berg and Mertz made an emphatic statement about recombinant DNA: "In the over three decades since adoption of these various regulations for conducting recombinant DNA research, many millions of experiments have been performed without reported incident. No documented hazard to public health has ever been attributable to the applications of recombinant DNA technology."[36] This statement seems perplexing at first; in a strict sense, Berg and Mertz may be correct. But the mutagenesis (leading to leukemia-like conditions) and deaths that occurred as a result of using recombinant retroviruses or adenoviruses as tools in gene therapy would not have been possible without construction of recombinant DNA. Nevertheless, biotechnology has become a global industry. An individual holding a brand-new PhD in genetic engineering can acquire fame and fortune in an environment that gives her the resources and freedom to work. The fourth event highlights virologists Ron Fouchier and Yoshihiro Kawaoka, who engineered the H5N1 virus, which causes lethal infection in birds, to be more transmissible between mammals.[37] As discussed in Chapter 9, Fouchier's team introduced three mutations into the H5N1 virus genome and then passed the mutant virus from the noses of infected ferrets to the noses of uninfected ferrets several times. Soon, the uninfected ferrets became sick. The genetically modified H5N1 virus had acquired the ability to transmit between ferrets. Kawaoka's team inoculated ferrets intranasally using a reassortant virus (H5N1; VN1203 and HIN1; PR8). Several ferrets were infected via aerosols. Many scientists worried that "if the new strains were accidentally or deliberately released, they could spark a deadly pandemic."[38] So why cross this line? Proponents would respond by stating that these gain-of-function studies promote public safety and disease prevention by paving the way for vaccines. Whether Kawaoka and Fouchier prompted David Evans and colleagues to create a synthetic chimeric horsepox virus, ruffling the feathers of some in the scientific and biodefense communities concerning the reemergence of smallpox, is unknown.[39] The fifth event emphasizes how scientists continually test the boundaries of science. This was a proposal by a collaborative team of synthetic biologists, system biologists, and a computer scientist to synthesize a complete human genome in a cell line.[40] In May 2016, an invitation-only meeting attended by approximately 150 scientists, lawyers, and ethicists was held at Harvard University Medical School to discuss this proposal. Finally, the sixth event is the claim by Jiankui He that he used CRISPR to carry out human embryo gene editing to make babies and their descendants resistant to HIV infection.[41] The motivation was to cure the sick: "He seemed quite sincere in his aim to engineer babies who would not suffer the illness and stigma that had plagued their HIV-infected father," said bioethicist Alta Charo of the Law School at the University of Wisconsin–Madison.[42] Additionally, *Xinhua*, China's government-run news agency, alleged that Jiankui He was hoping to gain personal fame for being the first to create CRISPR-edited babies.[43] It is possible that Jiankui He was prompted to do more because of the work of his fellow countryman Zhou, who edited mutated $\beta$-globin genes in tripronuclear zygotes and found off-target cleavage activity.[44] And, like a chain reaction, this led Mitalipov to study how to reduce the risk of off-target mutations and mosaics as he edited viable embryos to correct a mutation of the *MYBPC3* cardiac gene.[45] In short, the need to cure the sick is a powerful force (in addition to obtaining patents, making money, and garnering fame) that, for better or for worse, propels the new eugenics. For further explanation of why the new eugenics persists, consider Greely's argument:

> To me, the biggest likely change in our world from CRISPR/Cas9 and other genomic editing methods won't be in humans but in the non-humans we use the methods to modify. As it gets cheaper and easier to modify genomes, non-humans genomes offer freedom from a lot of regulation, liability, and political controversy, *while offering plenty of*

> *opportunities to improve the world, become famous, or make money — with combinations of all of the above. …* Want to end malaria? Come up with a modified version of *Aedes aegypti* that can't transmit yellow fever, dengue fever, or chikungunya viruses to humans and will outcompete and eventually eliminate the wild type. Want to make a really economical biofuel? Take an algae and modify its genome in thousands of ways to optimize it for producing hydrocarbon fuel. Want to bring back the passenger pigeon? Use Crispr-Cas9 to modify the genomes of existing band tail pigeons to march, more or less, the genomes sequenced from specimens of the extinct passenger pigeon. Want to corner the market in high-end gifts? Start playing around with horse genomes adding in bits and pieces from other species in an effort to produce actual unicorns. Want to make a splash as an artist? Use Crispr-Cas9 to make a warren of truly glow-in-the-dark rabbits.[46] (italics added)

As far as Jiankui He is concerned, there was widespread condemnation of his claim and calls for mechanisms to prevent germ line editing until this emergent technology matures. This invites the same question as before, with some modification: Why try to cross this line at all? Are there tangible medical benefits? Surprisingly, the statements of some bioethicists on tangible benefits read like the writings of those scientists who advocate a more aggressive research agenda in human embryo research. For example, one bioethicist argued that if genetic editing can eradicate birth defects, then it should be a moral imperative, while another believed that embryo hybrids are morally justified if they prevent premature death.[47]

Michael Specter relayed the description of a nightmare that Doudna recounted when he interviewed her:

> "I had a dream recently, and in my dream" — she mentioned the name of a leading scientific researcher — "had come to see me and said, 'I have somebody very powerful with me who I want you to meet, and I want you to explain to him how this technology functions.' So I said, Sure, who is it? It was Adolf Hitler. I was really horrified, but I went into a room and there was Hitler. He had a pig face and I could only see him from behind and he was taking notes and he said, 'I want to understand the uses and implications of this amazing technology.' I woke up in a

> cold sweat. And that dream has haunted me from that day. Because suppose somebody like Hitler had access to this — we can only imagine the kind of horrible uses he could put it to."[48]

Maybe all professionals in biotechnology who have to decide between medical benefits and risks should experience a nightmare like Doudna's in order to slow the momentum of genome modification. In this age of biotechnology fueling the new eugenics, with the motivation to prevent and cure all illnesses and make life better for all humans, it would be wise to slow down, continue to learn about the genome, and understand that humans do not determine what is better; that is for the environment to decide.

## Notes

1. Julian Huxley, *New Bottles for New Wine* (London: Chatto & Windus, 1957), 17.
2. Humanity+ "What Is Transhumanism?" accessed November 30, 2018, http://whatistranshumanism.org/.
3. Ibid.
4. Rainer Breitling, Eriko Takano, and Timothy S. Gardner, "Judging Synthetic Biology Risks," *Science* 347, no. 6218 (January 9, 2015): 107.
5. Humanity+ "What Is Transhumanism?" http://whatistranshumanism.org/.
6. Ibid.
7. Lee M. Silver, "How Reprogenetics Will Transform the American Family," *Hofstra Law Review* 27, no. 3, article 12 (1999): 651, http://scholarlycommons.law.hofstra.edu/hlr/vol27/1553/12.
8. Lee M. Silver, *Remaking Eden: Cloning and Beyond in a Brave New World* (New York, NY: Avon Books, 1998), 206–222.
9. Paul Knoepfler, *GMO Sapiens: The Life-Changing Science of Designer Babies* (Hackensack, NJ: World Scientific Publishing Co. Ptc. Ltd., 2016), 187.
10. Bradley J. Wilcox, Timothy A. Donlon, Qimel He, Randi Chen, John S. Grove, Katsuhiko Yano, Kamal H. Masaki, D. Craig Wilcox, Beatriz Rodriguez, and J. David Curb, "FOXO3A Genotype Is Strongly Associated with Human Longevity," *Proceedings of the National Academy of Sciences of the USA* 105, no. 37 (September 16, 2008): 13987–13992; Ying He,

Christopher R. Jones, Nobuhiro Fujiki, Ying Xu, Bin Guo, Jimmy L. Holder Jr., Moritz J. Rossner, Seiji Nishino, and Ying-Hui Fu, "The Transcription Repressor DEC2 Regulates Sleep Length in Mammals," *Science* 325, no. 5942 (August 14, 2009): 866–870; Katherine Harmon, "Rare Genetic Mutation Lets Some People Function with Less Sleep," *Scientific American* (August 13, 2009), http://www.scientificamerican.com/article/genetic-mutation-sleep-less/; Paul Knoepfler, *GMO Sapiens: The Life-Changing Science of Designer Babies* (Hackensack, NJ: World Scientific Publishing Co. Ptc. Ltd., 2016), 187.

11. Jim Kozubek, *Modern Prometheus: Editing the Human Genome with CRISPR-CAS9* (New York, NY: Cambridge University Press, 2016), 56–57; Knoepfler, *GMO Sapiens*, 187; Jennifer A. Doudna and Samuel H. Sternberg, *A Crack in Creation: Gene Editing and the Unthinkable Power to Control Evolution* (New York: Houghton Mifflin Harcourt, 2017), 230–231.
12. Kozubek, *Modern Prometheus*, 56–57; Knoepfler, *GMO Sapiens*, 187.
13. Harrison Wein, "Questions about HDL Cholesterol," June 18, 2012, https://www.nih.gov.
14. Kozubek, *Modern Prometheus*, 56–57; Knoepfler, *GMO Sapiens*, 187.
15. Knoepfler, *GMO Sapiens*, 187–88; Raven *et al.*, *Biology*, 349; Doudna and Sternberg, *A Crack in Creation*, 230–231; David Cyranoski, "Super-Muscular Pigs Created by Small Genetic Tweak," *Nature* 523, no. 7758 (July 2, 2015): 14.
16. Ibid., 13–14.
17. Ibid., 14.
18. Doudna and Sternberg, *A Crack in Creation*, 164–166; Knoepfler *GMO Sapiens*, 187; Kozubek, *Modern Prometheus*, 289–296.
19. Michael Specter, "The Gene Hackers: A Powerful New Technology Enables Us to Manipulate Our DNA More Easily Than Ever Before," *New Yorker* (November 8, 2015): 22.
20. Kevin M. Esvelt, Andrea L. Smidler, Flaminia Catteruccia, and George M. Church, "Concerning RNA-guided Gene Drives for the Alteration of Wild Populations," *Elife* 3, (July 17, 2014): 1–5, e03401; Knoepfler, *GMO Sapiens*, 190–196.
21. Andrew Hammond, Roberto Galizi, Kyros Kyrou, Alekos Simoni, Carla Siniscalchi, Dimitris Katsanos, Matthew Gribble, Dean Baker, and Eric Marois, "A CRISPR-Cas9 Gene Drive System Targeting Female Reproduction in the Malaria Mosquito Vector," *Nature Biotechnology* 34, no. 1 (December 7, 2015): 82; Kozubek, *Modern Prometheus*, 322–323; Doudna and Sternberg, *A Crack in Creation*, 148–151.

22. Ibid.
23. World Malaria Report (Geneva, SUI: World Health Organization, 2017), 32–41.
24. Humanity+ "What Is Transhumanism?" http://whatistranshumanism.org/; Knoepfler, *GMO Sapiens*, 181–185.
25. Kozubek, *Modern Prometheus*, 58–59, 339–340.
26. The National Academies of Sciences, Engineering, and Medicine, *Review of the Federal Strategy for Nanotechnology-Related Environmental, Health, and Safety Research* (Washington, DC: The National Academies Press, 2009), 5.
27. Nigel Williams, "Death of Dolly Marks Cloning Milestone," *Current Biology* 13, no. 6 (March 18, 2003): 209.
28. Paul Berg and Janet E. Mertz, "Personal Reflections on the Origins and Emergence of Recombinant DNA Technology," *Genetics* 184, no. 1 (January 2010): 15; Kozubek, *Modern Prometheus*, 128–130.
29. Berg and Mertz, "Personal Reflections," 15.
30. Ibid.
31. Robert L. Sinsheimer, interview by Shelley Erwin, May 30, 1990, and March 26, 1991, California Institute of Technology Oral History Project (Pasadena, CA: Caltech Archives, 1992), 53.
32. Ibid., 50–54; Kozubek, *Modern Prometheus*, 106–110.
33. Sinsheimer, interview, 50–54.
34. Kozubek, *Modern Prometheus*, 124.
35. Kozubek, *Modern Prometheus*, 124–129; Sinsheimer, interview, 57–58.
36. Berg and Mertz, "Personal Reflections," 16.
37. Colin Russell, Judith M. Fonville, André E. X. Brown, David F. Burke, David L. Smith, Sarah L. James, Sander Herfst *et al.*, "The Potential for Respiratory Droplet — Transmissible A/H5N1 Influenza Virus to Evolve in a Mammalian Host," *Science* 336, no. 6088 (June 22, 2012): 1541–1547; Masaki Imai, Tokiko Watanabe, Masato Hatta, Subash C. Das, Makoto Ozawa, Kyoko Shinya, Gongxun Zhong *et al.*, "Experimental Adaptation of an Influenza H5 HA Confers Respiratory Droplet Transmission to Reassortant H5 HA/H1N1 Virus in Ferrets," *Nature* 486, no. 4703 (June 21, 2012): 420–428.
38. Katherine Harmon, "What Will the Next Influenza Pandemic Look Like?" *Scientific American* (September 19, 2011), https: www.scientificamerican.com; Jocelyn Kaiser, "U.S. Halts Two Dozen Risky Virus Studies: One-Year 'Pause' to Develop New Policy on Controversial Research That Enhances Pathogens," *Science* 346, no. 6208 (October 24, 2014): 404.

39. Ryan S. Noyce, Seth Lederman, David H. Evans, "Construction of an Infectious Horsepox Virus Vaccine from Chemically Synthesized DNA Fragments," *PLoS ONE* 13, no. 1: e0188453; https://doi.org/10.1371/journal.pone.0188453.
40. Clyde A. Hutchison III, Ray-Yuan Chuang, Vladimir N. Noskov, Nacyra Assad-Garcia, Thomas J. Deerinck, Mark H. Ellisman, John Gill, *et al.*, "Design and Synthesis of a Minimal Bacterial Genome," *Science* 351, no. 6280 (March 25, 2016): 1–11; aad6253, http://dx.doi.org/10.1126/science.aad6253.
41. Jon Cohen, "What Now for Human Genome Editing? Claimed Creation of CRISPR-edited Babies Triggers Calls for International Oversight," *Science* 362, no. 6419 (December 7, 2018): 1090; Dennis Normile, "Government Report Blasts Creator of CRISPR Twins: He Jiankui Loses University Job After Preliminary Findings That He 'Dodged Supervision' and Violated Regulations," *Science* 363, no. 6425 (January 25, 2019): 328.
42. Cohen, "What Now for Human Genome," 1090.
43. Normile, "Government Report Blasts Creator," 328.
44. Puping Liang, Yanwen Xu, Xiya Zhang, Chenhui Ding, Rui Huang, Zhen Zhang, Jie Lv *et al.*, "CRISPR/Cas9-mediated Gene Editing in Human Tripronuclear Zygotes," *Protein Cell* 6, no. 5 (May 2015): 363–372.
45. Heidi Ledford, "CRISPR Fixes Embryo Error: Gene-editing Experiment in Human Embryos Pushes Scientific and Ethical Boundaries," *Nature* 548, no. 7665 (August 2, 2017): 13–14.
46. Henry T. Greely, "Of Science, CRISPR-Cas9, and Asilomar," Center for Law and the Biosciences Blog, *Stanford Law Review*, April 4, 2015, https://law.stanford.edu/2015/04/04/of-science-crispr-cas9-and-asilomar/.
47. Julian Savulescu, Jonathan Pugh, Thomas Douglas, and Christopher Gyngell, "The Moral Imperative to Continue Gene Editing Research on Human Embryos," *Protein Cell* 6, no. 7 (July 2015), 476–479; Maggie Fox, "Human-Cow Hybrid Embryos Made in Lab," Australian Broadcasting Corporation, April 3, 2008, http://www.abc.net.au/science/articles/2008/04/03/2206835.htm.
48. Michael Specter, "The Gene Hackers: A Powerful New Technology Enables Us to Manipulate Our DNA More Easily Than Ever Before," *New Yorker* (November 8, 2015): 21.

# Addendum

# COVID-19

When I began writing this book in 2017, the term "COVID-19" (previously known as the 2019 novel coronavirus) had yet to be coined. By the time I completed my second draft in late 2020, COVID-19 was a household word, even though most people could not articulate which virus the acronym was referencing. I wrestled with the idea of including COVID-19 in Chapters 9 and 15 as one of my key events in history that predicted the continued application of biotechnology on living organisms while ignoring the long-term risks because the benefits to science and human health (i.e., vaccines) outweighed the risks. But I decided not to include COVID-19 in detail because — as I worked on my book — the pandemic was current and not historical. Additionally, there were several theories and no definitive answers on the origins of the virus.

I did, however, write a short paragraph on COVID-19 in Chapter 9 relating to the funding of dangerous virus research. Specifically, I wrote that the National Institutes of Health terminated a $3.4 million grant, which was first awarded in 2014 and renewed in 2019, to the nonprofit EcoHealth Alliance that supported research into bat coronaviruses in China.[1] The impetus for terminating the grant stemmed from the Trump administration's belief, based on circumstantial evidence, that Wuhan Institute of Virology in Wuhan, China, released the coronavirus. Coincidentally, the institute employs Chinese virologist Shi Zhengli, who had received funding from the grant.

The official name for COVID-19 is severe acute respiratory syndrome coronavirus 2 (SARS-CoV-2), 2019.[2] The virus (and other strains of the coronavirus) was considered to have originated from bats and was transmitted to humans through intermediate hosts.[3] Although the direct source of the COVID-19 pandemic is still unknown, investigators have detected SARS-CoV-2 infections in animals and humans working or living on mink farms in western Europe.[4] Additionally, investigators have reported SARS-CoV-2 in horseshoe bats in China and in pangolins smuggled from South Asian countries, but none of these animals are directly the progenitor of SARS-CoV-2.

In July 2021, the director-general of the World Health Organization, Tedros Adhanom Ghebreyesus, urged China to increase its transparency about the early days of the COVID-19 pandemic to resolve and dispel conspiracies concerning the origins of the coronavirus outbreak.[5] In the summer of 2021, Ghebreyesus created an investigative team to aggressively probe two leading theories for the emergence of SARS-CoV-2 in Wuhan, China, in December 2019: 1) the virus made a "jump" from an unknown animal species into humans; 2) a researcher studying coronavirus found in animals was unknowingly infected and, subsequently, infected the Wuhan public after leaving the lab; 3) the virus was genetically engineered in the Wuhan laboratory.[6] The genetic engineered theory is extremely controversial and would add to my general theme that researchers emphasize the tentative benefits of biotechnology (i.e., viral vaccines) while ignoring the risks. However, there is absolutely no evidence of bioengineering. Moreover, the animal–human infection and the lab origin theories are the two most likely events that many researchers support. For instance, early cases of SARS-CoV-2 were identified in the animal markets in Wuhan, China; therefore, contact with contaminated uncooked food could be an important source of coronavirus transmission.[7] This information does not diminish the fact that researchers in the Wuhan lab study some of the most dangerous viruses in the world; therefore, virologists at the Broad Institute of MIT and Harvard would like an "audit" of Chinese labs.[8] What country would allow foreign scientists to audit their labs? This brings me to the main point of my book: *scientific advancement, if not guided responsibly and with public input, can be detrimental to public safety.*

# References

1. Science Magazine, "Nobel Awardees Decry Grant Halt," *Science* 368, no. 6494 (May 29, 2020), 921.
2. Caroline Ash, Gemma Alderton, Priscilla Kelly, Seth Scanlon, Valda Vinson, and Brad Wible, "A time to Reflect: Lessons from Two Years of the COVID-19 Pandemic," *Science* 375, no. 6585 (March 11, 2022): 1099.
3. Peng Zhou and Zheng-Li Shi, "SARS-CoV-2 Spillover Events," *Science* 371, no. 6525 (January 8, 2021), 120–122.
4. Ibid.
5. Jon Cohen, "WHO Chief Pressures China On Pandemic Origin," *Science* 373, no. 6553 (July 23, 2021), 378.
6. Ibid.
7. Ibid.
8. Ibid.

# Glossary

**antigen:** A protein-sugar complex on the surface of a cell. Foreign antigens introduced in the body will cause an immune reaction.

**antigenic drift:** Accumulation of mutations within the genes that code for antibody binding sites; new forms of antigens are created constantly so that a vaccine created today for one strain of a virus, for example, will probably not be effective six months from now, because the new strain will have different antigens and the immune system cannot recognize it. Without any previous contact with the strain, there will be no memory cells for an effective response. Viral RNA polymerase lacks proofreading ability (i.e., does not correct mistakes during replication). As a result, mutations accumulate constantly, changing surface antigens.

**antigenic shift:** Strains of two or more different microbes (e.g., viruses) combine to form a new subtype having a combination of the surface antigens of the original strains.

**antiserum:** Blood serum containing antibodies for specific antigens.

**archaic *Homo sapiens*:** Hominin fossils in this category have cranial capacities of 1,200 cubic centimeters on average, approaching the modern human average of 1,350 cubic centimeters. They have been dated at approximately 300,000 years old.

**assay (genetic):** Identifies changes in genes, amino acids, or proteins.

**bacteriophage:** A virus that infects bacterial cells.

**bionanotechnology:** Manipulation and exploitation of cells, molecules, DNA, RNA, and proteins.

**biotechnology:** The manipulation of the genome of biological organisms to prevent or cure diseases. In agriculture, crop plants are genetically modified to resist diseases and harsh climates. Biotechnology is also being used in artificial reproductive technologies (ART).

**biovar:** A variant prokaryotic strain that differs physiologically or biochemically from other strains in its species.

**blastocyst/blastula:** In vertebrates, the embryo stage after fertilization and before gastrulation, which is a hollow sphere of cells and an early preimplantation embryo.

**blastomere:** One of the cells of a blastula.

**branch migration:** The process by which base pairs on homologous DNA strands are exchanged at a Holliday junction.

**cell:** The basic structural, functional, and biological unit of all living organisms.

**chimeric organism:** A single organism that is composed of cells with distinct genotypes (i.e., two different zygotes involved in sexual reproduction).

**chromatin:** This material, consisting of DNA, RNA, and proteins, makes up eukaryotic chromosomes. Epigenetic modification, such as methylation and acetylation, occurs at the chromatin and alters gene expression.

**cleave:** To split or cut in biology.

**cleavage:** In vertebrates, a rapid series of cell divisions of a fertilized egg forming a hollow sphere of cells.

**clone:** A genetically identical organism produced naturally (e.g., human identical twins) or artificially (e.g., by merging the nucleus from one cell with an enucleated egg to produce an offspring that is genetically identical to the donor of the nucleus).

**complementary DNA:** DNA synthesized from messenger RNA (mRNA) or microRNA (both single stranded) templates by action of the enzyme reverse transcriptase. If a particular protein is absent in a cell, scientists can transfer complementary DNA that codes for the protein to the recipient cell. Retroviruses naturally produce complementary DNA and integrate the DNA copy into their host's genome, creating a viral genome (RNA genome).

**conjugation:** The temporary union of two bacteria or other unicellular organisms for exchange or transfer of genetic material.

**cosmid:** A type of hybrid plasmid that contains a Lambda phage (λ) cos sequence or genes; used to introduce DNA fragments into *E. coli*.

**CRISPR-Cas system:** A genetic editing technique based on how bacteria or other prokaryotic organisms rid themselves of infection. It is a family of DNA sequences derived from viral DNA fragments that have previously infected the bacteria but are subsequently used to detect and destroy DNA from similar viruses in future infections.

**cumulus cells:** A cluster of cells that surround the immature egg (oocyte) in the ovarian follicle and after ovulation.

**cyclin-dependent kinases (Cdks):** Any group of enzymes that control progress through the cell cycle. They regulate transcription and mRNA. These enzymes are only active when combined with cyclin (which initiates certain processes of mitosis).

**DAPI (4′,6-diamidino-2-phenylindole):** A fluorescent stain that binds strongly to adenine-thymine-rich regions in DNA. Used extensively in fluorescence microscopy.

**Darwinian evolution:** Individuals with adaptive traits for a *given environment* will survive, reproduce, and pass on the adaptive traits to their offspring. These traits will increase in frequency over time as long as the environment remains the same; the result is gradual change through time.

**deoxyribonucleic acid (DNA):** Codes for the biological characteristics of organisms. It is a double helix. It consists of phosphates, sugars, and four nitrogenous bases: adenine–thymine and guanine–cytosine. It is homoplasmic, or identical, in every cell.

**differentiated cell:** Differentiation is a process by which a less specialized cell becomes a more specialized cell type during the development of a multicellular organism.

**differential RNA sequencing (dRNA-seq):** A sequencing technique designed to examine the precise position of transcription start sites.

**double-strand breaks:** Both strands in the double helix are severed (which can lead to genome rearrangement, impacting cell function).

**ectoderm:** One of the three embryonic germ layers of early vertebrate embryos. It gives rise to the outer epithelium of the body (i.e., skin, hair, and nails) and to the nerve tissue (brain and spinal cord).

**embryonic stem cell (ES cell):** A stem cell derived from the inner cell mass of a blastocyst. ES cells can develop into different tissue in the adult organism (and in animals give rise to an adult organism when injected into a blastocyst).

**enucleated cell:** A cell that had the nucleus removed.

**endocytosis:** A cell takes in matter by creating a canal in its membrane.

**endoderm:** One of the three embryonic germ layers of early vertebrate embryos. It gives rise to the epithelium that lines internal structures (e.g., digestive and respiratory tracts).

**enzyme:** A substance produced by a living organism that is capable of speeding up certain biochemical reactions.

**epigenetics:** The study of heritable phenotype changes that do not involve alteration of the DNA; changes in gene expression but not the genome itself.

**erythropoietin (EPO):** A hormone produced by the kidney that promotes the formation of red blood cells by the bone marrow.

**eugenics:** The science of improving a human population by promoting the breeding of a desired group based on genotypic and phenotypic traits and restricting or preventing the breeding of another group deemed undesirable (based on the same traits). This process is controlled by the group in power. In early-20th-century America, involuntary sterilization of males and females was sanctioned by laws enacted by several states.

**eukaryotic cell:** A cell composed of a cell membrane with a cytoplasm. In the cytoplasm are many membrane-bound organelles, including a nucleus containing DNA packaged in chromosomes and proteins.

**fibroblast:** A flat, irregularly branching cell of connective tissue that produces collagen and other fibers.

**F plasmid:** A conjugated plasmid that contains genes that allow the plasmid DNA to be transferred.

**fullerenes:** Molecules composed entirely of carbon in the form of spheres, ellipsoids, or tubes. Named after the American architect Richard Buckminster Fuller.

**gamete:** A haploid female sex cell or haploid male sex cell. In humans, these reproductive cells normally each have 23 chromosomes.

**gametogenesis:** The process by which cells undergo meiosis to form gametes.

**$G_0$ (gap phase):** A period in the cell cycle during which cells exist in a quiescent state (cells are not actively dividing).

**gastrula:** The embryonic stage in vertebrates when the blastula with its single layer of cells transforms into a three-layered embryo made up of ectoderm, mesoderm, and endoderm.

**gastrulation:** In vertebrates, the process by which the blastula transforms into the gastrula.

**genetic engineering:** The application of science, technology, and engineering to deliberately modify the genomes of living organisms.

**genome:** The entire DNA sequence of an organism (e.g., *Homo sapiens* has 23 pairs of chromosomes).

**genome library:** Collection of the entire genome of an organism.

**global transposon mutagenesis:** A biological process that allows genes to be transferred to a host organism's chromosomes, which modifies the function of an extant gene on the chromosomes, resulting in a mutation.

**glycoproteins:** Proteins found in cell membranes that consist of sugars attached to a polypeptide chain.

**granulosa cells:** A thin layer of cells in the epidermis (skin).

**halophilic (halophiles):** Organisms that thrive in high salt concentrations, particularly those in the domain Archaea.

**hemagglutinin:** Glycoproteins that cause red blood cells to clump together (e.g. influenza virus strains **H1**N1, **H5**N1, etc.).

**hemoglobin S:** The sickle cell gene; sickled red blood cells are caused by a point mutation in the gene for the hemoglobin molecule.

**Hfr cell:** A high-frequency recombination cell (e.g., *E. coli*) due to integration of a conjugative plasmid (F plasmid) into the cell's chromosome.

**histone proteins:** These positively charged proteins are found in the chromosomes. The positively charged histones and negatively charged phosphates enable the DNA molecule to coil around a core of eight histone proteins.

**HNH motif:** A structural motif bearing the amino acid sequence H-N-H found in nucleases. It is capable of inducing a strand break.

**Holliday junction:** A branched nucleic acid structure that contains four double-stranded arms joined together.

**homeodomain:** Part of the protein that attaches to specific regulatory regions of the target genes.

**homing endonuclease:** An enzyme that can only cut particular sequences.

***Homo erectus*:** A prehistoric bipedal human ancestor that was the first hominin to migrate outside of Africa — dated at approximately two million to 200,000 years old.

**homologous recombination:** A type of genetic recombination in which nucleotide sequences are exchanged between two similar or identical molecules of DNA (used by cells to repair harmful breaks that occur on both strands of DNA).

**horizontal (lateral) gene transfer (HGT):** The movement of genes between unicellular or multicellular organisms (as opposed to transfer of genes from parents to offspring).

**human-animal hybrid:** An animal incorporating elements of both human and animal.

**induction:** Binding of an inducer to a repressor resulting in expression of a gene encoding an effect.

**introgression:** When genetic material is dispersed from one parental population into another via hybrids. Introgression of genetic material can sometimes result in genetic variation that can be advantageous for the survival of the parental taxa.

**in vitro:** A process that takes place outside a living organism (e.g., in vitro fertilization).

**in vitro mutagenesis:** A mutation produced in a segment of cloned DNA. The DNA is then inserted into a cell (or organism) to study the effects of the mutation.

**in vivo:** A process taking place inside a living organism (e.g., natural fertilization inside the body).

**Lamarckian:** Any characteristics acquired during the lifetime of an organism and then inherited by its offspring (e.g., culture or epigenetics).

**Laron syndrome:** Insensitivity to growth hormone resulting in short stature (i.e., dwarfism).

**lysogenic cycle:** The viral cycle in which the bacteriophage integrates its nucleic acid into a host bacterium's genome and is replicated during cell reproduction.

**lytic cycle:** The viral cycle that results in destruction of the infected cell. Bacteriophages that promote this cycle are virulent.

**macrorestriction map:** A linear depiction of the order and distance between the points on a DNA strand at which restriction enzymes cleave a chromosome.

**meganuclease I-*Sce*I:** A homing endonuclease found in the organism *Saccharomyces cerevisiae* (yeast). A rare cutting endonuclease that recognizes an 18-base-pair sequence (TAGGGATAACAGGGTAAT). Does not occur in a human or mouse genome.

**meiosis:** Sex cell (gamete) production; there is a meiosis I, during which variation is created, and a meiosis II. At the end of meiosis, haploid sex cells are produced.

**mesoderm:** One of the three embryonic germ layers of early vertebrate embryos. It gives rise to muscle, bone, and other connective tissue; the peritoneum; the circulatory system; the excretory system; and reproductive systems.

**mesophilic (mesophile):** An organism that grows best at moderate temperatures, between 20 and 45 degrees Celsius, such as some microorganisms.

**methyl group:** One carbon atom bonded to three hydrogen atoms ($CH_3$ or sometimes $H_3C$). It attaches to cytosine — a process known as methylation — changing the expression of the gene.

**MGAS:** Group A streptococcus strains.

**mitosis:** The production of adult cells (somatic cells); at the end of mitosis, diploid cells are produced.

**morula stage:** A small ball of cells resulting from division of a fertilized egg in the early stage of embryonic development.

**nanometer:** One billionth of a meter.

**nanomedicine:** Application of nanotechnology to molecular processes at the cellular level.

**nanotechnology:** The branch of technology that deals with dimensions less than 100 nanometers (i.e., manipulating atoms, molecules, cells, etc.).

**neuraminidase:** Enzyme present in many pathogenic microorganisms; in influenza virus strains, it enables the virus to spread its particles after entering a cell's genome (e.g., H1**N1**, H5**N1**, etc.).

**neurula:** The early stage of embryonic development in which neurulation (i.e., development of the nervous system in vertebrates from the ectoderm) occurs.

**nootropic drugs:** Cognitive enhancers or "smart" drugs that may improve memory, creativity, or motivation in healthy individuals.

**northern blot hybridization (RNA blot):** A method used in molecular biology to study gene expression by detection of isolated mRNA in a sample: RNA fragments are separated by size, using gel electrophoresis, and then blotted onto a sheet of filter paper.

**nuclease:** An enzyme that cuts or cleaves chains of DNA into smaller units.

**nuclear transplantation:** A method used in cloning in which the nucleus of a donor cell is transferred to a donor egg that has had its nucleus removed (an enucleated egg).

**nucleocapsid protein:** A genome (RNA or DNA) and protein coat of a virus.

**oligonucleotides:** Short DNA or RNA molecules that have many uses in genetic research.

**oogenesis:** The production or development of an ovum.

**oocytes:** Immature egg cells in ovaries. In fertilized, they complete meiosis II.

**open reading frame (ORF):** A continuous stretch of codons that begins with a start codon and ends with a stop codon — it can be translated.

**palindromic DNA sequence:** A nucleic acid sequence on double-stranded DNA or RNA where the 5′ to 3′ base sequence on one strand matches the 3′ to 5′ base sequence on the other strand (e.g., the DNA sequence ACCTAGGT is palindromic because its base-by-base complement is TGGATCCA).

**P1 phage:** A temperate bacteriophage that infects *E. coli* and other bacteria: when in the lysogenic cycle, the phage genome exists as a plasmid in the bacterium.

**plaque assay:** A way to determine viral titer as phage-forming units so that known amounts can be used to infect cells.

**plasmid:** A small, circular DNA strand in the cytoplasm of a bacterium or protozoan that can replicate independently of chromosomes.

**plasmid partition system:** A mechanism that ensures the stable inheritance of plasmids during cell division.

**polymerase:** An enzyme that synthesizes long chains of polymers (i.e., DNA or RNA).

**preimplantation:** The stage after fertilization of the egg and before implantation in the wall of the uterus.

**preimplantation genetic diagnosis (PGD):** This is a biotechnology under the category of artificial reproductive technology (ART). It is a procedure used prior to implantation to help identify genetic defects within embryos and prevent the transmission of the defective gene to the offspring. With continual improvement of CRISPR-Cas gene-editing technique, defective genes can be edited out before implantation.

**primitive streak:** During gastrulation in birds, reptiles, and mammals, a dorsal, longitudinal strip of ectoderm and mesoderm.

**prokaryotic cell:** A cell lacking a membrane-bounded nucleus or membrane-bounded organelles.

**protease:** An enzyme that breaks down proteins and peptides.

**protospacer adjacent motif (PAM):** A two-to-six-base-pair DNA sequence immediately following the DNA sequence targeted by the Cas9 nuclease in the CRISPR bacterial adaptive immune system. PAM is a component of the phage plasmid and not a component of the bacterial CRISPR locus; Cas9 cannot bind to or cleave the target DNA sequence if it is not followed by the PAM sequence.

**PstI:** A type II restriction endonuclease (isolated from gram negative bacteria), which cleaves DNA at the recognition sequence 5′-CTGCA/G-3′, generating fragments with 3′ cohesive termini (i.e., sticky ends four base pairs long).

**recombinant DNA:** Fragments of DNA from two different species (e.g., a bacterium and a mammal) spliced together in the laboratory.

**repetitive extragenic palindromic (REP):** A sequence that is 35 nucleotides long, including inverted repeat; the same DNA sequence from 5′ to 3′ (5′-AGCGCT-3′ and 3′-TCGCGA-5′) occurs on both complementary strands.

**replicon:** An RNA or DNA molecule that replicates; a nucleic acid molecule that replicates as a unit.

**restriction enzymes:** Enzymes produced by certain bacteria having the ability to cut DNA molecules at a specific sequence of bases.

**restriction sites:** Locations on a DNA molecule containing specific sequences of nucleotides recognized by restriction enzymes.

**retrovirus:** A virus that begins as RNA and then gets transcribed back into DNA using a reverse transcription enzyme.

**reverse genetics:** A process used to understand the function of a gene by analyzing phenotypic effects.

**reverse transcriptase:** An enzyme used to generate complementary DNA from an RNA template.

**ribonucleic acid (RNA):** A nucleic acid present in all living cells, characterized by the presence of sugar, ribose, and the pyrimidine uracil.

**RuvC:** One protein that is part of a three-protein complex that mediates branch migration and resolves the Holliday junction formed during homologous recombination in bacteria. Essential in bacterial DNA repair, it cleaves the Holliday junction.

**sequence motif:** A nucleotide or amino acid sequence pattern that is widespread.

**serum:** A gold-colored, protein-rich liquid that is visible when blood clots.

**sex selection:** The increased frequency of a trait, usually in males, because of the advantages it confers in winning mates. The females are attracted to these traits, and they will mate with the males possessing them.

**sialic acid:** A family of nine-carbon acidic monosaccharides — all influenza A virus strains need sialic acid to bind with cells.

**sialidases:** Neuraminidases cleave sialic acid residues from glycoproteins.

**sickle cell gene ($Hb^S$):** A random point mutation in the sixth position of the beta chain results in a sickled red blood cell. Individuals who have inherited the gene from one parent are heterozygote dominant and called carriers. Individuals who have inherited the gene from both parents are homozygous recessive and develop sickle cell anemia.

**sickle cell anemia:** A condition in which red blood cells are sickle shaped, which reduces their ability to bind oxygen and clogs blood vessels. Without enough normal red blood cells, anemia develops.

**SIRV virus:** *Sulfolobus islandius* rod shaped virus 1.

**SSV:** Simian sarcoma virus — a gamma retrovirus that infects primates.

**somatic cells:** These are body cells or adult cells with the full complement of chromosomes: twenty-three pairs (or forty-six total) in humans.

**Sop C region:** A position on the *E. coli* chromosome.

**southern blot hybridization (DNA blot):** A method used in molecular biology for the detection of a specific DNA sequence: DNA fragments are separated by gel electrophoresis, denatured into single-stranded DNA, and then blotted onto a sheet of filter paper where respective DNA sequences are located by probe hybridization.

**suicide plasmid:** A plasmid that cannot replicate because it does not have the sequence "origin of replication" (Or i) and will not replicate during cell division.

**TALENs (transcription activator-like effector nucleases):** Restriction enzymes that can be engineered to cut specific sequences of DNA. These enzymes can be introduced into cells for use in gene editing.

**Tay-Sachs disease:** A Mendelian disease caused by the accumulation of a lipid-sugar molecule ganglioside, an important constituent of cell membranes. Normally, gangliosides are broken down by the enzyme hexosaminidase A. In Tay-Sachs individuals (homozygote recessives), hexosaminidase A is not produced in the body and excess gangliosides accumulate in neuronal cells. An affected infant will not survive its third birthday (the child will experience loss of hearing and sight, paralysis, and become unresponsive to stimuli in the environment).

**telomere:** A specialized structure that caps each end of a chromosome.

**thermophilic (thermophile):** An organism that thrives at relatively high temperatures, between 41 and 122 degrees Celsius, such as some bacteria and many archaea.

**tissue plasminogen activator (TPA):** A human protein that causes blood clots to dissolve. As an enzyme, it catalyzes the conversion of plasminogen to plasmin (the major enzyme responsible for clot breakdown).

**transcription factors:** One set of proteins required for RNA polymerase to bind to a specific DNA sequence to begin the transcription process (conversion of DNA to messenger RNA).

**transduction:** A process by which foreign DNA is introduced in a cell by a virus via HGT (i.e., viral transfer of DNA from one bacterial cell to another).

**transformation:** The uptake of DNA directly from the environment; a natural process in some bacterial species that incorporate "free" DNA from another ruptured bacterial cell.

**transgenic organism:** An organism into which a gene has been introduced through genetic engineering. Transgenic organisms are used in research to help determine the function of an inserted gene.

**viral challenge:** The deliberate infection of bacteria with a virus in effect-on-function experiments.

**zinc-finger nucleases (ZFNs):** Artificial restriction enzymes that fuse a DNA-binding domain to a DNA-cleavage domain.

**zinc-finger protein:** A protein structural motif. Some of the most abundant proteins in eukaryotic genomes, in which a zinc ion (not involved in binding) serves to stabilize the protein. The "fingers" are secondary structures held together by the zinc ion.

# Index

Printed in the United States
by Baker & Taylor Publisher Services